Mollusk Shells in Troia, Yenibademli, and Ulucak

Mollusk Shells in Troia, Yenibademli, and Ulucak

An Archaeomalacological Approach to the Environment and Economy of the Aegean

Canan Çakırlar

BAR International Series 2051
2009

Published in 2016 by
BAR Publishing, Oxford

BAR International Series 2051

Mollusk Shells in Troia, Yenibademli, and Ulucak

ISBN 978 1 4073 0469 4

BAR Publishing is the trading name of British Archaeological Reports (Oxford) Ltd. British Archaeological Reports was first incorporated in 1974 to publish the BAR Series, International and British. In 1992 Hadrian Books Ltd became part of the BAR group. This volume was originally published by John and Erica Hedges Ltd. in conjunction with British Archaeological Reports (Oxford) Ltd / Hadrian Books Ltd, the Series principal publisher, in 2009. This present volume is published by BAR Publishing, 2016.

Printed in England

BAR
PUBLISHING

BAR titles are available from:

BAR Publishing
122 Banbury Rd, Oxford, OX2 7BP, UK
EMAIL info@barpublishing.com
PHONE +44 (0)1865 310431
FAX +44 (0)1865 316916
www.barpublishing.com

Praise the sea and stay on land.

Provençal Proverb

(Braudel 1972: 17)

Denizden babam çıksa yerim.

Turkish Proverb

Abstract

This study investigates the archaeomalacological assemblages from three primary sites in the Aegean: Troia, Yenibademli, and Ulucak. The principle aim of the study is to reveal the environmental factors, and cultural and economic stimuli that may have affected the temporal and spatial patterns in the role and nature of shellfishing as an economic activity and the role of shellfish as an element of human nutrition. Archaeological mollusk shells are additionally treated as palaeoenvironmental archives; enhancing and complementing the studies of past coastal environments in the Aegean. The strong relationship between the coastal environment and human subsistence strategies is demonstrated within a greater framework by reviewing the available data from other archaeomalacological studies from the Aegean.

Although the chronological focus of the study is the Bronze Age, results from Post-Bronze Age Troia and from Chalcolithic Ulucak are included in the dissertation, in order to add perspective to temporal and spatial patterns.

Methods included the chronological analyses of the proportion of mollusk remains within the faunal assemblages, analyses of calculations of relative abundance of mollusk taxa, taphonomic analyses, contextual analysis, statistical analyses of the morphometric properties of important mollusk species, periodic sampling and examination of modern mollusk populations, stable isotopic analysis, seasonality analysis of incremental shell growth, and employment of ethnographic analogies.

Prey species were identified, the environments exploited for mollusk foraging were reconstructed, and gathering and processing techniques were explained. The factors affecting the decisions concerning harvest times were discussed. The evidence for the “murex” dye production in Troia was evaluated.

The changes in the size of *Cerastoderma glaucum* shells in Troia and *Patella caerulea* in Yenibademli indicate chronological changes in mollusk exploitation modes and demonstrate the consequences of human exploitation pressure on mollusk populations. Changes in the demographic structure of *C. glaucum* populations in Troia support this view. Seasonal incremental analysis of *C. glaucum* populations in Troia revealed that harvesting of this species took place year-around, with increased activity during warmer months. Changes in the growth pattern of *C. glaucum* may be indicative of changes in local climatic conditions. A major morphological difference is observed between the archaeological and modern populations of *C. glaucum* in the Troad. This situation is attributed to environmental changes.

This study should be considered as a pilot study for the archaeomalacology of the Eastern Mediterranean. The palaeoeconomic and palaeoenvironmental implications that emerged from this study can only be tested and improved if further morphometric, stable isotopic and incremental growth analyses on the archaeomalacological material from the region are carried out in conjunction with regional research on extant mollusk populations.

Deutsche Zusammenfassung

Die vorliegende Arbeit behandelt die archäomalakologischen Reste aus hauptsächlich drei Fundstätten im ägäischen Raum: Troia, Yenibademli, und Ulucak. Das Hauptziel der Arbeit ist es, die Umweltbedingungen sowie die kulturellen und ökonomischen Faktoren herauszufinden, die die zeitlichen und räumlichen Muster der Bedeutung und Natur des Molluskensammelns als eine ökonomische Aktivität und die Rolle der Schalentiere als ein Element der Ernährung beeinflussst haben. Die archäologischen Molluskenschalen werden auch als Paläoumwelt-Archive behandelt. Sie komplementieren und verbessern die Forschungen zur Küstenpaläoumwelt der Ägäis. Der starke Zusammenhang zwischen Küstenumwelt und Subsistenzstrategien von Menschen wird, durch Einbeziehung von vorhandenen Daten aus anderen archäomalakologischen Studien des ägäischen Raumes, in einen größeren Rahmen gestellt.

Obwohl der chronologische Schwerpunkt der Arbeit auf der Bronzezeit liegt, wurden auch die Ergebnisse aus dem post-bronzezeitlichen Troia sowie aus dem chalkolitischen Ulucak in die Dissertation eingeschlossen, um die zeitlichen und räumlichen Muster in eine größere Perspektive stellen zu können.

Die angewendete Methodik umfasst chronologische Analysen der Anteile der Molluskenreste an den Faunenresten im Allgemeinen, Analysen der relativen Anteile der verschiedenen Molluskenarten, taphonomische Analysen, Analysen der archäologischen Fundkontexte, statistische Auswertungen der morphometrischen Eigenschaften der wichtigsten Molluskenarten, periodische Sammlungen und Untersuchungen der modernen Mollusken-Populationen, stabile Isotopie, Jahreszeitbestimmung der Wachstumsinkremente der Molluskenschalen sowie ethnographische Analogien.

Die gesammelten Arten wurden identizifiert, die für das Molluskensammeln genutzten Biotope rekonsturiert, und die Technik des Sammelns und Bearbeitens erläutert. Die Faktoren, die den Zeitpunkt der Sammelaktivität beeinflussten, wurden diskutiert. Die Hinweise auf die Purpur-Herstellung in Troia wurden geprüft und bewertet.

Die Größenveränderungen der *Cerastoderma glaucum*-Schalen in Troia und der *Patella caerulea*-Schalen in Yenibademli weisen auf chronologische Veränderungen in der menschlichen Nutzung der Mollusken hin und zeigen die Wirkung des menschlichen Ausbeutungsdrucks auf Molluskenpopulationen. Veränderungen in der demographischen Struktur der *C. glaucum*-Populationen in Troia unterstützen diese Ergebnisse. Die Jahreszeitbestimmung der inkrementellen Wachstumsraten der *C. glaucum*-Populationen in Troia zeigt, dass diese Art in allen Jahreszeiten gesammelt wurde, jedoch verstärkt während der wärmeren Jahreszeiten. Veränderungen in den Wachstumsraten der *C. glaucum*-Schalen weisen möglicherweise auf Veränderungen der lokalen klimatischen Bedingungen hin. Ein wesentlicher morphologischer Unterschied wurde zwischen archäologischen und modernen *C. glaucum*-Populationen in der Troas beobachtet. Diese Situation wurde auf umweltliche Veränderungen zurückgeführt.

Diese Arbeit sollte als eine Pilotforschung für die Archäomalakologie des östlichen Mittelmeerraums angesehen werden. Die päläoökonomischen und umweltgeschichtlichen Schlussfolgerungen dieser Arbeit können nur überprüft und erweitert werden, wenn weitere morphometrische, geochemische, und inkrementelle Analysen des archäomalakologischen Materials aus diesem Gebiet durchgeführt werden, in Zusammenhang mit regionalen Forschungen an heutigen Molluskenpopulationen.

Acknowledgments

Without the help and support of a number of people and institutions, this monograph, which is based on my doctoral dissertation, would not have come to fruition.

Foremost among them is my dissertation advisor Prof. Dr. Dr. H-P. Uerpmann. I am greatly indebted to him for the invaluable advice and support he provided over the last four years. The study also benefited enormously from the insightful comments and constructive criticism of my second reader Prof. Dr. J. Nebelsick. I am also grateful to Prof. M. Satır for his encouragement to carry out inter-disciplinary studies. I additionally owe warmest thanks to Dr. M. Uerpmann for her enduring support for my research on coastal resources since 2002.

The archaeomalacological material for the study was granted by the director of the Troia Project, late Prof. Dr. M. Korfmann, the director of the excavations at Yenibademli, Dr. H. Hüryılmaz of Hacettepe University, and the director of the excavations at Ulucak, Prof. Dr. A. Çilingiroğlu of Ege University. Their support was not limited to granting the study material, but was expanded to providing unpublished stratigraphical information and illustrations from project archives. I would like to thank them for their kind interest and support in this work.

In addition I would like to thank my colleagues at Tübingen University for helping with various aspects of the project. Dr. P. Jablonka, Dr. R. Becks, Dr. P. Pavuk, P. Hnila, and S. Blum provided the stratigraphy of most of the excavation units in Troia. C. Gündem analyzed the faunal material from Troia, Yenibademli, and Ulucak with me during 2004 and 2005. Dr. S. Münzel, Dr. Katleen Deckers, Dr. M. Haidle, Dr. S. Ünlüsoy, Dr. Ç. Çilingiroğlu, Dr. A. Bobokhjan, and P. Zidarov helped with various queries relating to my research and always provided stimulating discussions. R. deBeauclair corrected the German abstract. Research assistants A. Erdem and F. Dedeoğlu kindly helped to set up a comfortable work-place at the Ege University during my stay at Izmir to study the Ulucak material.

Dr. C. Wolkerdörfer of Freiberg University and Prof. Dr. İ. Kayan of Ege University helped the development of this project by sharing their knowledge on the geology and geomorphology of Troia and the Aegean coast. Dr. J. Göbel of Deutsches Museum, München, was never tired to answer e-mails about her work on the Trojan landscape. I would like to thank them for their readiness in correspondence.

Dr. C. Richardson of the University of Bangor, Dr. B. Öztürk of Ege University, P. Leontarakis of the Fisheries Insitute at Kavala, Dr. S. Mariani of Hull University, Dr. B. A. Gümüş from Isparta, T. Bodur of Akdeniz University, and Prof. Dr. A. Morales, of Universidad Autonoma de Madrid, assissted the project with their insights about Mediterranean molluscan habitats, taxonomy, shell geochemistry, and morphology. I would like to thank these people for sharing and supporting my excitement about archaeological shell material and enlightening me about biological approaches to mollusk shells. T. Theodoropoulou of the National Museum of Natural History in Paris, and C. E. Veropoulidou, of Aristotle University of Thessaloniki, doctoral candidates working on archaeomalacological and archaeoitchthyological material from Greece, N. Andreasen, of Cambridge University, who kindly provided me with his unpublished M.Sc. thesis, and Dr. D. S. Reese, of Peabody Museum at the Yale University, shared their knowledge and experience on the shells in Aegean archaeology. I would like to thank them for their colloboration. Dr. U. Önal of Çanakkale University, marine biologist O. Taşkın, excavation guard Ş. Taşkın, fishermen N. Doğan, G. Gün and A. Gün from Yeniköy, fisherman E. Cumalı from Tevfikiye, and Çanakkale University biology student E. Çaydan from aided sampling the local mollusk populations in the Troad. Beyond its scientific value, shellfishing with them was an experience immemorial. C. Karataş's help with winter accommodations in Çanakkale unexpectedly turned fieldworks a rather relaxed part of this project.

I. Yılmaz of Stuttgarter Kunstakademie helped shaping the illustrations. Laboratory work was aided by K. Langguth and Ö. Ekin. Working with them was delightful. Additionally, very special thanks go to B. Bevan, K. Revolinski and S. Avery who undertook the long and painful task of revising the language and structure of this work.

Many members of the Archaeomalacology Working Group gave support to this research. These people include Dr. C. Claassen of the University of S. Carolina, Dr. I. Quitmyer of University of Florida, and Dr. D. Bar-Yosef of Haifa University. I greatly appreciate their assistance and support with scientific matters.

Last, but not the least, I would like to spell out my appreciation for my friends A. Aksoy, L. Gaffaroğlu, A. Papakçı, D. Müller, and E. Rossberger, all the 'Kollegen' at Luginsland, and all the 'Kızlar' at Istanbul and Ankara. Without their emotional support, I would have never been able to carry out this research.

This research was financially supported by the Deutsche Forschungsgemeinschaft (DFG) Graduiertenkolleg "Anatolien und Seine Nachbarn" at Tübingen University, Troia Projekt, Institute for Aegean Prehistory (INSTAP), Landesgraduiertenförderung Program of Baden-Württemberg, and by the Wenner-Gren Foundation.

Preliminary results on the analysis of *C. glaucum* populations from Troia have been presented at a variety of conferences and invited lectures at several institutions.

This book is dedicated to my family: Zeki, Emine, Cem and Beyza Çakırlar.

Contents

Abstract III

Deutsche Zusammenfassung IV

Acknowledgments V

1. Introduction 1

1.1. Archaeomalacology 1
1.2. Archaeomalacology in the Aegean region 2
1.3. Research objectives 3
1.4. Organization of the study 4

2. Archaeology and environment 6

2.1. Location and current geography 6
2.1.1. Troia 6
2.1.2. Yenibademli 6
2.1.3. Ulucak 6
2.2. Palaeogeography 6
2.2.1. Geomorphic evolution of the Trojan and Beşik Bays/Plains 7
2.2.2. Geomorphic evolution of the Büyükdere Valley 10
2.2.3. Changes in the geomorphic features around Ulucak 11
2.2.4. Summary 11
2.3. Climate 11
2.3.1. Present climate 12
2.3.2. Climate of the Middle and Late Holocene 13
2.4. Aquatic environments 15
2.4.1. Troia 16
2.4.2. Yenibademli 17
2.4.3. Ulucak 17
2.4.4. Summary 17
2.5. Biosphere 17
2.5.1. Vegetation 17
2.5.2. Fauna 18
2.6. Archaeology 18
2.6.1. Archaeological research at Troia 20
2.6.2. Archaeological research at Yenibademli 25
2.6.3. Archaeological research at Ulucak 25
2.7. Subsistence economy 26
2.7.1. Subsistence economy of Troia 26
2.7.2. Subsistence economy of Yenibademli 28
2.7.3. Subsistence economy of Ulucak 29
2.8. Earlier work on the mollusk remains of Troia 29

3. Methods 32

3.1. Excavation techniques and sampling 32
3.2. Identification, description, and recording of specimens 33
3.3. Quantification 33
3.4. Morphometry 35
3.5. Incremental growth analysis 36
3.6. Sampling extant mollusk populations 38
3.7. Isotopy 39
3.8. Ethnographic analogy 42

4. Analytical results 43

4.1. Aquatic mollusks 43
4.1.1. Gastropods 43
4.1.1.1 *Patella* spp. 45
4.1.1.2. Haliotis tuberculata 45
4.1.1.3. Monodontinae 46
4.1.1.4. *Cerithium* spp. 46
4.1.1.5. Cypraeidae 47
4.1.1.6 Phalium granulatum 47
4.1.1.7. Charonia tritonis 47
4.1.1.8. Bolinus brandaris 47
4.1.1.9. Hexaplex trunculus 47
4.1.1.10. Fasciolaria lignaria 48
4.1.1.11. Columbella rustica 49
4.1.1.12. Conus mediterraneus 49
4.1.2. Bivalves 49
4.1.2.1 *Arca noae* 49
4.1.2.2. Glycymerididae 49
4.1.2.3. Mytilus galloprovincialis 49
4.1.2.4. *Pinna* spp. 49
4.1.2.5. Pectenidae 49
4.1.2.6. Spondylus gaederopus 50
4.1.2.7. Ostrea edulis 50
4.1.2.8. Acanthocardia tuberculatum 50
4.1.2.9. Cerastoderma glaucum 50
4.1.2.10. Solen marginatus 51
4.1.2.11. *Donax* spp. 51
4.1.2.12. Tapes decussatus 52
4.1.2.13. *Unio* spp. 52
4.2. Modern collections 52
4.2.1. Survey of modern mollusk populations in the Troad 52
4.2.2. Properties of modern *C. glaucum* populations in the Troad 55
4.2.2.1. Morphology 55
4.2.2.2. Description of incremental shell growth 55
4.2.2.3. Age profiles 58
4.2.2.4. Seasonal incremental growth pattern 58
4.2.3. Summary 62
4.3. Assemblages of Troia 63
4.3.1. Proportion of mollusk remains in faunal assemblages 63

4.3.2. Archaeomalacological assemblages 64
4.3.3. Taphonomic description of mollusk specimens 68
4.3.4. Archaeological context of Bronze Age mollusk remains 70
4.3.5. Properties of *C. glaucum* populations in Bronze Age Troia. 73
4.3.5.1. Morphology of Bronze Age *C. glaucum* populations 73
4.3.5.2. Description of incremental shell growth in Bronze Age *C. glaucum* populations 80
4.3.5.3. Age profiles of Bronze Age *C. glaucum* populations 83
4.3.5.4. Seasonal incremental growth pattern of Bronze Age *C. glaucum* populations 85
4.3.5.5. Isotopic compositions of Bronze Age *C. glaucum* specimens 88
4.3.6. Summary 91
4.4. Assemblages of Yenibademli 92
4.4.1. Proportion of mollusk remains in faunal assemblages 92
4.4.2. Archaeomalacological assemblages 93
4.4.3. Taphonomic description of mollusk specimens 93
4.4.4. Archaeological context of mollusk remains 95
4.4.5. Morphology of *P. caerulea* populations 97
4.4.6. Morphology of *C. glaucum* populations 98
4.4.7. Summary 98
4.5. Assemblages of Ulucak 100
4.5.1. Proportion of mollusk remains in faunal assemblages 100
4.5.2. Archaeomalacological assemblages 100
4.5.3. Taphonomic description of mollusk specimens 100
4.5.4. Archaeological context of mollusk remains 101
4.5.5. Morphology of the *C. glaucum* assemblages 102
4.5.6. Summary 102
4.6. Archaeomalacological data from other sites in the Aegean 102
4.6.1. Species spectra 104
4.6.2. Morphometric, incremental, and isotopic data 105
4.6.3. Summary 106
4.7. Comparative account of analytical results 107
4.7.1. Proportion of mollusk remains in faunal assemblages 107
4.7.2. Comparison of archaeomalacological assemblages 107
4.7.3. Comparative morphology of archaeological and modern *C. glaucum* populations 107
4.7.4. Comparative incremental growth pattern of archaeological and modern *C. glaucum* populations 111

5. Shellfish procurement and consumption 115

5.1. Food species 115
5.2. Shellfishing grounds 119
5.2.1. Shellfishing grounds in the Troad 119
5.2.2. Shellfishing grounds of Yenibademli 120
5.2.3. The origin of mollusk assemblages at Ulucak 121
5.2.4. Summary 121
5.3. Techniques of mollusk procurement: gathering and cultivation 122
5.4. Shellfish processing 123
5.5. Nutritional value of mollusks 124
5.6. Harvest time 125
5.7. Dye production in Troia 128

6. Variation and continuity in the archaeomalacological record 131

6.1. Species diversity 131
6.2. Proportion of mollusk remains in the faunal assemblages 132
6.3. Relative abundance of mollusk taxa 135
6.4. Size and age 139
6.5. Other morphological attributes 141
6.6. Incremental growth pattern in *C. glaucum* populations 142

7. Summary, conclusions, and prospects 144

7.1. Summary 144
7.2. Concluding remarks 146
7.3. Prospects for future research 147

References 149

Appendix 165

List of Figures

Figure 1. Map of the Aegean and the location of the place names mentioned in the text. 5

Figure 2. Evolution of the Trojan and Beşik Bays. 7

Figure 3. Geomorphology of the Büyükdere Valley. 10

Figure 4. Geographical features in the vicinity of Ulucak....11

Figure 5. Temperature and precipitation recorded at stations nearest to Troia, Yenibademli, and Ulucak....12

Figure 6. Sea surface temperatures off the coast of Çanakkale....15

Figure 7. Archaeological plan of the citadel area of Troia....22

Figure 8. Archaeological plan of the citadel and the lower town of Troia....23

Figure 9. Topographical plan of Yenibademli....25

Figure 10. Topographical plan of Ulucak....26

Figure 11. Descriptions of the morphological measurements on *C. glaucum* valves. 36

Figure 12. Description of the morphological measurements on *P. caerulea* shells....36

Figure 13. External lines on modern *C. glaucum* valves and size variation in the modern populations of the Troad....37

Figure 14. Laboratory equipment used for valve sectioning. 38

Figure 15. Sampling of live *C. glaucum* at the deltaic lagoons of the Karamenderes. 39

Figure 16. Location of sampling stations in the Troad. 40

Figure 17. Shells of different mollusk classes that frequently occur in archaeological contexts 43

Figure 18. Definition of bivalve (left) and gastropod (right) shell part terms used in this study. 44

Figure 19. Definition of aquatic ecological zones....44

Figure 20. Shells of *P. caerulea, P. rustica,* Monodontinae, and *H. trunculus*....46

Figure 21. Shells of *A. noae*, *M. galloprovincialis*, *F. glaber*, *O. edulis*, *S. marginatus*, and *T. decussatus*....48

Figure 22. A lagoon (Station 2) at the deltaic area of the Karamenderes in February 2006....53

Figure 23. SST (° C) and salinity (PPT) values at the time of collection. 54

Figure 24. Morphology of modern *C. glaucum* populations from the Troad (Distributions). 55

Figure 25. Comparison of the morphology of modern *C. glaucum* populations from the Troad by station (ANOVA)....56

Figure 26. Comparison of completed breadths of successive white and gray incremental zones in modern *C. glaucum* populations....57

Figure 27. Mean valve height (VH) achieved at each growth stage in modern *C. glaucum* populations. 58

Figure 28. Estimated age profiles of modern *C. glaucum* populations in the Troad. 59

Figure 29. Seasonal growth pattern of modern *C. glaucum* populations. 60

Figure 30. Growth pattern of modern *C. glaucum* specimens (1) 61

Figure 31. Growth pattern of modern *C. glaucum* specimens (2) 62

Figure 32. Proportion of mollusk remains in the faunal assemblages of Troia. 63

Figure 33. Changes in the relative abundance of species represented in the mollusk assemblages of Troia. 64

Figure 34. Taphonomy of *Patella* spp., *C. vulgatum, H. trunculus*, and *Pinna* spp. in Troia. 67

Figure 35. *C. glaucum* valves from Troia : Burnt, partiallly broken, and (naturally ?) holed specimens......69

Figure 36. Examples of bivalve shell corrosion in Troia......70

Figure 37. Example of a typical excavation Unit with high mollusk content from Troia 71

Figure 38. "Shell layer" at Quadrate D5 in Troia......71

Figure 39. Valve height of *C. glaucum* populations from Troia (Distributions). 74

Figure 40. Comparison of the valve heights of the *C. glaucum* populations in Troia by period and phase. 75

Figure 41. Solidity of *C. glaucum* populations from Bronze Age Troia (Distributions)......76

Figure 42. Comparison of solidity in *C. glaucum* populations in Troia by period (a) and phase (b) (ANOVA). 77

Figure 43. Curve of *C. glaucum* valves from Troia (Distributions) 78

Figure 44. Comparison of curve in *C. glaucum* populations in Troia according to periods and phases (ANOVA)......79

Figure 45. Comparison of breadths of white and gray incremental zones in archaeological *C. glaucum* (ANOVA). 80

Figure 46. Comparison of completed successive breadths of white incremental zones in *C. glaucum* populations by occupational phase in Troia. 81

Figure 47. Comparison of completed successive breadths of gray incremental zones in *C. glaucum* populations by occupational phase in Troia 82

Figure 48. Incremental growth pattern of sample *C. glaucum* individuals from Bronze Age Troia (1)......83

Figure 49. Incremental growth pattern of sample *C. glaucum* individuals from Bronze Age Troia (2)......84

Figure 50. Age profiles of archaeological *C. glaucum* populations from Troia. 85

Figure 51. Marginal incremental zones of archaeological *C. glaucum* from different occupational phases in Troia. 86

Figure 52. Marginal incremental zones of archaeological *C. glaucum* valves from different excavation units in Troia. 87

Figure 53. Incremental growth pattern of the *C. glaucum* specimens sampled for isotopic analysis. 89

Figure 54. Oxygen and carbon isotopic composition of incremental zones from two *C. glaucum* specimens from Troia. 90

Figure 55. The proportion of mollusk remains in the faunal assemblages of Yenibademli. 92

Figure 56. Changes in the relative abundance of mollusk species represented in Yenibademli (NIS). 93

Figure 57. Typical taphonomy of various mollusk specimens from Yenibademli......94

Figure 58. Glycymerididae specimens from Yenibademli. 95

Figure 59. Taphonomy of various mollusk specimens from Yenibademli......96

Figure 60. Live collected (Beşik Bay, 2004) *A. tuberculatum* specimen with abraded umbo......96

Figure 61. Morphology of the *P. caerulea* populations in Yenibademli (Distributions)......97

Figure 62. Morphology of the *P. caerulea* populations in Yenibademli by occupational phase (ANOVA)......98

Figure 63. Morphology of the *C. glaucum* populations in Yenibademli (Distributions)......99

Figure 64. Morphology of the *C. glaucum* populations in Yenibademli by chronological period (ANOVA). 99

Figure 65. Proportion of the mollusk remains in the faunal assemblages of Ulucak. 100

Figure 66. Changes in the relative abundance of mollusk species represented in Ulucak (NIS)......101

Figure 67. A *C. rustica* and a *C. glaucum* specimen from Ulucak. 101
Figure 68. Context CSP in Ulucak 102
Figure 69. Morphology of the *C. glaucum* specimens in Ulucak by chronological period (Distributions). 103
Figure 70. Comparison of the morphology of the *C. glaucum* specimens in Ulucak by chronological period. 103
Figure 71. Comparison of the valve heights of *C. glaucum* from Troia, Yenibademli, Ulucak, and the modern populations from the Troad 108
Figure 72. Comparison of the solidity of *C. glaucum* from Troia, Yenibademli, Ulucak, and the modern populations from the Troad (ANOVA). 109
Figure 73. Comparison of the curve of *C. glaucum* from Troia, Yenibademli, Ulucak, and the modern populations in the Troad (ANOVA). 110
Figure 74. The relationship between curve (y) and solidity (x) indices for *C. glaucum* populations from Troia, Yenıbademli, and Ulucak, and modern *C. glaucum* populations in the Troad. 111
Figure 75. Comparison of the demographic structures of modern and archaeological *C. glaucum* populations. 112
Figure 76. Comparison of shell growth achieved at the completion of each incremental zone in modern and Bronze Age *C. glaucum* populations in the Troad. 112
Figure 77. Comparison of the character of the marginal incremental growth zone in modern and archaeological *C. glaucum* populations. 112
Figure 78. Morphological differences between *C. glaucum* and *C. edule* 142

List of tables

Table 1. Global climate in the Holocene and its regional reflections in the eastern Mediterranean.......... 14

Table 2. Chronology of Troia.......... 21

Table 3. Chronology of Ulucak.......... 26

Table 4. Morphological properties of the modern *Patella* populations in Saros Bay.......... 45

Table 5. Relationship between valve height and age in modern *C. glaucum* populations in North Atlantic and the Mediterranean.......... 51

Table 6. Characteristics of *C. glaucum* collection stations in the Troad.......... 53

Table 7. Method of growth stage determination.......... 59

Table 8. Some special mollusk accumulations in archaeological deposits of Bronze Age Troia.......... 72

Table 9. Some examples of the differences between quantified archaeomalacological results and excavators' accounts of mollusk remains in Troia.......... 73

Table 10. Oxygen and carbon isotopic composition of carbonate samples from two Bronze Age *C. glaucum* specimens.......... 89

Table 11. Overview of archaeomalacological assemblages from other sites in the Aegean.......... 113

Table 12. Nutritional values of commonly represented mollusk taxa.......... 125

Table 13. Temperature and precipitation at the nearest stations to Troia, Yenibademli, and Ulucak.......... 165

Table 14. Sea surface temperatures off the Çanakkale coast.......... 165

Table 15. Systematic list of the mollusk taxa represented in the archaeomalacological assemblages.......... 166

Table 16. Scientific and common names of the mollusk taxa represented in the archaeomalacological assemblages.......... 167

Table 17. Morphology of modern *C. glaucum* populations in the Troad.......... 168

Table 18. Faunal assemblage of Troia (All occupational phases).......... 169

Table 19. Faunal assemblages of the Early Bronze Age Troia.......... 170

Table 20. Faunal assemblages of the Middle Bronze Age Troia.......... 171

Table 21. Faunal assemblages of the Late Bronze Age Troia.......... 172

Table 22. Faunal assemblages of the Troia VIIb and VIIa/b periods.......... 174

Table 23. Faunal assemblages of the Post-Bronze Age Troia.......... 175

Table 24. Mollusk assemblages of Troia (All occupational phases).......... 177

Table 25. Mollusk assemblages of the Early Bronze Age Troia.......... 178

Table 26. Mollusk assemblages of the Middle Bronze Age Troia.......... 179

Table 27. Mollusk assemblages of the Late Bronze Age Troia.......... 180

Table 28. Mollusk assemblages of Troia VIIb and VIIa/b.......... 181

Table 29. Mollusk assemblages of the Post-Bronze Age Troia.......... 182

Table 30. Morphology of *C. glaucum* populations from Troia.......... 184

Table 31. Faunal assemblages of Yenibademli.......... 187

Table 32. Mollusk assemblages of Yenibademli.......... 188

Table 33. Morphology of *P. caerulea* populations from Yenibademli 189

Table 34. Morphology of *C. glaucum* populations from Yenibademli 190

Table 35. Faunal assemblages of Ulucak. 191

Table 36. Mollusk assemblages of Ulucak 192

Table 37. Morphology of *C. glaucum* assemblages from Ulucak 193

List of abbreviations

° C	Celsius degree
AD	Anno domini
ANOVA	One-way analysis of variance
B	Breadth
BC	Before Christ
Beh.	Behälter
BP	Before present
ca.	Circa
cal.	Calibrated
cf.	Confer
cm	Centimeter
DIC	Dissolved organic carbon
DO	Dissolved oxygen
e.g.	Exempli gratia
EBA	Early Bronze Age
et al.	et alibi
g	Gram
H	Height
i.e.	id est
K	Kelvin
kcal g^{-1}	Kilocalories per gram
km	Kilometer
L	Length
l	Liter
LBA	Late Bronze Age
m	Meter
MBA	Middle Bronze Age
mg	Miligram
mm	Millimeter
MNI	Minimum number of individuals
N	North
NIS	Number of identified specimens
P	Precipitation
PBA	Post-Bronze Age
PDB	*Belemnitella Americana* (Pee De. Fm.)
per. com.	Personal communication with
per. obv.	Personal observation
pH	Potential hydrogen
SST	Sea surface temperature
T	Temperature
VH	Valve height
VI E	Early Troia VI
VI E/M/L	Early, Middle, or Late Troia VI
VI L	Late Troia VI
VI M	Middle Troia VI
VL	Valve length
Vol.	Volume
WIS	Weight of identified specimens

Chapter 1. Introduction

Shelled aquatic mollusks constitute one of the most abundant elements of marine and continental macrofauna. Extant archaeological evidence shows that humans exploited these animals as a food resource at least since the Middle Palaeolithic Age, ca. 130,000 – 30,000 BP (Bailey and Parkington 1988, Claassen 1998, Walter et al. 2000, Waselkov 1987). Numerous shell-bearing archaeological sites along the coasts and rivers of the world explicitly manifest the intensive human use of this resource during the Holocene (Bailey and Parkington 1988, Claassen 1998: 2, Erlandson and Fitzpatrick 2006, Yesner 1984).Part of the evidence concerning the human use of shelled aquatic mollusks in the Holocene comes from the Bronze Age (ca. 3000-1000 BC) habitation sites in the Aegean region.

The present study investigates the aquatic mollusk assemblages from archaeological excavations at three settlement mounds in the Aegean; namely Troia, Yenibademli, and Ulucak (Figure 1), by way of applying a series of analytical techniques to the sites' archaeomalacological record and evaluating the results in view of the existing scientific knowledge concerning the archaeology of these sites in specific and the Aegean Bronze Age in general.

The two main aims of this study are; (1) to explain the structure of shellfishing activity; and, (2) to elucidate the environmental (including climatic), cultural, and economic stimuli affecting change and continuity in the nature of shellfishing as an economic activity, and the role of shellfish as an element of human nutrition at the sites chosen for study. The results will contribute to the understanding of subsistence strategies and the coastal environments in the Aegean Bronze Age.

1.1. Archaeomalacology

The term "archaeomalacology" has become widespread rather recently. Therefore, it is necessary to explain what is meant by archaeomalacology before discussing the archaeomalacological record. In briefest terms, archaeomalacology is the study of archaeological mollusk remains (Bar-Yosef 2005b). In a similar manner to archaeozoology and archaeobotany, it combines the methods and approaches of archaeology and biology. In archaeomalacology, it is archaeology and the specialized zoological field of malacology[1], the study of mollusks, which are merged in order to conduct research with a broad array of goals.

Archaeomalacology was born from the necessity of specialization in the study of mollusks in archaeological contexts since other disciplines dealing with past cultures not only often marginalized the importance of invertebrates in creating the archaeological record but also fell short of the biological knowledge required to interpret archaeological mollusk remains. While archaeomalacology can be regarded as part of archaeozoology, as a discipline concerned with archaeological faunal assemblages, archaeozoology tends to place relatively more emphasis on the study of the characteristically more ubiquitous and abundant vertebrate remains. Mollusks, while constituting an integral element of archaeological faunal assemblages, are invertebrates and bear properties that are distinct from those of vertebrate remains. These properties include their differential biology and ecology, the chemistry and morphology of their shells, and the value and meaning of both their flesh and shells for human populations.

The burgeoning institutionalization of the field is evident in the scientific literature. Many specialized studies dealing with archaeological mollusks do not identify themselves with the term "archaeomalacology." In the classic series *Cambridge Manuals in Archaeology,* for example, Claassen's textbook about archaeomalacology appears under the title *Shells* (1998), while other volumes of the same series that pertain to organic remains in archaeology appear under titles such as *Zooarchaeology* (Reitz and Wing 1999) or *Vertebrate Taphonomy* (Lyman 1994). This and other similar situations reflect both the state of research about archaeological mollusk remains and the lack of a formalized corpus of scientific study that envelops this research.

The scope of recent and historical research on archaeological mollusks attests to the diversity of approaches embodied by the discipline. It is possible to summarize these approaches under two categories that occasionally overlap: (1) approaches dealing with the primary biological properties of mollusk remains; and, (2) others focusing on the secondary uses of the mollusk shells in human societies. Studies of intrusive and natural occurrences of terrestrial snails in archaeological deposits and the landscapes that surround them, as well as research focusing archaeological mollusk remains that represent past human consumption of the molluscan soft-body can be included in the first category (e.g., Evans 1972, Quitmyer et al. 1997). These studies aim at palaeoeconomic and palaeoenvironmental reconstructions of past human existence. Studies of culturally modified or imported mollusk shells with the aim of explaining the development of human cognition, technology, art, and trade can be included in the second category (e.g., Bar-Yosef 1999, Chilardi et al. 2005). Proceedings of recent archaeomalacological conferences demonstrate the wide array of subjects that fall under both of these categories (Bar-Yosef 2005a, Quitmyer and Szabo 2008).

It was, though, long before the recent designation and rapid development of archaeomalacology that the aquatic mollusk remains were acknowledged as valuable tools in illustrating past human activity and environment. The discovery, or

[1] The term is combined of the Greek words *malacos* for soft and *logos* for study. Malacology differs from conchology, which is the branch of zoology studying the shells of mollusks.

rather the initial documentation of "shell-bearing sites" (Widmer 1989 cited in Claassen 1998: 12) along the river banks and coasts of the world has led to the development of specific analytical techniques to explain the environment and the culture of the peoples who were responsible for the formation of these sites. Large-scale interdisciplinary studies conducted on the matrices of Mesolithic shell middens of Northern Europe stimulated archaeologists' interest in mollusk remains as early as the second half of the 19th century (Bar-Yosef 2005b: 2, Claassen 1998: 2-3, Dall 1877). The term "shell midden archaeology" is often used to denote the body of knowledge obtained from the investigations of such sites and their study. Although shell midden archaeology primarily focuses on the study of mollusk remains, its study material is not limited to mollusks.

The theories and applications of "New Archaeology" triggered the wide-range integration of scientific approaches into archaeology in the 1960s (Renfrew and Bahn 1996: 37). These theoretical and methodological developments stimulated a scientific paradigm that was maturing for the in-depth exploration of the potential of archaeological mollusk remains with interdisciplinary methods and approaches. In 1969, N.J. Shackleton published a short article summarizing the analytical techniques that can be used to reconstruct palaeoenvironmental and palaeoeconomic conditions from archaeological mollusk remains. In this milestone article, he introduced the idea and method of palaeotemperature reconstruction from the stable isotopic ($\delta\ ^{18}O$) compositions of aquatic mollusks from archaeological contexts (Shackleton 1969). In England, Evans' (1972) volume on the possibilities of interpreting terrestrial mollusks from archaeological contexts triggered the development of archaeomalacology in Anglo-Saxon countries.

Shell midden archaeology in Northern Europe, the British Isles, the Americas, Australia, Japan and South Africa has developed rapidly since the 1970s. Studies in these regions have focused on diverse research questions including the beginnings of human adaptations to coastal landscapes (e.g., Erlandson 2001); the impact of agriculture on human dependence on mollusk resources (e.g., Deith 1988); seasonality of shellfish gathering (e.g., Claassen 1983, 1986, Deith 1983a, 1983b, Deith and Shackleton 1986, 1988, Koike 1980, Milner 2001, Quitmyer et al. 1997, Shackleton 1973, 1974) and, indirectly, seasonality of site use (e.g., Deith 1989); modes of foraging behavior (e.g., Buchanan 1988); human impact on molluscan populations (e.g., Mannino and Thomas 2001, 2002, Yesner 1984, 1988); effects of changing coastal geomorphologies and sea-levels on the chronological patterns of shellfishing activity (e.g., Jerardino 1997); and coastal industries that relied on shellfish as a raw material (e.g., Henshilwood 2001, Reese 1980).

Coevally, analytical techniques applied on archaeomalacological material have been improved. Emphasis has been put on the importance of observations on modern mollusk populations to build analogies with ancient populations in seasonality studies and palaeoenvironmental reconstruction (e.g., Koike 1980, Claassen 1988, 1993, Quitmyer et al. 1997, 2005, Shackleton 1970). Ethnoarchaeological investigations have been successfully used as analogies to test hypotheses concerning the role of shellfishing in past societies (e.g., Meehan 1977, 1982, 1983, Thomas 2002). Researchers have started to pay greater attention to the taphonomy of mollusk remains to distinguish culturally modified specimens from specimens representing food refuse (e.g., Light 2005). Comparative biometrical investigations of ancient and modern mollusk shell material became more common and were primarily used to disclose the effects of human activity on mollusk populations (e.g., Cabral and Silva 2003, Mannino and Thomas 2001, 2002, Marelli and Arnold 2001, Yesner 1984). Today, researchers increasingly promote the necessity of multi-dimensional approaches to archaeomalacological questions and the use of a broad spectrum of analytical techniques (e.g., Claassen 1998, 2000, Jones and Quitmyer 1996, Mannino and Thomas 2001, 2002, Quitmyer et al. 1997).

The approach the present study adopts is to treat archaeological mollusk remains as elements of biogenic sediments filtered through human activity: they represent an environment common to human and mollusk populations alike. What is sought after is an understanding of environmental and cultural phenomena affecting and resulting from this mutual existence and the formulation of the structure of human-mollusk-environment relationships in the sites under focus. In this sense, the study should be included in the above mentioned biologically-oriented category of archaeomalacological approaches.

1.2. Archaeomalacology in the Aegean region

It would not be correct to claim that recent and past theoretical and analytical developments in the study of archaeological mollusks have had a significant influence on the archaeology of the Aegean. Despite their ubiquitous and often conspicuous presence, mollusk remains recovered in the archaeological sites along the Aegean coast or the immediate hinterland received negligible attention from the scientific community involved with the archaeology of the Aegean, often including those schools of research focusing on and prioritizing issues related to regional subsistence economies and palaeoenvironment.

This situation can be considered as part of a worldwide tendency to downplay the dietary and economic role of shellfish for prehistoric societies. Mollusks are regarded by some archaeologists as a universally low-priority resource (Moss 1993). This tendency coincides with the general trend to neglect the role of foraging activities in subsistence economies based on agriculture and animal husbandry (Deith 1985, 1988a). Influential works addressing Aegean prehistory using environmental approaches failed to consider mollusk remains as a useful part of the archaeological record. For instance, one of the most frequently cited works that handle settlement patterns in prehistoric Greece does not touch upon shellfishing as an element of the relationship between human settlers and the coastal environment in Aegean prehistory (Bintliff 1977). In Bintliff's work only fishing has been taken into consideration as a means of exploiting coastal resources (Bintliff 1977: 113-116).

On the other hand, archaeomalacological reports in the form of inventories are numerous in the literature dealing with Aegean prehistory. Most of them appear either as amendments to site reports (e.g., Buitenhuis 1995, Coy 1977, Higgs and Clench 1962, Lamb 1936: 215, Shackleton 1972) or as sections in site-specific archaeozoological studies (e.g., Boessneck and Schäffer 1986, Gejvall 1969, Peters 1983, and von den Driesch 1999). The few studies which were published as individual articles with archaeomalacological view points are restricted in content to quantities of specimens identified by taxon per archaeological context, without presenting an interpretation of the remains (e.g., Reese 2006). Nevertheless, the quantified accounts presented by these works often constitute the sole and important source of archaeomalacological data of the region.

Such reports and some unpublished data were compiled by Karali (1999) in an attempt to give a comprehensive account of the archaeomalacological evidence in Aegean prehistory. This monograph adopted a holistic approach to the archaeomalacology of the Aegean, paying equal attention to both culturally-modified and unmodified mollusk remains from archaeological deposits, as well as pictorial and plastic depictions of the mollusks in Aegean art and daily life.

The works of J.C. and N.J. Shackleton and Deith, on the other hand, mark an obvious methodological and conceptual divergence from general trends in Aegean archaeomalacology (Deith 1988a, Shackleton J.C. 1983, 1988a, 1988b, Shackleton N.J. 1968a, Shackleton and Deith 1988). When N.J. Shackleton reported on the shell assemblage from the Neolithic site of Saliagos in 1968, this became the first study in the region to ponder about palaeoeconomical issues by analyzing the morphometric features of archaeological mollusk remains. In 1970, the same researcher presented the results of isotopic ($\delta\ ^{18}O$ and $\delta\ ^{13}C$) analyses on a small sample of *Cerastoderma glaucum* from the Neolithic site of Nea Nikomedeia in Greece and suggested that the nature of coastal environments, pluvial regimes, and the time of shellfish gathering could be reconstructed from the shells of this species, if a sufficiently sized sample could be analyzed. J.C. Shackleton evaluated diachronic changes in the species spectrum of the shell assemblage of the Franchthi Cave in Greece together with the reconstructions of past shorelines in order to determine the environmental factors affecting shellfish gathering at this site (1983, 1988a, 1988b). In an appendix to J.C. Shackleton's monograph on the mollusk remains from Franchthi (1988a), Deith and N.J. Shackleton (1988) presented their interpretations on the seasonal patterns of shellfishing activity at the site using the isotopic compositions of archaeological and beach-collected shells. In addition to these complementary works, Becker (1986: 232-237) showed that temporal changes occurred in the relative abundance of *Unio* spp. and *C. glaucum* remains in Kastanas (Greece) and demonstrated that the valves of both taxa became smaller in length and width throughout the occupational phases of the settlement. Both of these observed changes in the archaeomalacological assemblages were related to environmental and economic factors, but space for a detailed interpretation of these results was not spared in this volume (Becker 1986).

Only a scattering of studies that employed methods other than species identification on mollusk remains followed these works. To the knowledge of the author, only one morphometric study was carried out in the 1990s on any archaeomalacological material from the Aegean. In his M.Sc. thesis, Andreasen (1998) attempted to reveal the relationship between the valve heights of the *C. glaucum* specimens from the Neolithic Makriyalos and past ecological conditions that prevailed in the past aquatic environments near the site. This work was never published. In addition, in 2005, Prummel assayed the impact of human forager populations on the coastal environment by comparing the morphometric attributes of archaeomalacological specimens from Greek sites.

As it can be easily deduced from the overview of the archaeomalacological studies in the Aegean presented above, there is a clear lack of attempts to explain the archaeomalacological record of the Aegean, either using advanced analytical techniques (microscopic, isotopic, taphonomic, or biometric) or using quantitative accounts based on species identifications of the mollusk remains. While studies of archaeomalacological assemblages with sundry research questions from both sides of the Aegean are in progress (Per. com. L. Karali, J. Peters, D. S. Reese, T. Theodoropoulou, and E. Veropoulidou), for the time being it should be admitted that the true potential of the Aegean archaeomalacological assemblages in reconstructing palaeoenvironmental and palaeoeconomic patterns remain largely unexplored.

1.3. Research objectives

After having discussed the methods and approaches of current archaeomalacological research in the world and the state of archaeomalacology in the Aegean region, it is now appropriate to explain the objectives of the present study in greater detail and to introduce the methodological approach adopted in this study.

As stated above, one of the main objectives of the present study is to explain the structure of shellfishing activity at the three sites under investigation. In order to reveal the structure of shellfishing activity, recurring patterns representing long-term trends, as well as the probable representation of short-term events and singularities are investigated in the archaeomalacological record. Shellfishing is considered as one of the several archaeologically recorded aspects of the relationship between human habitation and the surrounding environment that affects both realms.

The most obvious and tangible remnants of shellfishing activity are the remains of mollusk shells; the primary material of this study. The present study takes two basic principles into consideration:

(1) The modality of human exploitation to which animal populations is subjected will leave its traces on the archaeological representation of these populations (Ballbè 2005, Reitz and Wing 1999: 315).
(2) Archaeological mollusk shells are archives of the environments of human populations who exploited the mollusk populations (Claassen 1998: 122-142, Richardson 2001).

In order to reveal the signatures of human behavior and environment in the archaeomalacological method, a series of methods are used. The relative abundance of taxa in the faunal assemblages of the sites, their archaeological context, taphonomic condition of mollusk specimens, variations and continuing patterns in shell morphology, shell microstructure (i.e., incremental growth pattern), and the chemistry of mollusk remains are analyzed. Based on the results of these analyses, various facets of shellfish consumption and shellfishing such as: (1) the contribution of shellfish to the animal-based diet; (2) prey choice in shellfishing; (3) gathering locations; (4) methods of procurement and processing; (5) intensity of shellfishing activity; (6) the function of shellfishing in multi-pronged economies at Bronze Age Troia, Yenibademli, and Ulucak; and, (7) the annual schedule of cockle (*C. glaucum*) gathering activity at Troia are investigated. The results are evaluated in light of current knowledge about the palaeoenvironment and material cultures of these archaeological sites.

An understanding of temporal and spatial patterns in these facets of shellfish gathering and consumption will form the basis for deciphering the factors which affected them. Factors affecting the economic value of shellfish and the nature of shellfishing may include: (1) altering demographic structures of the individual sites and related cultural innovations; (2) organization and needs of local economies; (3) changes in coastal geomorphology and sea-levels; (4) fluctuations in regional and global climate; (5) seasonal pattern of environmental conditions as well as human activities; and, (6) human foraging pressure on shellfish populations. In addition, the palaeoenvironmental data extracted from archaeomalacological material will be utilized to complement the current understanding of past environmental conditions at the primary sites discussed in the study.

The results and interpretations of the study are reviewed in view of published archaeomalacological data from contemporary sites in the Aegean to attain a broader perspective of shellfish and shellfishing in the Bronze Age Aegean. Ethnographic accounts pertaining to human exploitation of aquatic mollusks bring additional valuable insights to the interpretation of the archaeomalacological record. The accuracy of the interpretations about shellfishing and the reconstructions of the environmental factors are calibrated by the ecological relationships observed in present-day natural phenomena. This calibration is consolidated by modern laboratory and fieldwork conducted on geographically coherent extant mollusk populations.

1.4. Organization of the study

Following the present chapter, Chapter 2 will contextualize the study by describing the archaeology and environment of Troia, Yenibademli, and Ulucak. Chapter 3 will give extensive information on the methods used in this study and the rationale involved in adopting them. Chapter 4 will provide the details of analytical results, compare the properties of the mollusk assemblages from the three selected sites, and review published archaeomalacological accounts from sites in the Aegean region. Chapter 5 will define the different aspects of shellfishing activity and the nutritional value of the different mollusk species identified at the primary sites under study. Chapter 6 will present explanations for the observed structure and conjunctures of the archaeomalacological record. Finally, Chapter 7 will review the results and interpretations of the study, evaluating them with reference to our understanding of Bronze Age economy and environment in the Aegean, and consider prospects for further studies of archaeomalacology in the region.

An appendix contains the extensive tables referred to in the text. The complete numerical data of morphometric measurements taken on *Cerastoderma glaucum* and *Patella caerulea*, as well as data obtained from the examinations of the incremental shell growth of *C. glaucum* have been published online. These and high-resolution formats of Figures 30, 31, 48, 49, and 53 are available for viewing and download at http://opencontext.org/projects/, "Aegean Archaeomalacology".

Figure 1. Map of the Aegean and the location of the place names mentioned in the text.

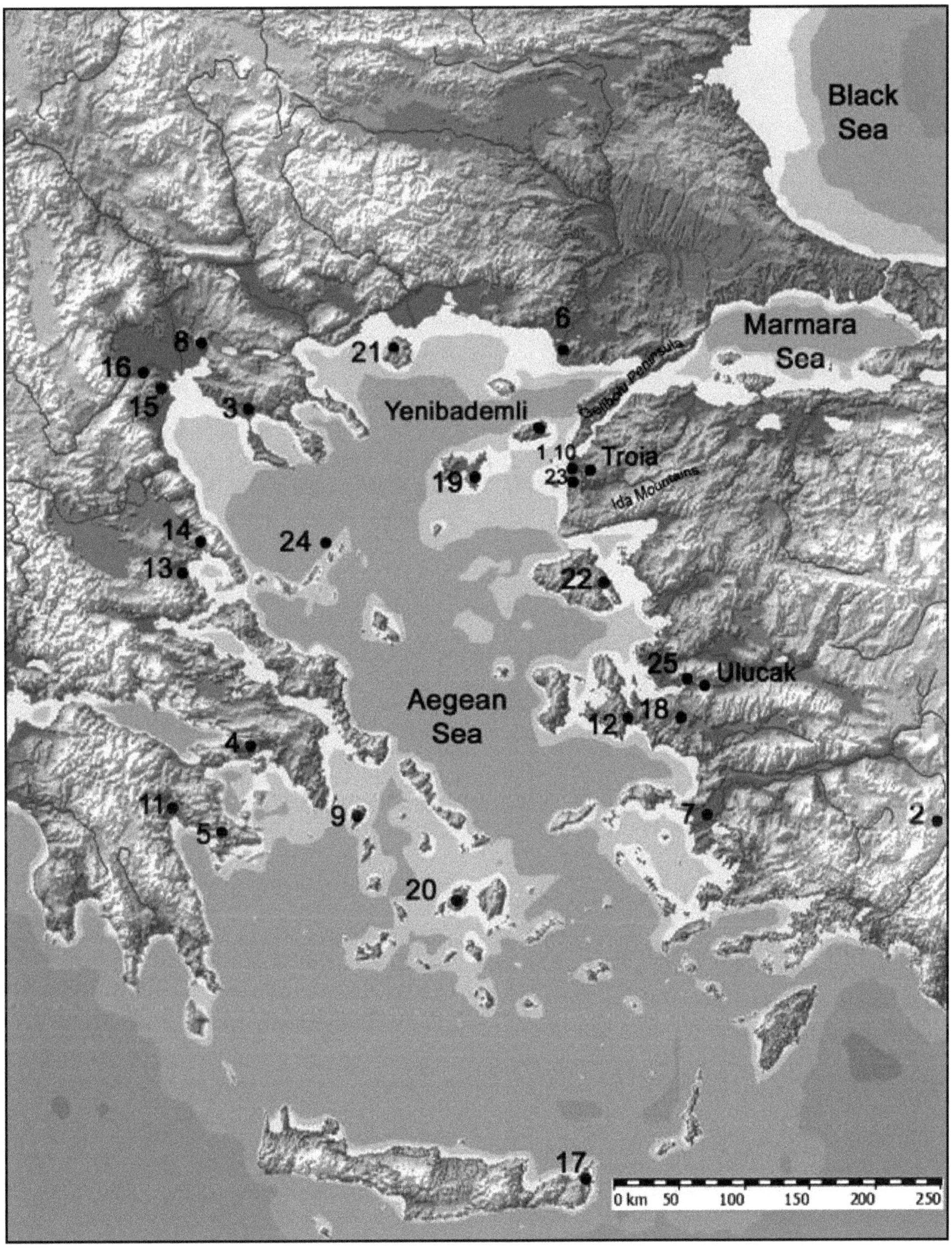

1. Álacalıgöl
2. Aphrodisias
3. Ayios Mamas
4. Eleusis
5. Franchthi
6. Hocaçeşme
7. Kalabaktepe
8. Kastanas
9. Kephala
10. Kumtepe
11. Lerna
12. Limantepe and Baklatepe
13. Magula Pavlina and New Halos
14. Magula Pevkakia
15. Makriyalos
16. Nea Nikomedia
17. Palaikastro
18. Panaztepe
19. Poliochni
20. Saliagos
21. Skala Sotiros
22. Thermi
23. Yassıtepe and Sivritepe
24. Youra
25. Zeytintepe

Chapter 2. Archaeology and environment

The aim of this chapter is to review the extant information concerning the archaeology and environment of the three sites that are investigated in this study. This review serves two purposes: firstly, it sets the primary material of the study in a geographical and archaeological context, and secondly, it provides the relevant climatic, palaeoenvironmental, cultural, and palaeoeconomic data, which will be later during the discussion of shellfishing and consumption (in Chapters 6 and 7) synchronized with the diachronic patterns of archaeomalacological record from these sites.

2.1. Location and current geography

The three primary archaeological sites that are investigated in this study are situated at diverse geographical surroundings in the northern Aegean. This situation offers the opportunity to exemplify the differences and similarities of human adaptive strategies in diverse geographical settings within a single study.

2.1.1. Troia

Troia (39° 58'N, 26° 13'E) is located in northwestern Anatolia in the Turkish province of Çanakkale (Figure 1). The landscape surrounding Troia is dominated by low plateau ridges, alluvial plains that lie between them, the Karamenderes (Skamander) and the Dümrek (Simoeis) rivers that feed these plains, the Strait of Dardanelles, and the Aegean Sea (Figure 2). The low plateaus consist of the Sigeon Ridge to the west, the Kumkale Ridge to the north, and the Troia Ridge, where the site is located at ca. 30-40 m elevation from the present sea-level. The Karamenderes River flows between the Sigeon and Troia Ridges, whereas the Dümrek lies between the Kumkale Ridge and Troia, flowing parallel to the Karamenderes River towards the north. The Aegean Sea falls to the west of the Sigeon Ridge. The southern end of the Dardanelles, the narrow marine bridge that connects the Black Sea and the Marmara Sea with the Aegean, lays ca. six km north of the Troia Ridge. This ridge is approximately 40-80 m high and runs on an east-west axis. The two ridges to the west and northeast of the Troia Ridge separate the joint Karamenderes and Dümrek flood plains from the Aegean and the Dardanelles respectively. Both the Sigeon Ridge and the Kumkale Ridge have steep slopes facing the sea, while they have steep gradients that gently let onto the Karamenderes-Dümrek Plain, forming indentations. The Ida Mountains constitute the most important geographical feature in the hinterland.

2.1.2. Yenibademli

Yenibademli (40° 13' N 25° 53' E) Höyük (i.e., settlement mound) is located in Northeast Gökçeada, formerly named Imbros, a volcanic island that lies off the coast of Gelibolu Peninsula (Figure 1). The mound is situated 1.7 km southwest of the present coastline, on the eastern bank of Büyükdere River, the largest river on the island (Figure 3). The floodplain of this river is flanked by steep limestone ridges to the east and west. The ancient settlement of Yenibademli was founded on a low outreach of the eastern limestone ridge. The mound proper covers an area of 130 m x 120 m, and rises about 18 m above the present day sea-level (Hüryılmaz 1998, Öner 2000, 2001).

2.1.3. Ulucak

Ulucak Höyük is situated about nine km east of the İzmir Bay (38° 27' N 27° 21' E), on the landlocked plain of Kemalpaşa, east of the Belkahve Pass, which separates it from the coastal plain of Bornova (Figures 1 and 4). The main water source of the Kemalpaşa Plain is the perennial Nif Stream (Kayan 1999). The plain is located in the east-west lying depression between the Spil Mountains to the north and the Nif Mountain to the south. The mound is ca. 221 m above sea-level and its surface area measures ca. 105 m x 300 m (Çilingiroğlu et al. 2004: 2, Çilingiroğlu and Abay 2005: 6).

As it is evident in the descriptions made above, the primary sites under study are marked by their geographical positions: Troia represents a coastal settlement on the Anatolian mainland; Yenibademli represents an island site; and Ulucak represents a site in the immediate hinterland of the Aegean coast. The geographical features in their immediate vicinity likewise differ in their particularities. Important to the purposes of the present study is how these geographical features have evolved in the course of the Middle and Late Holocene.

2.2. Palaeogeography

The features of the physical environment determine the elements and characteristics of the biological environment, which constitutes the resource base of subsistence systems. Thus, the economy of human groups has been significantly influenced by the stable and changing characteristics of the landscapes surrounding habitation sites and areas. The most important geographical factor that affects the general interpretation of the modes of shellfishing and shellfish consumption at the sites that are discussed in the present study is naturally their geographical position vis-à-vis the bodies of water used for resource procurement. Troia and Yenibademli represent two settlements that lie directly on the Aegean coast; both lacking any major obstacle to shellfishing grounds. Ulucak, on the other hand, is a shell-bearing habitation site located on a plain that had no direct contact with the coast during the Quaternary Period. Another very important factor that affects the interpretations is related to the character of the water bodies that were used for resource procurement.

Figure 2. Evolution of the Trojan and Beşik Bays (Modified after Kayan et al. 2003: 397, Figure 7 and Göbel 2005: 193, Abbildung 54).

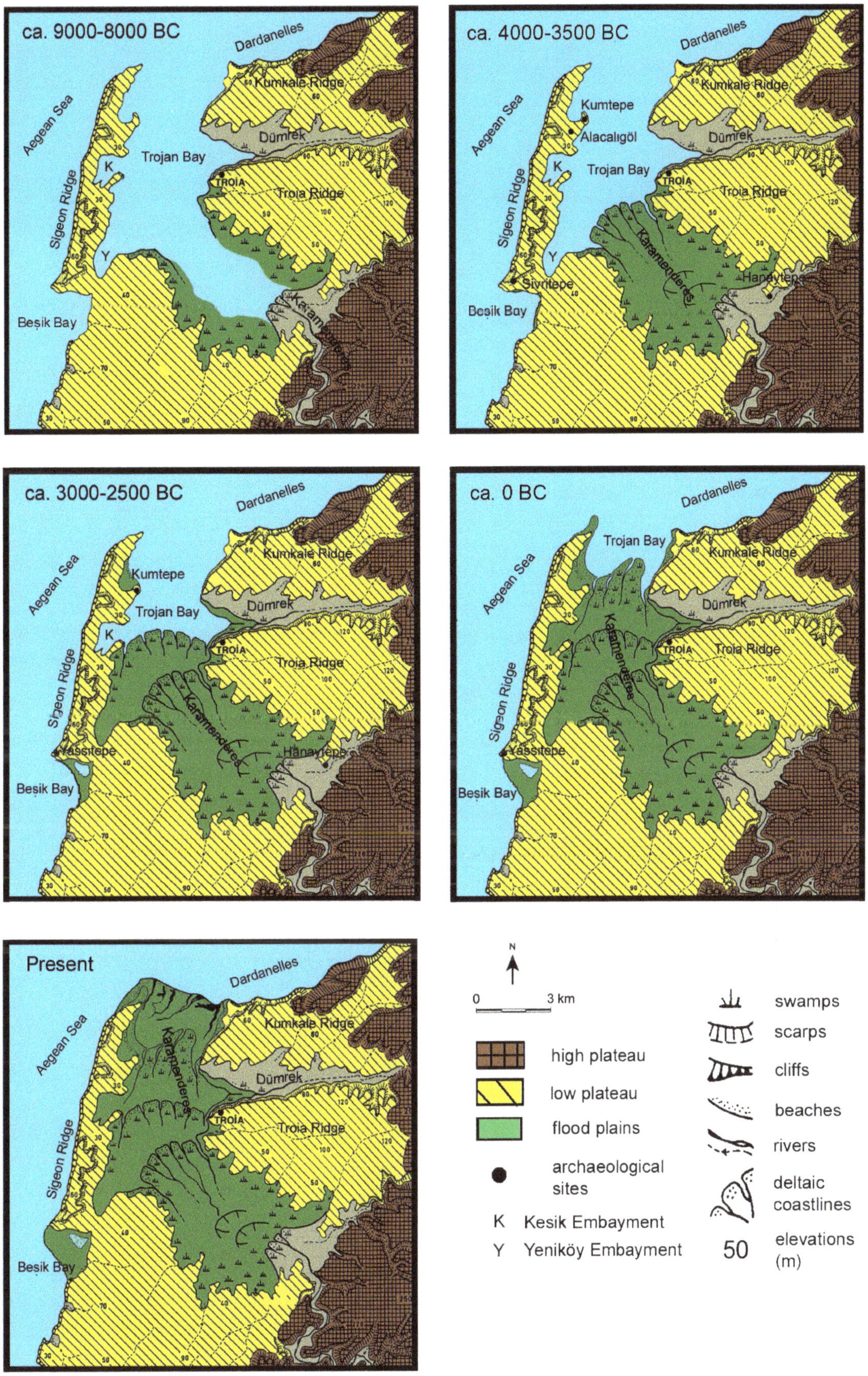

Changes in the geomorphology of river plains, the location of coastlines, and the character of bays and river courses have a direct influence on the available aquatic mollusk biomass, and therefore on shellfishing patterns.

The coastlines and river plains of the Mediterranean went through extensive changes in the course of the Holocene (Belknap and Mart 1999, Brückner et al. 2005, Erol 2003, Galili et al. 2005, Kayan 1988, Kelletat 2005, Kraft et al. 2003a and 2003b, van Andel and Shackleton 1988). Although each coastal plain in the Aegean has a biography of its own, the factors that took part in their formation process are similar (Kraft et al. 2003b): fluctuations in eustatic sea-levels (i.e., global changes in the volume of water or changes in the volume of ocean basins), vertical tectonic movements of the continental plates, and sedimentary processes are the main factors (Rapp and Kraft 1979). Since most of the former embayments in the Aegean and their environs were stages for human occupation from at least the Neolithic period onwards, rapid formation of the coastal plains is also thought to have been at least partially reinforced by land erosion caused by anthropogenic manipulation of the landscape through land clearance, terracing, abandonment of former fields, and herding activities (Butzer 2005, Roberts 1991, 2004).

Eustatic sea-levels, which had dropped down to ca. 120-130 m lower than present day levels during the Last Glacial Maximum (ca. 18000 BP), began to rise with the onset of warmer climate at the beginning of the Holocene at ca. 12000 BP (Flemming et al. 1998, van Andel and Shackleton 1988). As a result, seawater rapidly inundated the low-lying continental shelves and indentations, forming marine embayments and islands in the Aegean (van Andel and Shackleton 1988). Sea-level rise continued at a rapid, yet steady pace, causing a counter force against delta progradation due to sediment accumulation until about 6000 BP (Brückner et al. 2005, Kayan et al. 1991, 1997a, 2003). After this date, due to the stabilization of the eustatic sea-level rise at a slower paste, terrestrial deposits began to silt up the marine embayments more effectively, forming large coastal plains that expanded towards the Aegean (Brückner et al. 2005, Kayan 1991, 1996, Kayan et al. 2003). While some researchers are of the opinion that a general drop in the relative sea-levels (i.e., caused by vertical motion of the regional land mass) up to 4.5 m below present-day sea-level accelerated the rate of alluvial accumulation at the Aegean coastal plains ca. 4000-3500 BP (Kayan 1988, 1991, 1996, 1997b, Kayan et al. 2003, Kelletat 2005, Öner 2000, 2001), others maintain that the accumulation of terrestrial sediments in the coastal plains of the Aegean after 6000 BP was not affected significantly by a drop in the sea-level (Brückner et al. 2005). Additionally, while some geomorphological studies indicate that the nature of fluctuations in the relative sea-levels in the Eastern Mediterranean differ from one location to the other (Kayan 1997b), cumulative studies suggest that their rough synchronization in the Late Holocene should be further investigated in order to reach conclusions about a common nature of their origin (Kelletat 2005). Current evidence, however, suggests that the sea-levels of the Middle and Late Holocene Aegean should be handled individually for each local case.

The evolution of coastal landscapes in the Troad[2], the sedimentation process of the Büyükdere River at Yenibademli, and the changing extent of the Bornova Plain, as well as the location of shifting river beds of Nif (Nyphaion) Stream near Ulucak are issues of particular interest for interpreting the patterns observed in the archaeomalacological record of these settlements.

2.2.1. Geomorphic evolution of the Trojan and Beşik Bays/Plains

The palaeogeography of the Trojan Bay has been a concern of scholarly debate since antiquity, both as the potential site of the settlement's harbor and as an important feature of the landscape described by Homer (Strabo, *Geography*, Book 13, Chapter 1, Kraft et al. 1980). Contemporary scientific interest in this geography aims at reconstructing the altering Holocene coastal environment during the course of human occupation in the Troad and providing the physical setting for cultural history (Kayan et al. 2003, Kraft et al. 2003b).

At the time of the Last Glacial Maximum, when the sea-level was at its lowest, a river valley constituted the threshold between the Marmara Sea and the Aegean (Aksu et al. 2002). The Karamenderes was a trubitary of this river that was formerly located in the valley in what is now the Dardanelles (Kayan et al. 2003: 386). As the eustatic sea-level rose and the Strait of Dardanelles was formed, the seawater inundated the Karamenderes Valley. Until about 7000 BP, the coastline of what became a narrow estuary[3] continued advancing southwards, where it reached just north of Hanaytepe (Figure 2; Kayan et al. 2003). The sea-level remained stable at the height it had reached in ca. 7000 BP until around 6000 BP (Kelletat 2005). According to Kayan (1988, 1991, 1997b), a drop in the relative sea-level followed this stand and continued until 2000 BP. Between 6000 BP and 2000 BP, the deltas of the meandering Karamenderes and Dümrek shifted towards the Aegean, progressively pushing the coastline to the north towards its current location directly on the Dardanelles (Figure 2; Kayan 1995, 1996, Kayan et al. 2003, Kraft et al. 1980, Kraft et al. 2003a, 2003b).

This process created a very irregular deltaic coastline with coastal swamps, coastal barriers and lagoons as well as backswamps, floodplains, river channels and marshes, and an alluvial deposit as thick as 40 m at some locations (Kayan et al. 2003, Kraft et al. 1980, 2003a, 2003b, Rapp 1982). The river channels, the locations and characters of lagoons, and other coastal features shifted their locations and changed their characters frequently (Kraft et al. 2003a). The complexity, the thickness, and the fine-grained character of the alluvial deposits make it impossible to delineate single

[2] The Troad is the region of Troia, bounded by the Dardanelles to the northwest and separated from the rest of Anatolia by the Ida Mountains.
[3] The term "estuary" is here used to denote an arm or inlet of the sea at the lower end of a river.

morphological features and their chronology (Kayan et al. 2003, Kraft et al. 2003a,).

Since the habitats of benthic mollusk taxa are restricted to distinct environments, identifying the possible locations of shellfishing grounds largely depends on the knowledge of the former aquatic environments, including the benthic types that were in existence and their locations. Extensive geomorphological investigations in the Trojan Plain delineated the spatial and temporal alterations in the benthic characteristics of the former Trojan Bay:

During the early phase of the Holocene, until about 7000 BP, when the effect of the alluvial sedimentation was lower than the effect of marine water inundating the Karamenderes Valley, the bottom of the Trojan Bay consisted of coarse lag deposits, even at places near the location of the former delta at the southern end of the valley. Fine-grained and homogenous marine mud overlays this initial Holocene deposit forming the lowest substrate of the estuarine bay. (Kayan et al. 2003: 389). During the Middle Holocene, after ca. 7000-6000 BP, the sediments in the geomorphologic cores taken to the south of Troia lose their homogeneity and become variable, indicating the formation of a transitional zone between terrestrial and marine environments. In this transitional zone to the south of the site, coarse sandy bottoms imply water flow, while muddy bottoms with high organic content indicate swampy or seasonally wet environments, as well as deltaic distributary channels. Marine muds and coarse sandy coastal sediments dated to ca. 5000-4500 BP indicate that the sea reached directly to the north of the Troia Ridge at the time (Kayan 1996, Kayan et al. 2003). This marine environment was later (ca. 4500 BP) filled by colluvium from the Kumkale and Troia Ridges and alluvium from the Dümrek River, forming a strip of dry land suitable for human use (Figure 2; Kayan 1996). The sedimentation rate, which was very slow until this point in time (Kayan 1995, 1996, 1997a, 1997b, Kraft et al. 2003a), accelerated during the Late Holocene due to a drop in the relative sea-level, allowing alluvial deposits to cover the area of the Trojan Bay until ca. 2000 BP, when the deltaic coastline shifted to the position indicated in Figure 2 (Kayan et al. 2003).

The southerly two of the indentations of the Sigeon ridge to the west, the Yeniköy and Kesik embayments, were also invaded by seawater until about 5000 BP (Figure 2). Between 5000 BP and 3500 BP, wide areas of these small embayments gradually became dry land. According to Kayan et al. (2003), a ca. 2 m drop in the relative sea-level was the main cause of this development. ^{14}C dates indicate that marine conditions prevailed in the Kesik embayment up until 3500 BP (Kayan et al. 2003: 395). The marshy and swampy (freshwater) conditions prevailed in the Kesik embayment until the 1960s when it was reclaimed by anthropogenic activity and turned into arable land (Kayan et al. 2003: 395).

The general picture of the Trojan Bay was that of an estuary with marine characteristics prevailing towards the Dardanelles and with a quiescent character towards the delta areas of the Karamenderes and Dümrek Rivers of the time. The area of the estuarine bay progressively diminished; slowly during the Middle Holocene and faster during the Late Holocene, leaving its place to the joined floodplain of the Karamenderes and Dümrek Rivers. Today the only relicts of this river-deltaic lagoons-estuary system are the minute distributary channels of the Karamenderes and Dümrek Rivers and the lagoons they create. These brackish features are located directly on the coast of the Dardanelles.

The Beşik Bay, located to the southwest of the Sigeon Ridge and separated from the Trojan Bay by a low threshold, was another coastal area available for shellfishing in the Troad (Figure 2). The low plateau north of the bay has been host to two prehistoric settlements, Yassıtepe and Sivritepe; the former dating to ca. 3000-2600 BC and the latter to ca. 4600-4500 BC[4] (Figures 1 and 2; Korfmann 1984, 1985, 1986b, 1988, 2000, Seeher 1985). The Beşik Bay has long been considered to have been used as a natural harbor for Troia (Korfmann 1984, 1986a). Today this former bay is filled with sandy sediment, forming a plain about two meters higher than the present sea-level (Figure 2; Kayan 1991: 82-83). Unlike the former Trojan Bay, instead of river alluvium, colluvium coming down the slopes of the bay along with rainwater has been the source of sedimentation here (Kayan 1991: 86). At the time of maximum transgression (ca. 6000 BP), the sea inundated the embayment almost until the bottom of the sandstone slopes surrounding the bay (Kayan 1991: 88, Plate 6). The bay had its widest area between 7000-6000 BP, during the time of human settlement at Sivritepe (Kayan 1991: 89). Because of a previously formed sand barrier lying across the bay on a north-south axis, the eastern part of the bay became a lagoon ca. 5200 BP (Figure 2; Kayan 1991: 88, Plate 7). Black, silty, and clayey mud formed the substrate of this lagoon. Located to the west of the lagoon was a shallow marine bay. The bottom of this side of the bay consisted of coarse-sandy and gravelly sediments (Kayan 1991: 89). With the drop in the relative sea-level, beaches formed both at the eastern and western sides of the sand barrier between 5000 BP and 3500-3000 BP. Colluvial activity accelerated with time and both the lagoon and the marine bay gradually became narrower (Kayan 1991: 90, Plate 8). The present coastline is almost aligned with Yassıtepe to the north (Figure 2).

In summary, terrestrial sediments (alluvial and colluvial) gradually overtook the marine environments of the continental shelves in the Troad during the Middle and Late Holocene. This was a process which started ca. 5000-4000 BC and accelerated between 2500 BC and 0 BC. Around 5000-4000 BC, both the Trojan and Beşik Bays had their widest areas with coarse benthic types. By the beginning of human occupation in Troia, the southeastern part of the Trojan Bay was filled with alluvial sedimentary deposits and the deltaic river mouth of the Karamenderes was located directly to the southeast of the settlement. To the north of the settlement marine environments prevailed. Coevally, the

[4] Yassıtepe additionally contains Archaic layers, whereas Sivritepe is entirely covered with a Hellenistic tumulus (Korfmann 1984). The location is also the site of a Late Bronze Age cemetery contemporary with Late Troia VI and Troia VIIa (Basedow 2000).

Kumtepe embayment turned from a marine to a brackish environment, subsequently becoming dry land over the course of time between 5000 BC and 3500 BC. A lagoon formed at the Beşik Bay ca. 3200 BC at the latest. By 2500-2000 BC, a narrow strip of land had formed at the northern foot of the Troia Ridge. From about this date onwards, sedimentary processes accelerated. While the Yeniköy embayment gradually became dry land between 3000-1500 BC, the Kesik embayment retained its marine conditions until about 1500 BC. By 0 BC, the deltaic coastline of the Trojan Bay was ca. 3-4 km north of Troia and the lagoon and the marine embayment of the Beşik Bay had considerably diminished in area.

However, two problems are potentially embedded in the chronological interpretation of the evolution of geomorphological features in the Troad. Firstly, the time line of the geomorphic alterations is based on the radiocarbon dating of the shells of marine invertebrates; mostly shells of *Cerastoderma glaucum* (per. com. İ. Kayan). The results of the radiocarbon analyses are calibrated but they have not been corrected for the marine reservoir effect. It is well known that the reservoir age of the Mediterranean changed over time (Manning et al. 2002). Although the effect of the marine reservoir on shell material in the Aegean has been revealed for the Early Holocene (Facorellis et al. 1998), there are currently no measurements of the marine reservoir effect in the Aegean for the period between 3800 BP and 100 BP (Manning et al. 2002). Moreover, detecting the marine reservoir effect on the carbon component of *C. glaucum* specimens can be particularly complicated because of the nature of the species, which adapted to a great variety of habitats where marine and hard (fresh) waters mix (See Section 4.1.2.9, per. com. B. Kromer).

Secondly, an unknown number of shells that have been used for radiocarbon analyses may not be representing primary geomorphological deposits; i.e., they may belong to individuals that died in another location and considerably earlier than the formation of the deposit in which they were recovered. Transported shells are certainly not representative of the deposit they are found in and therefore not suitable to date or otherwise characterize these deposits. This phenomenon is analogous to the "old tree" effect known from problems occurring in the radiocarbon dating of carbonized architectural wood remains. It is, however, very difficult to determine whether a fossil or sub-fossil element is in its *in situ* or transported position in drilled cores, which constitute the sole analytical material of geomorphological research program in the Troad. Therefore, the timeline of the evolution of both the Trojan Plain and the Beşik Bay should be approached with caution, when correlation with diachronic archaeological patterns is necessary.

2.2.2. Geomorphic evolution of the Büyükdere Valley

Yenibademli is located in the Büyükdere Valley (Figure 3). The valley consists of a coastal alluvial plain formed by the largest river on Gökçeada Island, the Büyükdere River, and the limestone ridges flanking this plain. The capacity of the Büyükdere River has been largely diminished by a modern dam built in its upper part (Hüryılmaz 2004).

Figure 3. Geomorphology of the Büyükdere Valley (Modified after Öner 2000, courtesy of the Yenibademli Excavations).

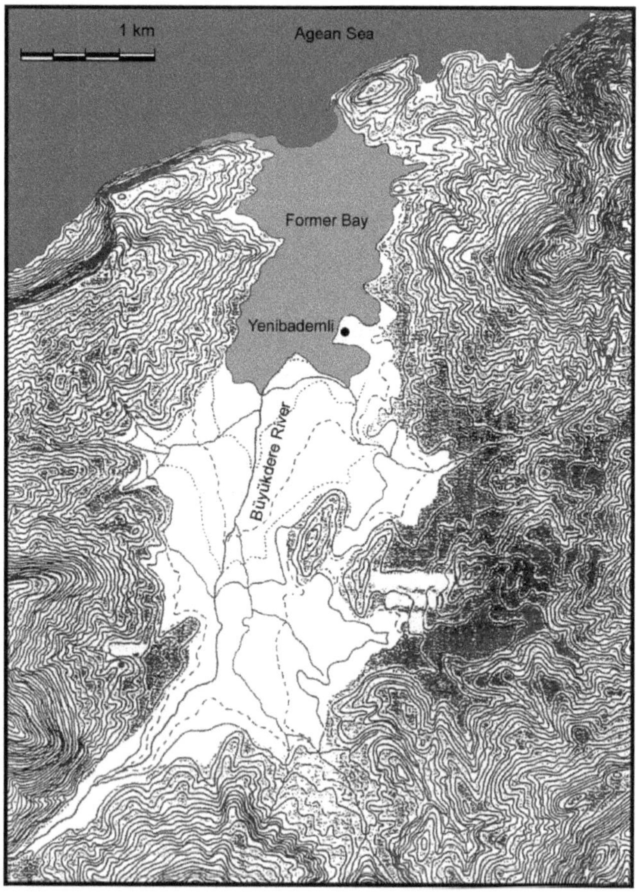

During the Early Holocene, as a consequence of marine transgression, the valley was filled with marine water (Figure 3; Öner 2001). Geomorphological research in the Büyükdere Valley and on the settlement proper has indicated that a marine bay dominated the northern part of the valley at the time of the beginning of settlement at Yenibademli, around 5000 BP (Figure 3; Öner 2000, 2001). No radiocarbon dates are available from the geomorphological cores. The date cited above was inferred from the relationship of the vertical position of the deltaic sediments with cultural layers. To the north and west of the mound, the Oligocene bedrock beneath the cultural layers of Yenibademli reaches directly to the muddy marine sediments (Öner 2001: 787, Figure 6). According to Öner (2000, 2001) the deltaic marsh at the mouth of the Büyükdere was located immediately to the southeast of the mound during the formation of cultural layers at Yenibademli. Adopting Kayan's (1988, 1991, 1997b) "marine regression" model of ca. -2 m for the Troad, Öner (2000, 2001) suggests that the sedimentation rate of the Büyükdere Valley accelerated between 5000-4000 BP. (Öner 2000, 2001). To the north and west of the mound, the Oligocene bedrock beneath the cultural layers of Yenibademli reaches directly to the muddy marine sediments (Öner 2001: 787, Figure 6). A wide and open bay ca. 1.7 km north of Yenibademli is all that remains from the former coastal landscape (Figure 3).

2.2.3. Changes in the geomorphic features around Ulucak

Ulucak, as mentioned above (cf. Section 2.1.3.), never had direct access to the sea coast during the Quaternary. Today it is located ca. 13 km east of the present coastline (Figure 4), the Bornova Plain and the Belkahve Pass constituting the two major obstacles between the site and the Aegean Sea. The development of the Bornova Plain and the former extension Bornova Plain are yet to reach marine sediments, a shallow muddy-sandy substratum for the former eastern extension of the bay is most likely (per. com. İ. Kayan). Subsequently, this extension was slowly silted by flood water running from the high ridges surrounding the embayment of İzmir. In the case that the regional land mass was not simultaneously moving upwards at the time of the maximum marine transgression (ca. 7000-6000 BP), the coastline of the İzmir Bay may have been located at its easternmost point about this time.

Figure 4. Geographical features in the vicinity of Ulucak (Modified after Çilingiroğlu et al. 2004, courtesy of the Ulucak Höyük Excavations).

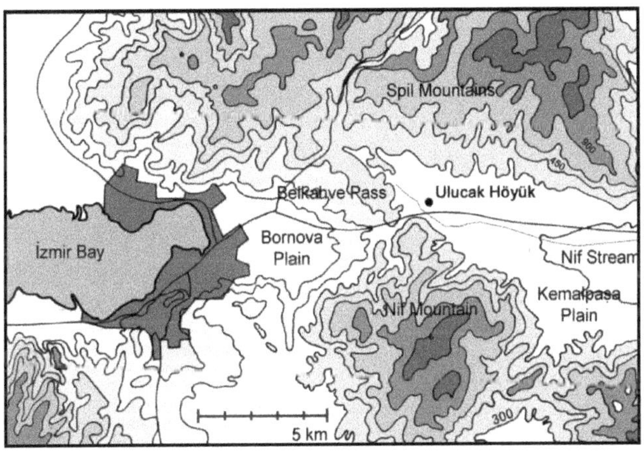

The only freshwater source located near Ulucak is the Nif Stream (Figure 4). Preliminary results of geomorphologic surveys around Ulucak have revealed that during the development of the mound no significant change in the course of the stream occurred (Çilingiroğlu et al. 2004: 8, Kayan 1999). The Nif remained as a low energy river causing no catastrophic floods throughout the occupational history of Ulucak. (Çilingiroğlu et al. 2004: 8).

2.2.4. Summary

The continental shelves of the coastal areas that are of primary concern to the present study, namely the Trojan and Beşik Bays in the Troad, the Büyükdere Valley on Gökçeada, and the İzmir Bay near Ulucak were gradually filled (partially or completely) with sedimentary deposits after ca. 7000-6000 BP, during the Middle and Late Holocene. The bays were of differential characteristics,

and character of the İzmir Bay to the west of this plain are of interest here as the most probable source for the marine mollusk assemblages recovered from Ulucak.

The Bornova Plain began to form during the Pleistocene (Kayan 2000). At the easternmost end of the İzmir Bay, the sedimentation process was slow due to the absence of a large river reaching the sea (Figure 4), while the alluvial plains north of the inner part of the İzmir Bay have been formed by the former deltas of the Gediz River and other rivers that were located on this coast earlier during the Holocene. Although on-going geomorphological investigations at the

embodying various benthic features, which continuously altered through time. Regardless of how the local sea-levels affected the sedimentation processes temporarily, the coastlines generally continued shifting towards their present position. The current interpretation of the time line of the fluctuations in the relative sea-levels and related changes in the sedimentary rates should be approached with caution.

2.3. Climate

Like geomorphology, climate is an important component of the physical setting of archaeological sites. It has direct influence on human societies on both short and long time scales. Moreover, in a similar manner to geomorphic features, climate is not a fixed constant and it is subject to rapid change. Despite these, archaeology has paid differential attention to the impact of climate and its fluctuations on human cultures and subsistence systems. While Palaeolithic archaeology has perceived climate as a determining factor in initial human adaptations, scholarly disciplines studying with the complex societies of the Holocene –especially of the Middle and Late Holocene– have largely neglected its influence (Fagan 2004: xiv, Issar and Zohar 2004: 13, Uerpmann 2003). This academic tendency partially stems from the fact that, until recently, the Holocene was thought to be a stable warm period (Carpenter 1966: 1-2). Only recent advances in deciphering fine-grained Holocene palaeoclimatological proxy-data and interdisciplinary research could generate serious consideration of climatic conditions as the main impetus or at least as one of the factors influencing the cause-effect relationships observed in the archaeological record of complex societies (Bintliff et al. 1982, Cullen et al 2000, Dalfes et al. 1997, de Menocal 2001, Fagan 2004, Issar and Zohar 2004).

Below, first, recent and modern climatic conditions around the sites in question will be reviewed, and then the present data on past climatic conditions will be discussed, thereby drawing the framework for two main questions of the present study. Firstly, the information about the recent climatic conditions resolved at short intervals (monthly and seasonally), constituting the sole source of information about regional intra-annual climatic cycles, will form the basis for explaining seasonality related questions of this study, i.e., the annual schedule of the mollusk harvest. And secondly, the synchronization of the cumulated knowledge about periodic climatic events with the apparent chronological patterns in the archaeomalacological data will form the essential basis to detect the climatic factors that may have had an impact on the former ecological and economic relationships inferred from the archaeomalacological record.

2.3.1. Present climate

Troia, Yenibademli, and Ulucak are latitudinally located within the warm-temperate climatic zone. The sites are located on or near the Aegean coast where the climate is typically Mediterranean. The Mediterranean type of climate can be characterized by its seasonality, in the most basic terms, by winter rains, caused by westerlies, summer droughts, caused by subtropical high, and a strong soil water deficit in summer (Bolle 2003: 6). Troia and Yenibademli are directly exposed to the mild humid coastal weather and maritime winds, whereas the influence of coastal humidity and maritime winds on Ulucak is depleted by the high ridges flanking the passage between the coast and the settlement.

Today, the mean annual precipitation totals in the western Anatolian coast reach 700 mm according to values covering the period between 1930 and 2000 (Türkeş 2003). Seasonality of precipitation deviates from the winter rain/summer drought pattern towards the north (Türkeş 2003: 192, Figure 6), where rainfall is distributed slightly more evenly throughout the months of the year (Figure 5). Monthly oscillations in mean temperature and precipitation measured in meteorological stations near the sites in question are listed in Table 13 and illustrated in Figure 5.

In the Aegean region of Turkey, there has been a significant shift from humid conditions in the 1960s to sub-humid or semi-arid conditions that prevailed during the 1990s. This trend may be related to both natural (geomorphic, pedologic, and global-climatic changes) and anthropogenic factors, such as the miss-use of cultural lands, especially the drainage of marsh areas in order to prevent malaria and create more arable land, the building of dams, and forest fires. (Türkeş 2003). This present phenomenon shows that caution must be taken while interpreting modern climatic trends and attests to the diachronic variability of regional climatic conditions. It also underlines the fact that one of the major challenges of palaeoclimatic and climatic research, as in all environmental research pertaining to the recent and earlier Holocene, is to distinguish between the natural and anthropogenic causes of environmental phenomena.

Figure 5. Temperature and precipitation recorded at stations nearest to Troia, Yenibademli, and Ulucak. The exact locations of the stations are given in Table 13 (After www.weatherreports.com).

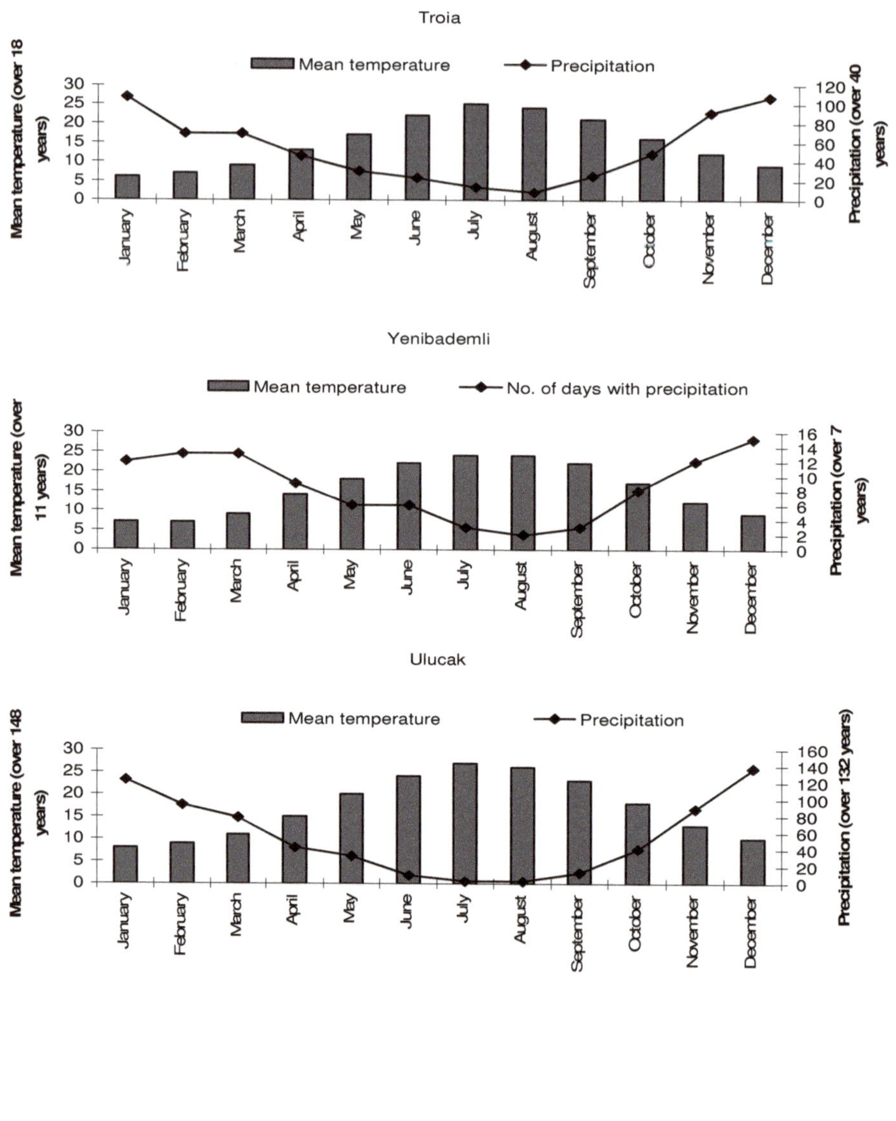

2.3.2. Climate of the Middle and Late Holocene

The Holocene represents an on-going interglacial period. While interglacial conditions were established soon after 10000 BP, it has been recently clarified that abrupt shifts to cool intervals took place within the warm climate of the Holocene (Carpenter 1966: 1-2, Roberts 1991: 62). Cyclical

widespread cooling events, the amplitude of which has not been as strong as those that occurred during the last glacial, have interrupted the continuity of patterns in this unit globally (Table 1; e.g., Bond et al. 1997, 1999, 2001, de Menocal 2001: 668, Issar and Zohar 2004, Roberts et al. 2001). These global events have the potential to influence precipitation, temperature, vegetation, zoogeography, and human societies to varying degrees on a regional scale. Since climatic evidence from terrestrial sediments such as the palynological record is potentially masked by anthropogenic interference, such as intensive land use resulting from population growth, particularly during the Middle and Late Holocene, here, current interpretations of globally distributed oceanographic and geological climatic proxy data and their correlation with the regional records from the eastern Mediterranean will be reviewed.

Global climate variability in the Holocene is regarded as a response to perturbations in solar activity, variations in the Earth's orbit, and volcanism (Bond et al. 2001, Mayewski et al. 2004). Astronomical changes, even in small magnitudes, control the amounts and courses of ice drift on ocean surfaces. The glacier extent, in return, directly affects the global climate (Bond et al. 2001). The events on the ocean surfaces, coupled with atmospheric events, create two global systems that control the climate: the Southern Oscillation (ENSO) and the North Atlantic Oscillation (NAO) (Issar and Zohar 2004: 15). While these global systems affect the earth's climatic system as a whole, the complexity and variability of regional proxy records pose difficulties in determining whether local signatures of climatic changes represent responses to global events or if they represent regional events (Mayewski et al. 2004, Roberts 1993). Recent observations on the present climate indicate that the NAO has direct influences on the climate of the eastern Mediterranean region (Cullen et al. 2002). It is, therefore, highly expected that the trends observed in the regional proxies from the Eastern Mediterranean are reflections of the NAO. Indeed, high-resolution regional proxy-data from the Middle East generally correlate with the isotopic record from deep sea cores in the North Atlantic.

It is widely accepted that cooling events occurred globally roughly between 9000-8000, 6000-5000, 4200-3800, 3500-2500, 1200-1000, and 600-150 BP (Bond et al. 1999, 2001, Mayewski et al. 2004). The synchronization, the geographical influence range, and the magnitude of regional responses to these events are issues which are still debated.

Of interest to the present study are the changes in the global climate as a framework and how the Mediterranean weather and environment responded to these changes in particular. Recent palaeoclimatological research in the eastern Mediterranean was partially able to determine the ways in which global climatic changes were recorded in regional proxies (Table 1).

The end of the last glacial period (Terminal Pleistocene) can be characterized by an unsteady process of deglaciation, spotted by at least two readvances of the glaciers, first one around 10800 BP (the Older Dryas) and the second one (the Younger Dryas) occurring between 11000 BP and 10000 BP (Roberts 1991: 50, 52-53, Rossignol-Strick 1999). The impact of the Younger Dryas on the eastern Mediterranean climate was extreme aridity (Rossignol-Strick 1999). Following the Younger Dryas, a milder and moister climate prevailed in the Mediterranean region. Pollen and foraminiferal records from the last sapropel[5] layer in the eastern Mediterranean, lake levels, and savanna-like conditions prevailing in the Sahara (Roberts 1991: 84-88, Roberts et al. 2001, Rosignol-Strick 1999) demonstrate that more humid conditions were present during the period between ca. 10000 BP and 6000 BP. By synchronizing proxy records from a wide range of geographical locations, Mayewski et al. (2004) found out that a cool episode caused the temperatures to decrease over much of the northern hemisphere roughly between 9000-8000 BP. Based on the increasing abundance of cold-water foraminifera species in the latest sapropel layer in the southern Aegean, Rohling et al. (2002) suggested that the SST of the Aegean became 2-4 ° C cooler during this period due to increased influence of winter precipitation over this region. Correlation between decreased δ ^{18}O and δ ^{13}C values of the speleothem deposits in Soreq Cave (Israel) were interpreted as signatures of increased precipitation (Bar-Matthews et al. 1999).

The four subsequent intervals of rapid climatic change brought cooler climates to higher latitudes and dryer climates to lower latitudes (Mayewski et al. 2004), causing regional droughts (de Menocal 2001). A relatively short-lived decline in moisture is demonstrated in the δ ^{18}O and δ ^{13}C profiles of speleothems from the Soreq Cave in Israel ca. 5200 BP, and then again between 4600 BP and 4000 BP (Bar-Matthews and Ayalon 2005), roughly corresponding to the 6000-5000 and 4200-3800 BP rapid climatic change episodes detected by Mayewski et al. (2004). Rohling et al. (2002) found evidence for two cooler episodes in the foraminiferal record of the Aegean around 6500-6000 BP and 3500-3000 BP. These episodes are roughly synchronized with other records from the North Atlantic (Rohling et al. 2002 and Table 1). Rohling et al. (2002) suggest that these cooler episodes should also be in the order of ca. 2-4 ° C. The chemical composition of a thin ash horizon from Tell Leilan and Abu Hgeira in northern Syria, immediately postdating the collapse of the complex societies that flourished here during the 3rd millennium BC, correlates well with an ash layer from a marine sediment core in the Gulf of Oman that indicated coeval arid conditions around 4150-3900 BP (Cullen et al. 2000). This phase is correlated with the 4200-3800 BP cool episode (Mayewski et al. 2004).

[5] Sapropel is the term used for organic rich, often laminated marine sediments (Charisi et al. 1998).

The eruption of Thera around 1650 BC may also have left its signature on the region's climate. A tree-ring series from Central Anatolia shows anomalously favorable growth around 1650 BC (Manning et al. 2001). Although the volcanic ash layer from this eruption is present in deep sea cores in both the Aegean and the polar regions (Rohling et al. 2002, Figure 1), the sediments that were deposited immediately after this layer do not bear significant traces of regional or global climatic fluctuations. At present, the anomalous tree-ring record from the dendrochronological series in Anatolia represents the only proxy-data that may be related to the Thera eruption. The next cool episode that left its signature in various records across the globe starting from 3500 BP and lasting until about 2500 BP (Mayewski et al. 2004) were also recorded in the proxy data from the eastern Mediterranean region. Fluctuations in sea and lake levels, as well as isotopes from stalagmites and lake sediments from the region indicate that the climate abruptly became colder and more humid around 1200 BC, bringing about increased precipitation (Issar and Zohar 2004: 27-28, Figures 3 and 3a, 159-161).

Table 1. Global climate in the Holocene and its regional reflections in the eastern Mediterranean.

Source	Mayewski et al. 2004	De Menocal et al. 2001	Rohling et al. 2002	Bar-Mathews and Ayalon 2005
Proxy	Multi-proxy	Multi-proxy	Foraminifera	$^{18}O_{speleothem}$
0 BP...				
	Cooler			
1000	Cooler	Droughts		
2000				
		Droughts		
3000	Cooler		Cooler SST in the Aegean	
		Droughts		
4000	Cooler			
				Arid in the Levant
5000				Arid in the Levant
	Cooler	Droughts		
6000			Cooler SST in the Aegean	
7000				
8000	Cooler		Cooler SST in the Aegean	

Climatic proxy-data of geological origin (evidence least likely to be affected by human activities) from locations closer to the archaeological sites of concern to this study are scarce. A recent study of a series of δ ^{18}O values from carbonate concretions sampled from drill holes sunk into the sediments around Troia indicated that no significant climate or temperature fluctuations took place in the region during the occupational history of Troia, but that temperatures were in general ca. 2-4 ° C lower than present (Göbel 2005: 174). Since the occupational history of Troia spans over more than 4000 years and it is exactly the climatic fluctuations throughout these 4000 years that need to be detected and correlated with past archaeological and ecological phenomena, the relevancy of Göbel's results for the purposes of this study is negligible.

To summarize the data above, it appears that three periods of rapid global climate change occurred during the period of concern for the present study: 6000-5000, 4200-3800, and 3500-2500 BP. These cool spells seem to have brought extensive droughts to most low latitude geographies. While suggested cooling of SST values in the Aegean and aridity in the Levant roughly correspond to the rapid climatic change period between 6000 and 5000 BP, the synchronization of these events are rather weak (Table 1). Although the effect of the following, 4200-3800 BP cooling event could not be demonstrated in the foraminiferal proxy, i.e., SST record of the Aegean, contemporary fluctuations in the proxy climatic data from various locations in the Middle East could be related to this event. The third, 3500-2500 BP period seems to have brought a cold and humid climate to the Middle East, the effects of which could also be detected in the foraminiferal record of the Aegean (Table 1). The impact of the Thera eruption on the climate of the Aegean is yet to be demonstrated unambiguously in regional multiple proxy records. The existing local palaeotemperature record from the Troad, on the other hand, needs chronological refinement.

In result, global climatic records provide a framework for the potential climatic causes of local palaeoenvironmental and archaeological phenomena. Self-evidently, the large time-spans of rapid climatic events and the synchronization with regional climatic data should be refined with future collaborative work. For the time being, while the extant climatic evidence can be used as a reference point in the evaluation of various lines of archaeobiological data from the sites that are in question in the present study, in the absence of high-resolution climatic record covering the Middle and Late Holocene periods from the North Aegean, climatic interpretations of the archaeobiological data need necessarily be inconclusive.

2.4. Aquatic environments

The water bodies around Troia, Yenibademli, and Ulucak represent the most essential resource for the past and present human subsistence at these locations. They also constitute the host environments for the aquatic mollusks that are discussed in this study. An understanding of the mollusk sector of the settlements' subsistence economies and accurate interpretations of the ecological relationships represented by the archaeomalacological record necessitates a close look at the present and former characteristics of the hydrogeographical components at their vicinity.

Figure 6. Sea surface temperatures off the coast of Çanakkale (After Anonymous 1982: 1809, Tables 2 and 3).

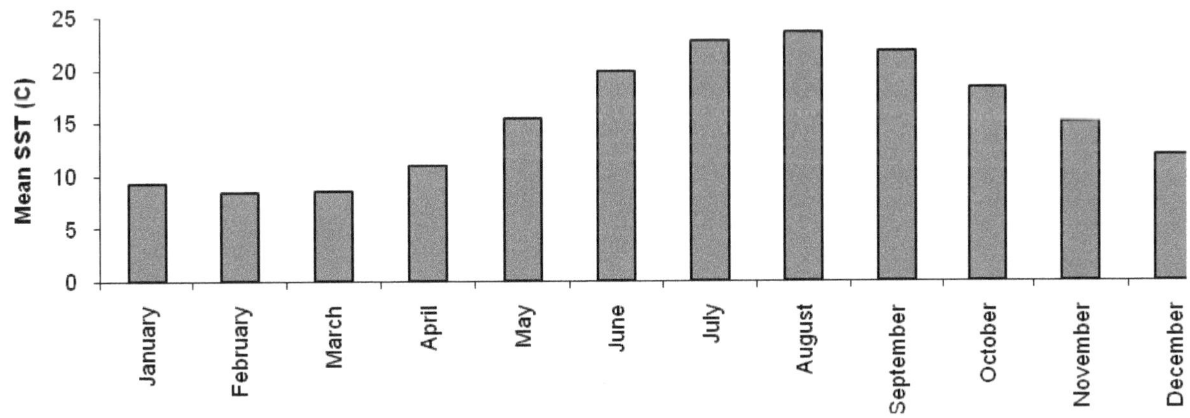

The aquatic mollusks that are dealt with in this study originate from the Aegean Sea, the Dardanelles, and the rivers that directly or indirectly discharge into these marine waters. The Aegean Sea can be described as a confluence zone between the Black Sea –via the Bosporus, the Marmara Sea, and the Dardanelles– and the eastern Mediterranean. It is divided into two geographical provinces along the 38th parallel (just below Ulucak). Sea temperature and salinity increase towards the south. According to the observations made between 1984 and 1997, average sea surface temperatures (SST) vary from 12 ° C in the winter to 23.5 ° C in the summer across the basin, with surface salinities averaging between 35.8 and 39.2 ‰ (Uçkaç 2005). The Dardanelles brings the colder and less saline water of the Black Sea into the North Aegean. The rivers are mostly perennial, influenced by the seasonality of climate, draining parallel to the precipitation curve (cf. Figure 5).

2.4.1. Troia

As mentioned above (Sections 2.1.1 and 2.2.1), the water bodies that dominate the Troad consist of the Aegean, the Dardanelles, the joint river systems of the Karamenderes and the Dümrek, and formerly the Trojan Bay. The beds of the Karamenderes and Dümrek, and the Trojan Bay went through extensive geomorphological changes (cf. Section 2.2.1). Today this bay is entirely silted with alluvial sediments from the Karamenderes and the Dümrek, whose deltaic mouths reach the Dardanelles forming very small lagoons and estuaries (Figure 16).

Both the Dardanelles and the North Aegean demonstrate high variability in their physical and chemical characteristics. The Dardanelles controls the water exchange between the Marmara Sea and the Aegean. In simplest terms, an upper, colder and less saline current originating from the Black Sea penetrates the Marmara Sea; mixed with Marmara seawater, it flows southward into the Aegean Sea, through the Dardanelles. Simultaneously, a deeper, warmer, and more saline current flows northward from the Aegean Sea into the Marmara Sea. The Black Sea water ranges between 5-15 ° C in surface temperature and 17-20 ‰ in salinity (Aksu et al. 2002). The salinity of the Black Sea is controlled largely by precipitation and river discharges. Evaporation is a less significant factor affecting water temperature and salinity (Aksu et al. 2002). Peak river discharges to the Black Sea occur during April and May (Özsoy et al. 1995, 1996). The surface salinity of the Marmara seawater can vary between 20 ‰ to 36 ‰ (Aksu et al. 2002: 142, Figure 19). The salinity of the North Aegean ranges between 33 ‰ and 38 ‰, whereas SST values vary between 12-25 ° C (Uçkaç 2005, Zodiatis 1996). Off the Aegean coast of Çanakkale salinity ratios as low as 30 ‰ have been recorded (Demirsoy 1996: 364). Monthly variations in the SST off the coast of Çanakkale are shown in Table 14 and illustrated in Figure 6.

The oxygen isotopic composition of the Black Sea and the Marmara Sea strongly correlate with the seasonally varying surface salinities of these waters. In the Marmara water mass, ca. 36 ‰ salinity corresponds to ca. +1.80 $\delta\ ^{18}O_{Water}$ ‰ PDB and ca. 22 ‰ salinity to -1.80 $\delta\ ^{18}O_{Water}$ ‰ PDB, whereas in the Black Sea ca. 22 ‰ salinity corresponds to -1.80 $\delta\ ^{18}O_{Water}$ ‰ PDB and ca. 17 ‰ salinity to -3.50 $\delta\ ^{18}O_{Water}$ ‰ PDB (Aksu et al. 2002: 142, Figure 19). While the surface waters of the Black Sea and the Marmara Sea contain enough dissolved oxygen (DO) for aquatic life forms to survive (ca. 5.0 mg/l in the Black Sea and ca. 7.0-8.0 mg/l in the Marmara Sea), it has been observed that the thickness of the surface waters containing sufficient DO levels have reduced from about 0 to -75 m to about 0 to -25 m in the course of the last 25 years, indicating increased pollution (Artüz 2002).

The estuaries and the lagoons of the Karamenderes and the Dümrek Rivers constitute a highly unpredictable transition zone influenced by freshwater coming from the river system, seawater penetrating into the estuaries and lagoons from the Aegean and the Dardanelles, and meteoric water. Evaporation is also an important factor affecting the temperature and salinity of these shallow waters. It can be inferred that the less saline and cooler upper current dominated by the Marmara Sea water has a greater impact on these brackish waters than the Aegean waters that flow through a lower level. While the impact of seawater is restrained by the coastal sand dunes that developed between the sea and the brackish water today, earlier in the Holocene, when the Trojan Bay was still in existence, the Dardanelles water could enter the bay (estuary) without meeting an obstacle. Formation of sand barriers, as was the case in Beşik Bay, has not been suggested for the palaeogeography of the Bronze Age Trojan Bay (Kayan 1991, Kayan et al. 2003, Kraft et al. 2003a, 2003b). The water in the bay became less saline from north to south, where the influence of freshwater was dominant.

There has been little modern ecological monitoring on the estuaries and lagoons of the Karamenderes and the Dümrek. One study reports SST and salinity values measured in May 2001, from an imprecisely described area along the coast around the river mouth of the Karamenderes. In one estuary, salinity measured 17 ‰ while the SST was 19.1 ° C, and in a lagoon, salinity was 10.9 ‰ while the SST was 19 ° C. Based on the high pH level (8.39 mg/l) measured in one of the lagoons, the same publication manifests concern about pollution in the deltaic river mouths, probably caused by agricultural toxic waste. (Tunçer and Erdemir 2001).

More information on the properties of the waters in the lagoons and estuaries at the mouth of the Karamenderes was collected during the field research conducted for this study. These will be explained in Chapter 4.

The exact locations of the meandering courses of the Karamenderes and the Dümrek Rivers during the occupation history of Troia have not been reconstructed (Kraft et al. 1980). However, it is certain that they were slow and shallow at the lower plain, carrying only fine grain sediments (Kayan et al. 2003). Their velocity must have been regulated seasonally, increasing with the beginning of September due to increasing precipitation, peaking during the spring due to snowmelt, and diminishing during the summer due to low precipitation rates and high evaporation. This natural pattern of drainage has largely been altered by irrigation canals and the Bayramiç Dam on the upper Karamenderes. Today, the water flow of the Karamenderes River is still greatest between January and May (Özcan 2005), but the Bayramiç Dam collects the excess water of the Karamenderes during the winter, both preventing flooding and reserving water to be redistributed to the fields on the lower plain during spring. The water drained from the fields is then redirected to the Dümrek River. The consequence of these hydraulic programs has been a considerable decrease in the amount of freshwater influx into the deltaic lagoons and alluvial sedimentation on the plain in the 20th century. The Late Bronze Age knowledge of hydraulic technical know-how in the Aegean region could have been used to regulate sedimentation and irrigation in the flood plain in front of Troia (Zangger 2003). But, at present there is no archaeological evidence to verify this suggestion. It is certain that during the Roman times, both the velocity and the capacity of the Karamenderes were reduced by an aqueduct directing water from the Kemerdere River, a tributary of the Karamenderes, to the water querries south of the settlement (Wolkersdorfer 2006). Additional hydraulic projects may have been pursued on the lower flood plain of the Karamenderes during this period, in order to redirect freshwater from the river bed to the agricultural fields located on the alluvial plain.

Some other relevant information is available about the present physical and chemical characteristics of freshwater resources around Troia: the mean temperature of the water sampled from the Karamenderes, the Dümrek, and their adjacent artificial canals is 23.3 ° C during July and August (mean of 25 values taken between 2001-2003, reproduced from Wolkersdorfer et al. 2004) and 17.3 ° C in April (mean of 4 values, Virchow 1879, cited in Wolkersdorfer et al. 2004). The salinity of the surface waters in the Karamenderes basin increases between January and May. This trait is attributed to the sea spray from the Aegean due to strong winter winds. DO in the same surface waters range between 1.98 and 12.08 mg/l throughout the year, with higher ratios between November and May. Alkalinity fluctuates between 6.44 and 8.8 pH, remaining mainly above 8.0 between November and May (Özcan 2005). Most aquatic animals require 6.5-8.0 pH for survival.

Additionally, proxy-data about the isotopic composition of the meteoric water that was active during the sedimentation process in the Troad is provided by the analyses of carbonate concretions sampled from sediment cores taken from the flood plain. The inorganic carbonate of calcium concretions originate from dissolved carbonate of the underlying terrestrial environment in interaction with meteoric waters (Göbel 2005: 176-178). $\delta\ ^{18}O$ PDB of the $CaCO_3$ samples from calcium concretions range between -8 ‰ and -4 ‰, with a mean of -5 ‰. The $\delta\ ^{13}C$ PDB range of inorganic carbon from the same samples is between -9‰ and -2 ‰, averaging at -6 ‰ (Göbel 2005: 116-117). While the majority (all except one) of the isotopic analyses of the inorganic carbonate samples indicate the dominating effect of freshwater during their formation, because the effect of the recyristallization on isotopic fractionation is unknown, there is no isotopic equilibrium between the analyzed carbonate and the meteoric water. Therefore, the value of the isotopic compositions of these carbonates samples to describe the chemical characteristics of the meteoric waters in Troia is low. Moreover, the isotopic compositions do not show significant fluctuations through the different layers of the sediment cores, i.e., through time.

2.4.2. Yenibademli

Gökçeada is one of the richest Aegean islands in terms of freshwater supplies. There are four large artificial water reservoirs on the island. One of these, the Zeytinliköy Dam Lake, is situated on the upper course of the Büyükdere River. This dam degrades the river's drainage capacity and the volume of its bed. While the velocity of the Büyükdere River has not been studied, it is known that the river is perennial (Hüryılmaz 2004). Under natural circumstances, the amount of water in the river valley must have been regulated seasonally. Geomorphological studies around the site have shown the presence of marine waters in the Büyükdere Valley during the formation of the mound (Öner 2001, 2002). Chronological fluctuations in the alluvial and colluvial sedimentation rates and the timing of the processes that led to the filling the former bay, to the extent that it was no longer suitable for the human exploitation of its aquatic resources, are not known.

2.4.3. Ulucak

The İzmir Bay is the most probable source of the marine mollusk remains yielded from the Ulucak excavations. The properties of the İzmir Bay water are influenced by the Gediz River discharge, Aegean Sea water and by the strong evaporation occurring during summer time. The bay is divided into three basins: the Outer Bay, the Middle Bay, and the Inner Bay (Sayın 2003). The Inner Bay is the narrow 60 km long easternmost section west of the Bornova Plain, and is therefore of particular interest. The mean depth of the Inner Bay is ca. 10 m. SST values, which range between 16 ° C in the winter and 24 ° C in the summer in the greater bay area, show stronger seasonal oscillations in the Inner Bay. Average salinity of the Inner Bay is ca. 38.0 ‰, a lower value than those in the Outer Bay where salinities can reach up to 41 ‰. The lower salinity of the Inner Bay results from anthropogenic waste water discharge of metropolitan İzmir. (Sayın 2003). Freshwater is discharged to the bay from the Gediz River, whose mouth was situated more southeasterly earlier in the Holocene (Çilingiroğlu et al. 2004: 81; Figure 1). The Nif, a tributary of the Gediz River, is highly polluted by industrial and agricultural agents.

2.4.4. Summary

In this section, existing data about the physical and chemical characteristics of the water bodies in the immediate surroundings of the archaeological sites in question was described. All three sites are associated with the northern Aegean basin, but the characteristics of the waters nearest to the sites show differential characteristics. Both the real and proxy data about these characteristics (SST, salinity, DO, pH, DIC, and stable isotopic compositions) are far from complete. Current data sets show predictable (e.g., seasonal) and unpredictable variability, mainly due to the transient character of the northern Aegean and the Dardanelles, as well as the dynamic nature of the continental shelf waters such as bays, estuaries, and lagoons. The prominent influence of anthropogenic activities on the present properties of these water bodies calls for caution when using the real data as analogies for palaeoenvironmental conditions.

2.5. Biosphere

The components of the biosphere shaped by the geographical and climatic characteristics of the landscape form the subsistence base of human societies. The present landscapes surrounding the archaeological sites of Troia, Yenibademli and Ulucak have been strongly modified by human activity. This has had severe consequences for the biosphere. The Aegean landscape is densely spotted by towns and villages of various sizes; plains are covered with crop fields (e.g., vegetables and cash crops such as cotton), whereas the low plateaus are used for extensive horticulture (e.g., olive orchards). Animal husbandry is based on sheep and goats, and to a lesser extent on cattle. Both artisanal and industrial fishing is practiced. Tourism and other industries heavily exploit the historical and natural landscape.

The understanding of the characteristics of the past biosphere depends mainly on archaeobiological and other palaeoenvironmental studies, whereas the components of the present biosphere in comparison with the palaeoenvironmental reconstructions highlights how the biological landscape was altered throughout the Holocene. In this section, both the actual data about the present biosphere and the implications of archaeobiological and other palaeoenvironmental studies around the sites in question are presented.

2.5.1. Vegetation

'Natural' vegetation survives only in patches in the Troad. Maquis, phrygana (i.e., scrubland vegetation such as lavender, rosemary bushes), and steppe-type plants dominate the low plateau, whereas moisture-dependent grasses, rushes, and sedges are found on the Trojan Plain (Riehl 1999: 52-53). The mountain slopes are covered with pine and juniper woods in reserve areas. Present vegetation in Gökçeada is similarly characterized by maquis, phrygana and steppe-type plants (Seçmen 1977). Patches of pine and juniper woods are also present on the island (Oybak-Dönmez 2005). The Kemalpaşa Plain, where Ulucak is located, is assigned to agriculture and industry. Otherwise, maquis, pine and oak constitute the remaining patches of natural vegetation around Ulucak (Çilingiroğlu et al. 2004: 5).

Palynological research has indicated that during the Middle Holocene much of the Mediterranean was dominated by temperate deciduous woodland (van Zeist and Bottema 1991, Roberts et al. 2004). Pine (*Pinus* spp.) started to dominate the forest landscape ca. 3000 BP, as a consequence of cumulating human impact (Roberts et al. 2004: 351). Palynological records from sediment cores from the Marmara region and the Dardanelles are characterized by moisture indicating species such as fir (*Abies*) in the lower layers and by higher percentages of pine (*Pinus* sp.) in the upper layers, indicating a trend towards drier and warmer conditions (Oybak-Dönmez 2005). Unfortunately, these records have not been dated. Early analyses on pollen samples from Early Bronze Age Troia indicated the presence of pine (*Pinus* sp.), oak (*Quercus* sp.), and pistachio (*Pistacia* sp.) (Gennett and Gifford 1982, Huber and Rapp 1987). According to Riehl's (1999: 83) interpretation of the macrobotanical crop and wild plant remains from the New Excavations in Troia and the nearby excavations in Kumtepe (Figure 2), land clearance activity started affecting the landscape around Kumtepe already before the Early Bronze Age. Combined multi-proxy data from wood charcoal, crop and wild plant analyses from Kumtepe and Troia shows that, during the Early Bronze Age, woodland still dominated the landscape, but maquis started showing an increasing presence due to agricultural and grazing activity (Riehl and Marinova 2008). The landscape was opening. The maquis cover recuperated during the Middle Bronze Age, only to be severely diminished in the latter half of the 2nd millennium BC (Riehl 1999: 84). Marsh vegetation covered the lower plain of the Karamenderes and the Dümrek Rivers throughout the Middle-Late Holocene and stayed up until 1960s (Aslan et al. 2003: 204, Figure 37, Kayan et al. 2003, per. com. with locals).

There has been no study of the vegetational history around Yenibademli and Ulucak so far. Palynological analyses of the geomorphological cores in the vicinity of the sites in question, as well as systematic studies of charcoal samples from archaeological deposits are of urgent necessity to build improved local vegetational histories.

2.5.2. Fauna

Seventeen species of terrestrial wild mammals were identified in the faunal assemblage of the New Excavations in Troia, mainly representing the hunted portion of the wild mammal populations that inhabited the vicinity of the settlement during the Middle-Late Holocene and, to a smaller extent, natural intrusions to archaeological deposits. These species include the European hedgehog (*Erinaceus europaeus*), beaver (*Castor fiber*), hare (*Lepus* spp.), wolf (*Canis lupus*), red fox (*Vulpes vulpes*), brown bear (*Ursus arctos*), weasels (Mustelidae), Eurasian badger (*Meles meles*), lynx (*Lynx lynx*), lion (*Panthera leo*), wild boar (*Sus scrofa*), fallow deer (*Dama dama*), red deer (*Cervus elaphus*), roe deer (*Capreolus capreolus*), wild cattle (*Bos primigenius*), wild goat (*Capra aegagrus*), and wild sheep (*Ovis orientalis*) (Fabiš 1996, 2003, Uerpmann 1992, 2003, 2006, Uerpmann and Uerpmann 2001). Of these species, only *E. europaeus, C. fiber, Lepus* spp., *V. vulpes,* Mustelidae, *M. meles, and S. scrofa* are still encountered in the flood plains and the lower plateaus of the Troad (per. obv.), the area which probably served as the primary food procurement area of the ancient settlements in the area. *B. primigenius* populations became extinct world-wide in the 17th century AD (Uerpmann 1999). A few *D. dama* populations still survive in enclosed reserves in western Turkey, but reserves in the Çanakkale province do not maintain *D. dama* herds (Demirsoy 1996: 580, Table 21). A field survey conducted between 2003 and 2006 reported *C. lupus, U. arctos, and C. capreolus* specimens in the Kazdağı National Park, situated ca. 100 km southwest of Troia in the Ida Mountains (Figure 1; Yiğit et al. 2006).

Achaeozoological research at Yenibademli revealed the former presence of *Lepus* spp., *V. vulpes*, *S. scrofa*, *D. dama*, *C. capreolus,* and *C. aegagrus* on Gökçeada. Wild carnivores larger than the red fox (*V. Vulpes*) are not represented in the studied portion of the faunal sample.[6] The Middle Holocene terrestrial wild mammal life on the island has been largely destroyed by anthropogenically induced reduction of suitable habitats and human exploitation pressure. Today *S. scrofa, D. dama, C. capreolus,* and *C. aegagrus* are no longer present on the island (Anonymous 1982). The wild mammal spectrum of the faunal assemblage from Ulucak consists of *Lepus* spp., *V. vulpes, S. scrofa, D. dama, C. elaphus, C. capreolus, B. primigenius* and *O. orientalis.*[7] Comprehensive studies of the present wild mammal populations of Gökçeada and the Kemalpaşa Plain are currently unavailable.

Troad is home to more than 156 bird species. This number amounts to one-third of all the bird species in Turkey. The remaining wetlands in the Troad constitute an important stop for white stork (*Ciconia ciconia*), little egret (*Egretta garzetta*), great grey shrike (*Lanius excubitor*), red-footed falcon (*Falco vespertinus*), and black-crowned night heron (*Nycticorax nycticorax*). There are very few cranes (*Grus grus*) in the area. Eastern imperial eagle (*Aquila heliaca*), greater spotted eagle (*Aquila clanga*), and lesser kestrel (*Falco naumanni*) have been recorded at the mountain slopes around the upper Karamenderes. (EURONATUR 2005).

The ancient spectrum of the avian fauna of the Troad has been reconstructed from the archaeological bird remains (from the excavations conducted between 1989 and 1995). The avian spectrum consisted mainly of species which prefer habitats in or near the water (e.g., *Anas* spp.) and some species whose habitats are open vegetation or woods (e.g., *Otis tarda* and *Grus grus*) (See Krönneck 1996 and 2003 for a complete species list of the studied avian remains from Troia). Remains of species whose occurrences are not restricted to certain habitats, such as *Aquila* spp. (eagles) and *Falco* spp. (falcons), were also present in the faunal assemblage (Krönneck 2003).

The Gediz Delta is an important bird area around İzmir. The ornithological record around the Nif basin –if extant– has not been accessible during the course of this research. The same is true for Gökçeada. Neither the bird remains of Yenibademli nor of Ulucak have been studied. Only three bird bone fragments were found in the studied portion of the archaeozoological assemblages of Ulucak. Eight bird bone fragments were recorded from Yenibademli. The limited number of bird bones yielded from these sites is unlikely to lead to an accurate reconstruction of the ancient avian fauna in these areas. However, because of its geomorphological and hydrological similarity to the Karamenderes and the Dümrek river valleys, although considerably smaller in area, it may be assumed that the Büyükdere Valley on Gökçeada was also host to a variety of aquatic birds, up until the time when it was completely silted and the water capacity of the Büyükdere River was reduced with hydrological projects.

[6] The author and C. Y. Gündem studied a random portion of the faunal remains of Yenibademli during the summers of 2004 and 2005.

[7] The author and C. Y. Gündem studied faunal remains from Chalcolithic and Early Bronze Age levels of Ulucak in the summer of 2004.

The Marmara Sea is inhabited by 196 species of fish. This number increases up to 346 species in the Aegean Sea. (Bilecenoğlu and Taşkavak 1999). Fifty-five of these species are economically important in the Çanakkale province (See Tekinay et al. 2002 for a full list of commercially important fish in Çanakkale). The majority of fish landings are (in order of importance) of sardine (*Sardina pilchardus*), bluefin tuna (*Thunnus thynnus*), and anchovy (*Engraulis encrasicolus*) (Tekinay et al. 2002). This list demonstrates the importance of the Dardanelles as a habitat for migrating and schooling pelagic fish and as a commercial fishing ground. Of the other economically important fish species in the North Aegean and the Dardanelles, gilt-headed bream (*Sparus aurata*), grey mullet (*Mugil cephalus*), and sea bass (*Dicentrarchus labrax*) are estuary dependent marine fishes. The silverside, sand-smelt, or Atherine (*Atherina* spp.) resides in estuaries, and eel (*Anguilla anguilla*) is catadromous. The Karamenderes is home to seven freshwater fish species: brown trout (*Salmo trutta macrostigma*), European chub (*Leuciscus cephalus*), bitterling (*Rhodeus amarus*), Danube bleak (*Chalcalburnus chalcoides*), barbel (*Barbus tauricus escherichi*), gudgeon (*Gobio gobio*), and loach (*Cobitis fahirae*) (Sarı et al. 2005).

Demir (2003) identified 362 mollusk species in the Marmara Sea and 546 shelled mollusk species in the Turkish waters of the Aegean Sea. Only a few littoral species are commercially exploited in Turkey today: *Sepia officinalis* (cuttlefish), *Loligo vulgaris* (squid), *Octopus vulgaris* (octopus), *Mytilus galloprovincialis* (common Mediterranean mussel), *Ostrea edulis* (European oyster), *Pecten jacobaeus* (pilgrim scallop), and *Ensis ensis* (razorclam) (Tekinay et al. 2002). As stated before (cf. Section 2.4.1), little ecological monitoring has taken place in the coastal waters in the Troad. The few bivalve species that were recorded from the deltaic areas of the Karamenderes and the Dümrek Rivers are *Cerastoderma glaucum* (lagoon cockle), *Venus verrucosa* (warty venus), *Chamelea gallina* (venus shell), *Tellina* spp. (tellin), *Solen* sp. (razorclam), *Donax* spp. (coquina), and *Tapes decussatus* (carpetshell) (Tunçer and Erdemir 2001). But since Tunçer and Erdemir (2001) did not indicate the exact sampling locations in their report, the reliability of this list is highly questionable.

More information on the present molluscan fauna of the deltaic distributary channels of the Karamenderes and the Dümrek, as well as that of the Beşik Bay, was assembled in the course of the present study. The details of the collected information will be given in Chapter 4.

The present molluscan fauna of Gökçeada coast has not been studied systematically.

According to quantitative assessments based on recent monthly samplings of the benthic mollusk fauna in the inner part of the İzmir Bay, *Corbula gibba* dominate the mollusk populations at all sampling stations. *C. gibba*, a small filter-feeder living in soft bottoms, is considered to be a pollution indicator. Other shelled mollusks are only very rarely present in the Inner Bay (Palaz 2005). The present mollusk fauna of the Inner Bay cannot be regarded as a valid modern analogue for the pre-industrial molluscan species spectrum of these waters.

2.6. Archaeology

An understanding of the cultural and historical context of the mollusk assemblages discussed in this study is essential to this study. The relevant results of archaeological research at Troia, Yenibademli, and Ulucak will be reviewed in this section, without attempting a comprehensive discussion of interdisciplinary research histories.

2.6.1. Archaeological research at Troia

The site, also named Hisarlık in Turkish, was first investigated by Heinrich Schliemann in the 1870s (Schliemann 1881). Schliemann's discoveries in Troia attracted public interest and stimulated a great deal of scientific debate. Wilhelm Dörpfeld continued the excavations at the site in the 1890s. It was Dörpfeld who developed the chronological system of Troia I to IX (Dörpfeld 1902). The second phase of investigations at the site was undertaken by a team from the University of Cincinnati under the direction of Carl Blegen during the decade that preceded the Second World War (1932-1938) (Blegen et al. 1950, 1951, 1953, 1958, Blegen 1963). Results of early archaeological investigations at the site have been summarized and discussed in various articles and monographs (e.g., Easton 2002, Sazcı 2003).

Fifty years after this initial phase of archaeological investigations, excavations in Troia were resumed by an international team under the direction of late Manfred Osman Korfmann in 1988. The Post-Bronze Age remains of the site have been investigated by a sub-team from University of Cincinnati under the direction of Brian Rose. The mollusk remains that are evaluated in this study were recovered during these "New Excavations" between 1989 and 2003. Excavations continue at present under the directorship of Ernst Pernicka (for the results of most recent seasons, see *Studia Troica* forthcoming no. 18).

Interdisciplinary studies on the Bronze Age site and its vicinity revised the site's relative and absolute chronology, its stratigraphy, and revealed new aspects of its cultural relationship with the contemporary Aegean and Anatolian realms. Geomorphological, archaeobiological, hydrographical, epigraphical, and geological studies, as well as geomagnetic prospection, and intensive and extensive surface surveys on and around the mound have accompanied and continue to accompany the excavations. The results of scientific studies related to Troia are published in the annual journal of *Studia Troica*. The most recent archaeological understanding of the site and its environs is outlined below.

Occupational layers in Troia cover a period spanning more than four millennia between the beginnings of the 3rd millennium BC and the end of Byzantine settlement (Table 2, Figures 7 and 8).

Prehistoric human settlement in the Troad is not restricted to Troia. Current research dates the beginnings of occupation in the area to the beginning of the 5th millennium BC (Gabriel 2000). This earliest known phase of human occupation in the area is represented by the lower levels of Kumtepe (Kumtepe A) (Figures 1 and 2). Kumtepe was located on a low promontory directly on the northwestern shore of the Trojan Bay (Kayan 1996). Occupational horizons earlier than Troia contain one more layer (Kumtepe B), which followed Kumtepe A after a hiatus, dating to ca. 3300-3100 BC. (Gabriel 2000). Both of these occupational phases correspond to the Chalcolithic Age of the Anatolian chronology, and to the Late Neolithic of the Aegean chronology. Additionally, material parallel to the inventory found in the Bronze Age levels in Troia was found in Kumtepe (Kumtepe C), indicating that the site was still in use as a neighboring village of Troia (Gabriel 2000). Other sites which present archaeological evidence older than the earliest layers in Troia include Hanaytepe, Sivritepe, and Alacalıgöl (Figures 1 and 2). The first occupational phase at Hanaytepe (Hanaytepe A), located 7 km southeast of Troia, is roughly contemporary with Kumtepe B (Schachner 1999). The site is identified with ancient Thymba and, apart from finds parallel to the Kumtper B layers, includes finds which can be dated up to 1200 BC, thus showing parallelism with Troia I, II and VI (Schachner 1999). The Sivritepe Tumulus to the east of the Beşik Bay has been constructed with the material of a 5th millennium BC site that partially remains underneath this Early Hellenistic tumulus (Figures 1 and 2; Seeher 1985, Korfmann 2000). Although the site has been largely disturbed by the later building of the tumulus, fragmented architectural remains and numerous pottery fragments have been excavated from its remains (Seeher 1985). Alacalıgöl, near Kumtepe, was discovered in 2003 by local farmers. Ceramic finds from this site are dated to the 5th millennium BC (Figures 1 and 2; Gabriel et al. 2004). Today, the occupational deposits at this site amount only up to 70 cm. The site has been severely destructed by agricultural activity and could not be excavated.

Other than these sites, scanty remains of a burnt layer without architectural remains pre-dating the first architectural phase in Troia were found in Troia itself. These remains were located in the southern section (mainly Quadrate D5) of the trench left over from Schliemann's excavations (the so-called "Schliemann's Trench"), which bisects the site in a north-south axis almost in the middle (Figures 7 and 8). The finds from this layer do not differ much from the finds associated with the earliest architecture at the site (Korfmann 1992). Earliest architectural layers in Troia date to ca. 2900 BC.[8] The absolute and relative chronology of Troia has been summarized in Table 2.

[8] Although the chronology mentioned here is expected to remain largely intact with the completion of these dissertations, the stratigraphical and chronological interpretations of single excavation units are subject to constant and rapid change, especially in transitional phases.

Table 2. Chronology of Troia.

Conventional archaeological periods		Conventional ages	Attributed eras (Korfmann 1996)	Absolute dates (Kromer et al. 2003, Rose 2006)
Troia I		Early Bronze Age	Maritime Trojan Culture	3100-2350 cal. BC
Troia II				3050-2150 cal. BC
Troia III				2500-1950 cal. BC
Troia IV		Middle Bronze Age	Anatolian Trojan Culture	2200-2020 cal. BC
Troia V				1950-1750 cal. BC
Troia VIa	Early Troia VI (Pavuk 2006)	Late Bronze Age	Trojan High Culture	2100-1100 cal. BC
Troia VIb/c				
Troia VId	Middle Troia VI (Pavuk 2006)			
Troia VIe/f				
Troia VIg	Late Troia VI (Pavuk 2006)			
Troia VIh				
Troia VIIa				
Troia VIIb		Iron Age	Balkan influenced Trojan Culture	1500-800 cal. BC
Troia VIII		Archaic/Hellenistic Periods		700-85 BC
Troia IX		Roman Period		0 BC-400 AD
Troia X		Byzantine Period		1050-1300 AD

The first three occupations –the overlapping phases of Troia I and II, and the succeeding Troia III phase– have been in unison called the "Maritime Troia Culture" by Korfmann (1996).[9] The phrase was coined to emphasize the characteristic material culture of the time period that shows parallelism with contemporary settlements along the coast of the Aegean and the Sea of Marmara (Korfmann 1996). The Maritime Troia Culture lasts until about 2200 BC and corresponds with the Early Bronze Age II of the Aegean chronology. The chronological term used for this period in this study is the "Early Bronze Age."

Troia I represents a small settlement (less than 100 m in diameter) encircled by a defensive wall that has gone through various stages of repairs and rebuildings between 2900 and 2350 BC (Figure 7). Long single-spaced houses with antechambers dominate the settlement structure. In the middle of these houses, a free-standing building constitutes an exception with its size and orientation. (Ünlüsoy 2006). Designated as "megaron", the size and the artifactual inventory of this building and its desendents in Troia represent palaces or cultic places. (Korfmann 1992, 1993, 1994, 1995, Sazcı 2005, Ünlüsoy 2006). Both the megaron and the defensive wall suggest that already at the beginning of settlement in Troia, the inhabitants were organized around a central (religious or political) body and they probably had a security concern.

The Troia II period, formerly designated as a separate cultural phase by Blegen, represents the expansion of the Troia I settlement outside of the citadel, ca. 2550 BC (Sazcı 2005, Ünlüsoy 2006). This period chronologically overlaps with the Troia I period. Results of the New Excavations indicate that Troia II is a continuation of the Troia I culture, not a new cultural era (Korfmann 1996). During this period, the defensive walls are expanded and reinforced with rebuilding. The citadel now covers an area of 11 000 m^2 enclosed in a ca. 350 m-long defensive wall. (Figure 7; Ünlüsoy 2006). Megarons replace the smaller, single-roomed houses of the Troia I period in the citadel. Buildings with a domestic character have been moved to the outer side of the citadel wall. They are located in the lower town extending southwards. The lower town is enclosed in a monumental timber fortification set into the bedrock (Figure 8; Korfmann 2000, Ünlüsoy 2006, Jablonka 2006). The citadel and the lower town of Troia II together occupy an area of ca. 90,000 m^2. The early layers of this period (Early Troia II) were destroyed by a catastrophic fire. This fire, however, does not cause an archaeologically visible break or decline in the cultural organization: in the following period (Middle Troia II), the megarons and the citadel wall become "monumental" and adjacent to them; a multi-celled structure is built (Ünlüsoy 2006). Pits are dug inside and outside all architectural structures. Archaeologists postulate that these pits have originally contained large storage containers and imply "insecure times." These insecure times, likewise, end with a large fire. After this fire, during Late Troia II, in strong contrast to the previous centuries, irregularly planned multi-roomed houses are built in the citadel. Only the largest megaron is still in use. The citadel wall is extended once more. Like in the previous layers of Troia II, these buildings were also destroyed by fire. (Ünlüsoy 2006). Fast-wheel pottery in Troia first appears in the deposits of Troia II (Blegen et al. 1950). Otherwise, observations on the material culture of the period suggest that the community residing in

[9] At the time of writing the relavency of term was being questioned by other members of the Troia Projekt who study the material culture of Troia I-III layers.

Figure 7. Archaeological plan of the citadel area of Troia (Courtesy of P. Jablonka, Troia Projekt).

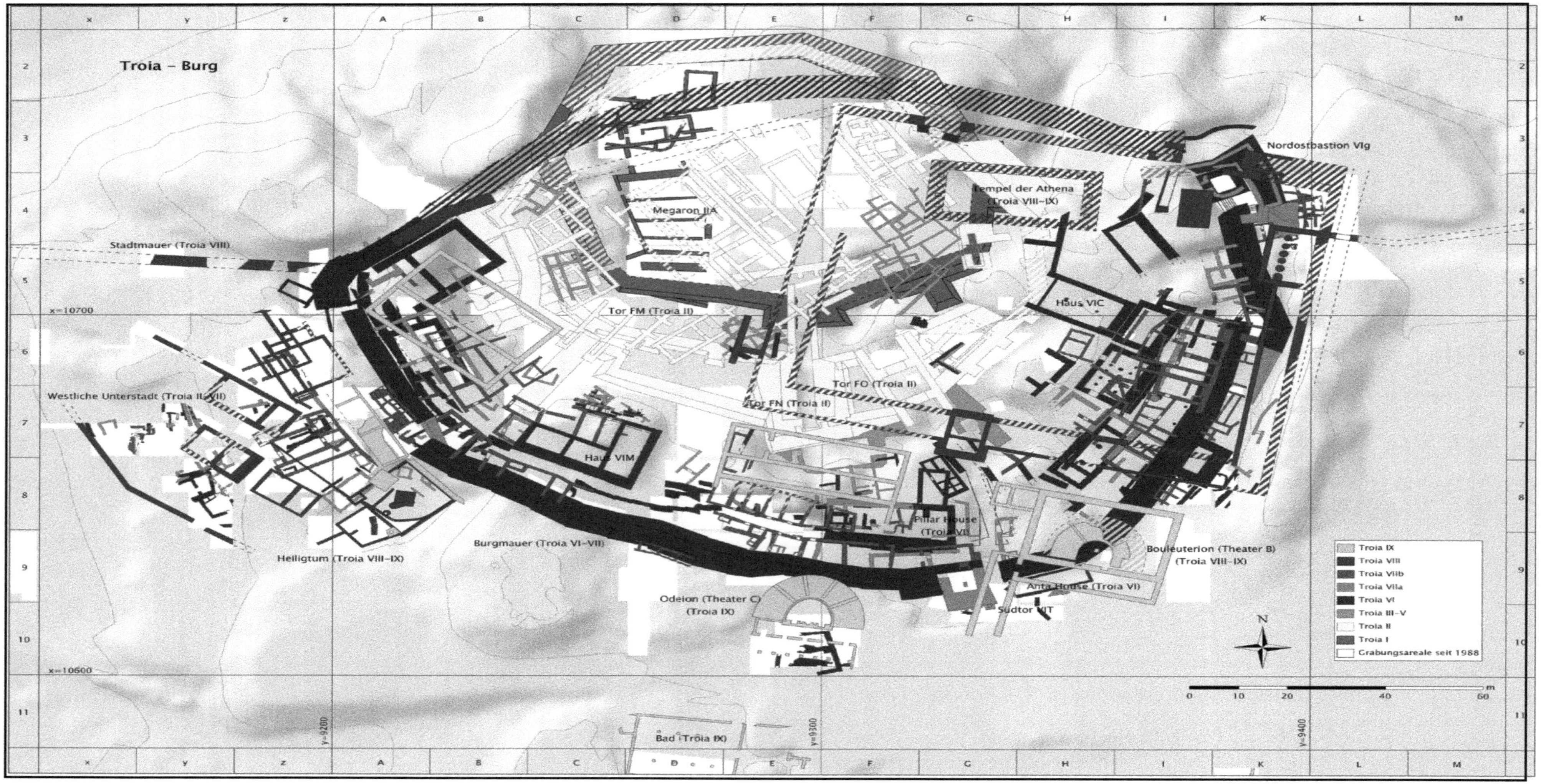

Areas excavated between 1988 and 2007 are shown in white.

Figure 8. Archaeological plan of the citadel and the lower town of Troia (Courtesy of P. Jablonka, Troia Projekt).

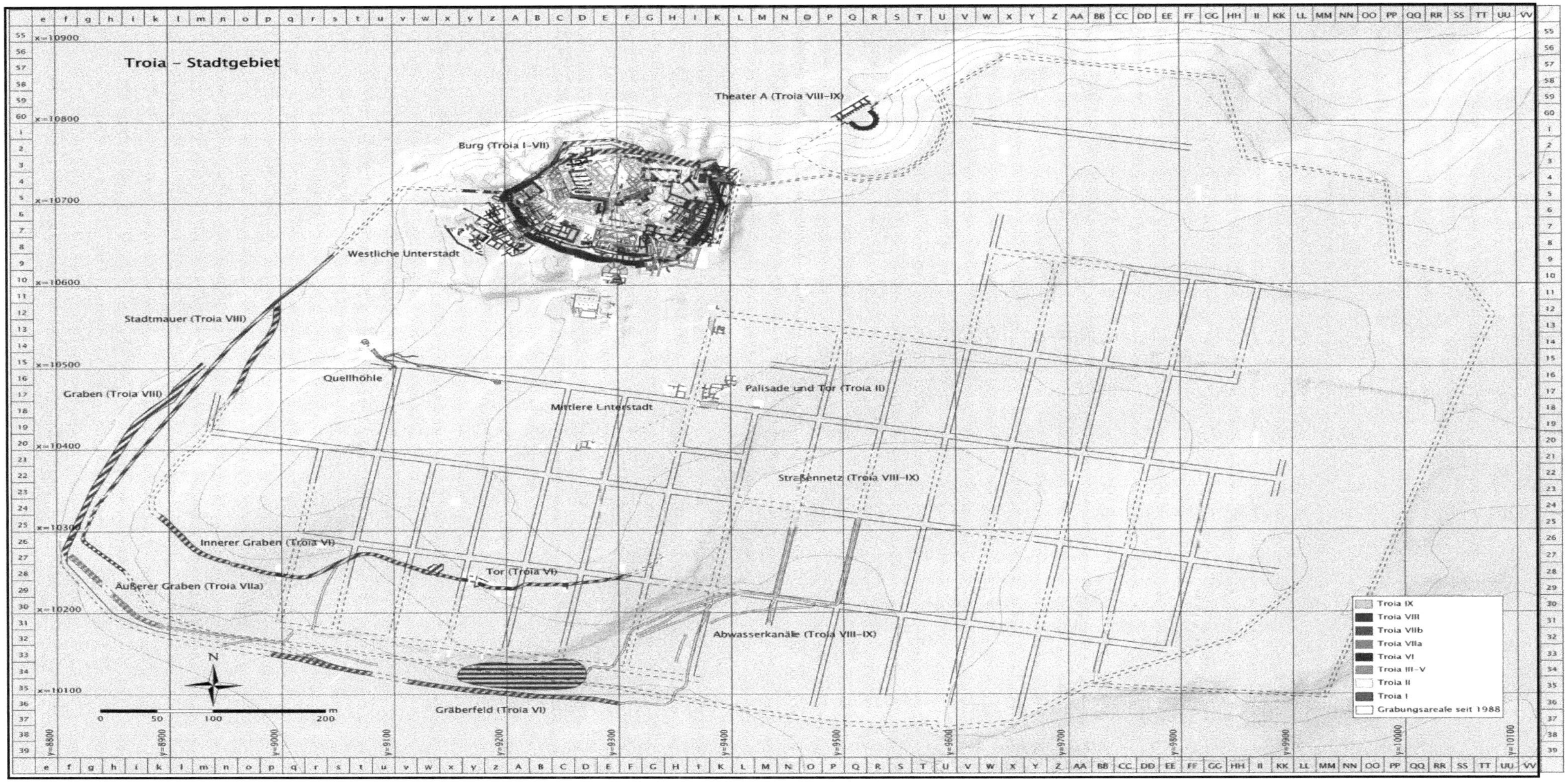

Areas excavated since 1988 are shown in white.

Troia between 2550-2250 BC was culturally identical to the Troia I population. The separation of residential areas from the citadel area until the last phase of the Troia II period implies the presence of a hierarchal society. The change in the architectural plan of the citadel during the last phase is perhaps an indication of a conflict between the lower town and the citadel, and a change in the structure of the social order. The introduction of fast-wheel pottery indicates further specialization in local economic activities.

The succeeding Troia III period is considered to begin around 2250 BC and lasts for only about 50 years (Korfmann 1996). During this short period three megaron houses and densely built small houses that surround them dominate the architectural plan of the citadel (Figure 7; Ünlüsoy 2006). The density of domestic structures within the citadel indicates continuous population growth. At least one of the four architectural periods of this phase was destroyed by a fire (Korfmann 1996).

The prosperity of the Early Bronze Age settlement and its significance within the Aegean-Anatolian cultural realm have been attested by the "treasures" that have been associated with the remains of this period, as well as numerous artifacts demonstrating contacts with contemporary settlements across the Aegean basin and beyond (Korfmann 2003, Sazcı 2005). The water quarries, which were in use during the Bronze Age and Post-Bronze Age periods in Troia, were built inside the bedrocks south of the settlement sometime during the Troia I-II periods (Frank et al. 2002).

Korfmann labeled the following settlements Troia IV and V together as the "Anatolian Troia Culture," based on the similarities of architectural schemes and some technological innovations with the material cultures of inland Anatolia (Korfmann 1996). Although it is true that many new elements indicating cultural influences from Anatolia are introduced in Troia during this period, influences from the Aegean region still dominate the material culture (Blum 2006). The period is dated from ca. 2200 to 1750 BC. It is contemporary with the regional designations of Early Bronze Age III and the Middle Bronze Age Periods. In this study, this phase of occupation in Troia is regarded as the "Middle Bronze Age."

Troia IV begins with no visible break after Troia III and Troia V is a direct continuation of Troia IV. During Troia IV, the settlement becomes almost double the size of the Troia II settlement (Blum 2006). Dense architecture is composed of house-complexes separated by small alleys (Figure 7; Blum 2006). Domed ovens, an Anatolian element, are introduced in Troia pointing at new food-preparation customs (Korfmann 1995, Blum 2006). This inference is partly supported by changes in the subsistence economy (cf. Section 2.7.1 below). The subsequent Troia V period is marked by a reorganization of the settlement plan. The citadel walls of Troia IV and V periods were larger than both the preceding and following occupational periods. The artifactual repertoire of the Middle Bronze Age Troia demonstrates social continuity from Troia III, with clear additions to the older traditions (Blum 2006).

The archaeological evidence indicates that at this time Troia may now be not only a well-organized society that contains sectors inhabiting different neighborhoods, but also practicing different customs, perhaps even representing groups with different origins. Regardless of the social structure of the Middle Bronze Age settlement, the population appears to have increased since the Early Bronze Age. The population increase and the introduction of new elements to the material culture imply a more complex society than those represented by the older settlements in Troia.

The succeeding period with its distinctive cultural-historical features is the so-called "High Troia Culture" (Korfmann 1996). This period includes the three phases of Troia VI (early, middle and late) and the Troia VIIa phase, which lasted for about a century. The periods coincide with the latter part of the Middle Bronze Age and the Late Bronze Age chronologies elsewhere in the region. In this study, Troia VI and VIIa periods are together termed the "Late Bronze Age." The three occupational phases of Troia VI are distinguished by the construction of new citadel walls (Figure 7). Troia VI ends with an earthquake. Troia VIIa comes to an end evidently as a consequence of warfare (Becks 2006).

One of the major aims of the New Excavations in Troia is to determine the extent of the lower town during the Late Bronze Age (Jablonka 2006). Partially for this reason and partially because the Late Bronze Age architecture within the citadel has been severely destroyed by later building activity in antiquity, most of the Troia VI-VIIa remains explored during the New Excavations come from outside the citadel walls (Figures 7 and 8). Together with the lower town, the entire city is thought to have reached a surface area of ca. 350,000 m^2, with an estimated population amounting to 5000 to 10000 inhabitants (Figures 7 and 8; Becks 2006).

The architectural plans of the Early Troia VI period suggest that a completely new population settled on the site (Korfmann 1995). Late Bronze Age Troia –with its houses and fortifications, lower city and the defensive ditches that surround it, streets and monumental buildings, and fountains and water systems, represent a flourishing city. The palace, which should be located in the middle of the citadel, was completely destroyed by the building of the Athena Temple during 3rd century BC (Figure 7; Rose 2006). Different architectural styles observed throughout the site, such as residential houses with apsidal, megaron, and L-formed ground plans, and the post buildings of the lower city, are indicative of a socially-differentiated population (Figures 7 and 8; Becks 2006, Jablonka 2006). Imported ceramics and other artifacts show close contact with the contemporary Mycenaean and Hittite realms (Becks 2006).

The architectural layer formed after the violent destruction of the Troia VIIa period buildings, namely Troia VIIb, is dated between ca. 1150 and 950 BC. This period correlates with the Early Iron Age, but it is conventionally either included in the Late Bronze Age or treated as a separate chronological unit. In the present study, Troia VIIb will be treated as a chronological unit separate from the Late Bronze Age, but the conventional denomination "Post-Bronze Age" used by the Troia Project for all occupational layers following this period will be kept.[10] The Early Iron Age represents a strong contrast with the previous Late Bronze Age imperial system of the Hittites and the influential cultural realm of Mainland Greece (Becks et al. 2006, Dickinson 1994). The archaeological record of the Early Iron Age is scarce in both Anatolia and the Aegean. In Troia, the material culture of this period indicates the migration of a new people with their origins probably in the Balkans or the northwestern Black Sea (Korfmann 1996, Becks et al. 2006). Mycenaean elements continue to influence the handicraft industries of the town. The settlement structure follows the pattern already established in Troia VIIa. The separation between the citadel and lower town persists with the continued use of the citadel wall. Houses in the citadel consist of one or two rooms, whereas those located in the lower town consist of multi-roomed complexes. (Figure 7; Becks et al. 2006). The mammal sector of the subsistence economy, however, demonstrates patterns that are clearly distinguishable from the earlier Troia VI and Troia VIIa periods (cf. Section 2.7.1, per. com. H.-P. Uerpmann). The Troia VIIb settlement collapses with a fire (Becks et al. 2006).

The two centuries following the end of the Troia VIIb settlement are regarded as the "Dark Ages". Until about 750 BC, archaeological remains in Troia are sporadic. The following periods VIII, IX and X are together termed the Post-Bronze Age in Troia and contain finds belonging to the Archaic (VIII), Classical (VIII), Hellenistic (VIII), Roman (IX) and Byzantine (X) settlements. Troia served as a pilgrimage center during those centuries as the scene of the legendary Trojan War. The population declined from the effects of two destructive earthquakes in 6th century AD. The settlement was not rebuilt until late Byzantine times when Troia became the residence of a bishopric. (Rose 2006).

2.6.2. Archaeological research at Yenibademli

Archaeological excavations at Yenibademli commenced in 1996 and continue at present. Work at the site is conducted by a team from Hacettepe University in Ankara under the direction of Halime Hüryılmaz. The archaeological project of Yenibademli aims to contribute to the island's cultural history by delineating the successive settlement structures on the mound and elucidating the nature of the contacts the island maintained with mainland Anatolia, Southeast Europe, continental Greece, and other Aegean islands (Hüryılmaz 1998). According to surface surveys, initial human occupation on Gökçeada dates back to the Neolithic period (Harmankaya 2001).

Yenibademli represents a settlement contemporary with Troia I and the Early Bronze Age II of the Aegean, dating to ca. 2900-2600 BC (Hüryılmaz 1998, 2002). Altogether seven architectural phases were recognized by the excavators. These are labeled with Roman numerals in chronological order, from I to VII; the youngest represented by I (per. com. H. Hüryılmaz).

Results of the archaeological investigations show that the site was densely settled in the Early Bronze Age and was enclosed in a defensive wall. The remains of this defensive stonewall were uncovered in Quadrate E9 in 1996 (Figure 9; Hüryılmaz 1998). A rampart located at the southeast of the mound provided access to the walled settlement (Hüryılmaz 1999). Some locally produced Mycenaean ceramics were encountered on this rampart, indicating that the rampart was used during the Late Bronze Age (Hüryılmaz 2000). Work in other excavation areas yielded elongated rectangular architectural units consisting of stone foundations and mudbrick superstructures with roofs supported by wooden posts (Hüryılmaz 2001). These structures are of domestic character (Hüryılmaz 2003). Artefactual finds do not demonstrate major differences between architectural units. The buildings are occasionally furnished with open hearths and flagstones (Hüryılmaz 2003).

With these finds, Yenibademli represents a small island settlement enclosed in a wall.

Figure 9. Topographical plan of Yenibademli (Courtesy of H. Hüryılmaz, Yenibademli Excavations).

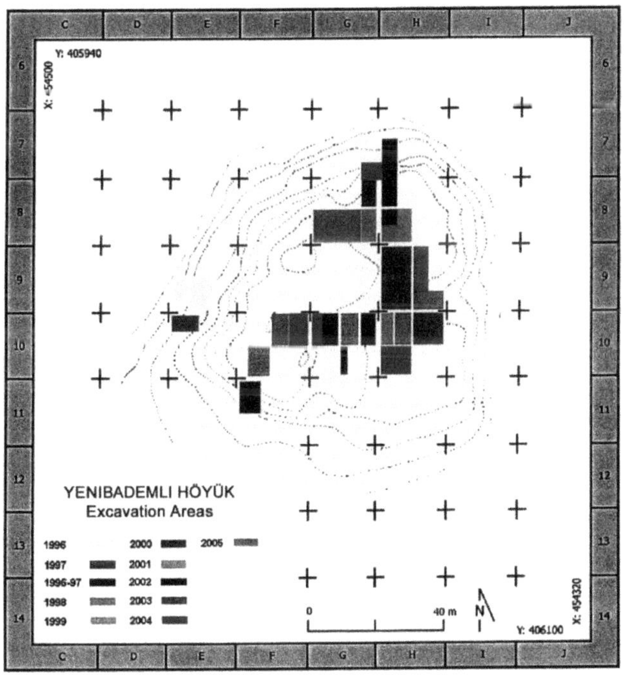

2.6.3. Archaeological research at Ulucak

Excavations at Ulucak started with a research perspective oriented towards the Neolithic period, a time period barely known from the Anatolian coast of Aegean (Çilingiroğlu et al. 2004: 1). The site represents one of the few prehistoric settlements investigated in the eastern Aegean and therefore presents critical information regarding the understanding of

[10] Both the chronology of the transitional period between the Late Bronze Age and Early Iron Age at Troia and the end of the "Dark Ages" at the settlement were subject to fierce discussion at the time of writing (per. obv. at the Troia Workshop in Tübingen, 28-30.10.2006).

the Neolithization of western Anatolia and the subsequent cultural contact with the rest of the Aegean. However, cultural layers later than the Neolithic period were also discovered at the site. Archaeologists labeled four main occupational periods in the stratigraphy of this mound. Their relative and provisional calendar dates are given in Table 3 (Çilingiroğlu et al. 2004). A topographical map of the excavations areas is represented in Figure 10.

The earliest occupational layers of the mound date to the Late Neolithic Period (Table 3). The period's ceramic assemblage is similar to those from the Lake District Neolithic in inner Anatolia. Radiocarbon analyses of charcoal remains date the period to the first half of the 6th millennium BC. Rectangular buildings constructed with wattle-and-daub, pisé and mudbricks, with and without stone foundations characterize the architectural remains. Numerous hearths and ovens were found in interior spaces. (Çilingiroğlu et al. 2004, Çilingiroğlu and Abay 2005).

The subsequent occupational phase (Level III) dates to the Chalcolithic Age, which has cultural parallels in Kumtepe A. The architectural remains of the period were poorly preserved only at one section of the excavated area. Stone foundations made of small field stones represent the main architectural feature of this period. (Çilingiroğlu et al. 2004).

Figure 10. Topographical plan of Ulucak (Courtesy of the Ulucak Project).

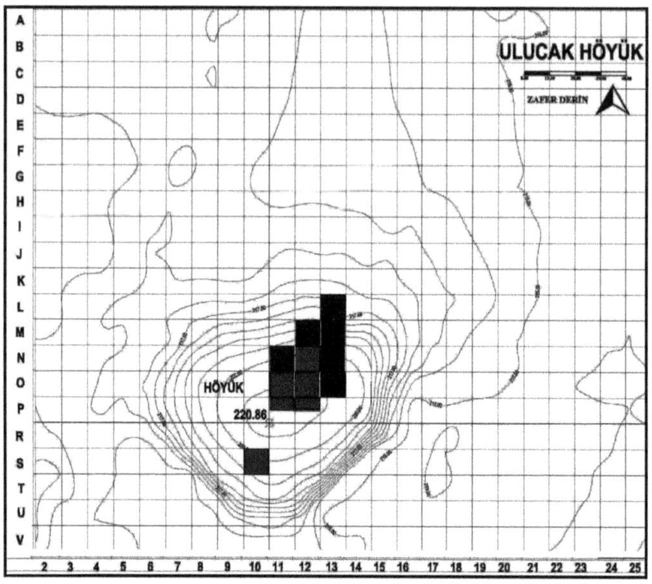

Above the remains dating to the Chalcolithic Age, remains of an Early Bronze Age II culture showing similarities with the Troia II material culture were found (Level II). At the time of writing, no radiocarbon dates were available for these levels. The stone architecture of the period has domestic characteristics. During this period the mound was terraced and was possibly surrounded by an enclosure wall. Because the Early Bronze Age layers were largely destroyed by later building activity, not much information is available about the archaeology of this settlement. At the time being, the Early Bronze Age settlement appears to represent an agricultural village, which may or may not have been enclosed in a wall. Although an Early and Middle Bronze Age cemetery was found close to the settlement, cultural layers corresponding to the Middle Bronze Age cemetery could not be located on the mound proper. (Çilingiroğlu et al. 2004).

Table 3. Chronology of Ulucak.

Occupational levels	Conventional cultural period	Calendar dates
IV	Late Neolithic Period	6000-5500 cal. BC
III	Chalcolithic Age	?
II	Early Bronze Age II	2550-2250 BC
I	Late Roman-Early Byzantine	?

The remains of the Late Roman and Early Byzantine settlements have been severely damaged by recent agricultural activity. The recovered architecture of these settlements represents domestic structures and a few storage pits. (Çilingiroğlu et al. 2004).

Apart from the Neolithic phases, absolute dates are not available for the site. Excavations and related interdisciplinary research continue.

2.7. Subsistence economy

The understanding of the subsistence economies of the sites that are investigated in the present study is shaped by the results of archaeobotanical and archaeozoological investigations. Although some information about the subsistence economy of each site is available, there is great disparity in the quality and quantity of information about the subsistence economies of the different sites and their chronological periods.

2.7.1. Subsistence economy of Troia

The subsistence economy of Troia has been investigated since the beginnings of archaeological research at Troia, at the end of the 19th century. Schliemann and his co-worker Virchow, a naturalist who assisted Schliemann in what would today be called environmental archaeology, attempted to draw a picture of the economic activities of the site's inhabitants based on observations of the faunal and botanical remains recovered at Troia. Virchow attended the excavations in the Troad and collaborated with Berlin scientists on special biological fields, such as malacology. This collaborative research made it –already then– clear that the ancient inhabitants were engaged in agriculture, animal husbandry, hunting, fishing, and shellfishing, exploiting a variety of ecological niches available in the landscape. (Virchow 1879, 1881, von Martens 1879, 1880).

Early in the 20th century, Gejvall's (1938, 1939, 1946) work provided details on the various properties of domesticated animal remains from Troia. Blegen and his coworkers catalogued archaeobiological remains found per archaeological context in their monographic series on the Cincinnati excavations at Troia, but they did not provide quantified accounts of these remains in absolute or relative amounts (Blegen et al. 1950, 1951, 1953, 1958).

Earlier work on the mollusk remains of Troia are discussed in detail later in this chapter (Section 2.8).

At the beginning of the New Excavations in Troia, H.-P. Uerpmann launched an archaeobiological project with the aim of illuminating the evolution of ecology and subsistence through successive occupational phases at Troia and its neighbors, based on the quantitative archaeobiological evaluation of the organic remains retrieved from the New Excavations (Uerpmann 1992). Research on animal husbandry practices, hunting, fishing, agriculture and land use, and the present study constitute various parts of this archaeobiological program.

The practice of plant cultivation and land use in Bronze Age Troia have been illuminated by archaeobotanical research: during the Early Bronze Age (Troia I-III), hulled wheat (*Triticum monococcum*) and fig (*Ficus carica*) were important components of the agricultural diet, whereas barley (*Hordeum vulgare*) and crop legumes were present but had a less important role (Riehl 1999: 83). Riehl reasoned that in the Early Bronze Age cereal cultivation took place on the lower plateau –more or less at the same elevation as the settlement (Riehl 1999: 55). Intensive small-scale cultivation was probably the norm of plant production during Troia I (Riehl 1999: 73).

The plant spectrum is most diverse during the Middle Bronze Age (Troia IV-V), representing the use of a wide variety of ecological habitats (Riehl 1999: 86). The frequent use of coastal habitats is best reflected by the samples dating to the Middle Bronze Age. This situation suggests intensive, garden-type cultivation mainly in the delta area, but also a generally risk-buffering form of plant cultivation (Riehl 1999: 57). Free-threshing wheat, (*Triticum aestivum/durum*), garden pea (*Pisum sativum*), and flax (*Linum usitatissimum*) were introduced to the settlement during the Middle Bronze Age (Riehl 1999: 84). It appears that olive (*Olea europaea*) was cultivated at all periods except the Middle Bronze Age, when flax probably substituted for it (Riehl 1999: 54). Riehl (1999: 56) suggests that emmer (*Triticum dicoccum*) may have been grown close to the coast, whereas barley may have been cultivated close to the settlement but also in the valley during the Middle Bronze Age.

During Troia VI, agricultural yield was dominated by barley, while chickpeas (*Cicer arietinum*) and crop legumes were of secondary importance (Riehl 1999: 84). The archaeobotanical assemblage of Troia VI indicates extensive cultivation practices and specialization in certain cereals (Riehl 1999: 73). Some previously well established cereals disappear in the record during this period. Emmer and einkorn, accompanied with rye (*Lolium* spp.) are the most abundant seeds in the macrobotanical assemblage of Troia VIIa. Chickpeas become important in Troia VIIa suggesting that the species may have been used as a source of protein in this period (Riehl 1999: 56). During Troia VIIb bitter vetch (*Vicia ervilia*) is one of the main crops beside barley (Riehl 1999: 85).

Archaeobotanical studies show that cereals and legumes were the most important agricultural products during the Bronze Age. Agriculture evolves from small-scale garden-type cultivation in the Early Bronze Age to specialization on certain cereals in the Late Bronze Age. Archaeobotanical evidence also indicates that risk-buffering agriculture may have been the mode of plant cultivation during the Middle Bronze Age. In general, moisture indicating species become less frequent throughout the Bronze Age periods (Riehl 1999: 85). The increase in weed species throughout time indicates the effects of human influence on the landscape (Riehl 1999: 84).

Analyses of animal bone remains have demonstrated the continuous importance of cattle (*BOS*)[11] from the Bronze Age up through the Hellenistic periods in the animal economy of Troia (Fabiš 1996, 2003, Uerpmann 2003, Uerpmann 2006). The proportion of cattle remains increases with the growing population in time. Remains of sheep (*OVIS*) and goats (*CAPRA*) follow cattle in weight of identified specimens (WIS). Primary and secondary uses of cattle, sheep, and goats in Troia still need to be interpreted in detail by way of analyzing culling ages and sex profiles of the ancient populations represented by their bones. Pigs (*SUS*) constitute the third most important sector of meat gain from domestic animals (Uerpmann 2003).[12] Domestic dogs (*CANIS*) were always present at the settlement. Analyses of butchery marks on their bones indicate that they were eaten, and some closed contexts –including intact skeletons– point to their value as sacrificial animals (per. com. D. Etkin).[13] Donkey (*ASINUS*) and horse (*CABALUS*) were introduced to the settlement during Troia VI (Uerpmann 2003, Uerpmann and Uerpmann 2001). Fallow deer (*D. dama*) and wild boar (*S. scrofa*) hunt was one of the important economic activities throughout the occupational history. An increase in fallow deer remains in contexts dating to Troia VI and VIIa periods may indicate both the increasing significance of hunting as a social practice and the increasing areas with open vegetation in the vicinity of the settlement (Uerpmann 2003). Fallow deer was granted a cultic role already in Troia VI (Uerpmann 2006). The relative abundance of equids and game animals such as the fallow deer decreases abruptly in the faunal assemblages coming from Troia VIIb contexts (per. com. H.-P. Uerpmann). In the subsequent Troia VIII phase, contrary to Troia VIIb, there is a marked increase in the proportion of fallow deer (*D. dama*) remains in the faunal assemblages, indicating an emphasis on hunting (Fabiš 2003). It is important to note that the faunal assemblage of the Troia VIII period originates mainly from ritual contexts (Fabiš 2003).

[11] The terminology for domestic animals follows the nomenclature suggested by Uerpmann 1993.
[12] Animal husbandry and hunting practices in Early Bronze Age Troia constitute part of C. Y. Gündem's doctoral dissertation. This work was in preparation at the time of writing.
[13] D. Etkin is the author of a master's thesis about the dog (*CANIS*) remains in Troia. This work was in progress at the time of writing.

Red deer (*C. elaphus*) and roe deer (*C. capreolus*) are only rarely represented in the faunal assemblages, indicating that the Ida Mountains and their slopes were not included in the primary hunting ground for Bronze Age Troia. The faunal assemblages of the Post-Bronze Age settlements have been studied by a small team under the direction of M. Fabiš. Since the subsistence economy and palaeoenvironmental reconstructions of these periods deserve a study in their own right, a comprehensive account of the Post-Bronze Age animal economy awaits publication.

Trojans' interactions with domestic animals and hunted ruminants seem to have deviated from the general Bronze Age pattern during Troia IV. Small domestic ruminants (i.e., sheep and goats) are underrepresented in the faunal assemblages from Blegen's excavations of the Troia IV and V settlements, whereas the proportion of fallow deer in the same assemblage is significantly high. There is a general lack of uniformity in the animal bone assemblages belonging to these periods, both from old and New Excavations, with high percentages of wild animals in certain samples, and accumulations of prematurely slaughtered pigs. (Uerpmann 2003).

Bird hunting is mirrored in the bird bone assemblage of the site. The variety of hunted birds has been discussed above in Section 2.5.3.

The role fishing played in the subsistence economy is reflected by the composition and amount of fish bones recovered from Troia. In the New Excavations at Troia, similar to the majority of other faunal remains (cf. Section 3.1), the fish remains have been habitually yielded through hand-collecting methods. Consequently, their value for expressing fishing activity in Troia is limited. Fish bones tend to be smaller than the remains of most other economically important animals and their state of preservation is generally relatively poor (Colley 1990). This situation compels the employment of dry and wet sieving techniques for a more complete recovery of fish remains, which will allow more accurate descriptions of fishing practices in ancient settlements, both coastal and inland (Clason and Prummel 1977, Rose 1994: 388, van Neer et al. 2005). In Aegean archaeology, the common method of sampling faunal remains is by hand-collecting. The restricting effect of hand-collecting methods on the descriptions of prehistoric fishing in the Aegean and in the greater Mediterranean region is great. Although most scholars share the opinion that the production level of the Mediterranean Sea is not sufficient to allow coastal settlements to have subsistence economies based on fishing (Purcell 1996, Rose 1994: 32-34), it is also agreed that the current understanding of the role of fishing in the subsistence strategies of the region is far from accurate (van Neer et al. 2005). Moreover, even if the Mediterranean is generally not a reliable fishing ground, areas such as the littorals and channels, like the Dardanelles and the Gibraltar, host large fish populations, at least seasonally. Opinions about the role of fish in ancient Mediterranean diet are further influenced by textual sources that degrade fishing as a lower occupation and address fish as an unpalatable food item (Purcell 1996). The simple truth is that the role fishing played in the ancient subsistence strategies of the Mediterranean represents a large gap in the knowledge of past Mediterranean economies.

In Troia, out of 439 fish bones that have been studied by Uerpmann and van Neer (1998, 2000), the majority belongs to the sea brass (Sparidae) family. Mackerel and tuna fish (Scombridae) follow this family in terms of frequency. In addition, sardines (*Sardina pilchardus*), bluefish (*Pomatomus saltator*), mullet (Mugilidae), sea basses and groupers (Serranidae) are among the most commonly represented marine species. Freshwater fish carp (Cyprinidae) and catfish (Siluridae) make their first significant appearance in the fish bone assemblage after the end of the Bronze Age (van Neer and Uerpmann M. 1998, 2000). Based on the chronological patterns of the species spectrum represented by the fish bone assemblage, fishing grounds utilized by the ancient inhabitants of Troia can be described as follows: throughout the occupation of Troia, the Trojan Bay provided most of the fish that were found in the fish assemblage obtained from the New Excavations. With the progradation of the Karamenderes and the Dümrek deltas, the riverbeds became important fishing grounds starting from Troia VII. The open sea fishing was more frequently practiced during the Early Bronze Age than during the Middle Bronze Age and Troia VI. In Troia VII, the use of open waters became important once again. Although the importance of open waters as fishing grounds appears to persist during the Post-Bronze Age, the limited amount of fish bones dating to the period does not allow satisfactory conclusions. (van Neer and Uerpmann 2000: 165).

2.7.2. Subsistence economy of Yenibademli

Archaeobotanical research on the macrobotanical remains from Yenibademli has shown that at least nine species of crop plants were cultivated by the inhabitants of the site, consisting of three species of cereals, a broad range of legumes, and grapes (*V. sylvestris*). Cereals included einkorn, emmer and barley. Charred cereal remains were recovered from grain containers located in domestic buildings. Remains of legumes have been recovered in greater amounts at the site and include bitter vetch (*Vicia ervilia*), faba beans (*Vicia faba*), grass pea (*Lathyrus sativus*), Spanish vetchling (*Lathyrus clymenum*), lentil (*Lens culinaris*) and clover (*Trifolium*). (Oybak-Dönmez 2005).

In 1999, a zoologist from Hacettepe University, Z. Ayaş, examined the faunal remains of the site and reported briefly about the taxonomic groups he identified (Hüryılmaz 2001). The archaeozoology of the site has been integrated into a regional archaeozoological program in 2004, as a result of an initiative taken by H.-P. Uerpmann in 2001.

The animal bone assemblage of Yenibademli is dominated by the typical Mediterranean domesticates; sheep, goat, cattle, and pig. Cattle constituted the main meat supplier for the settlement. Pig was kept for its meat and slaughtered young. The limited amount of dog remains does not provide evidence for their dietary consumption. Fallow deer was the most important and virtually the sole hunted animal, representing a key source of protein. Other game animals such as wild boar, hare, and wild goat are rare.[14] Only five fragments of fish bones were recorded from the studied sample of the faunal assemblage in Yenibademli. One would expect a higher representation of fish bones to have been recovered at a coastal site where animal bones are generally well preserved. Had dry or wet sieving techniques been employed during the excavations and the residual soil (heavy fraction) been investigated for faunal remains, the amount of fish bones could have been much larger. The restricted amount of fish bones does not permit any quantitative interpretation of fishing practices employed at the site.

2.7.3. Subsistence economy of Ulucak

Little is known about the subsistence economy of the site. Preliminary work on a few archaeobotanical samples (nine altogether) revealed the presence of einkorn (*Triticum monococcum*) cultivation in the Neolithic, and the presence of some wild fruits and nuts such as the acorn. Although pulses were absent in these samples, this situation is thought to be a consequence of the restricted sample size. (Megaloudi 2005). Plant cultivation in Neolithic Ulucak is now the subject of a doctoral research conducted by Aylan Erkal.

Archaeozoological investigations on a limited (604 specimens) sample of animal bones from the site attested to the presence of fully domesticated ungulates, primarily sheep and goats, and some hunting. This animal bone sample was selected from various contexts distributed between different prehistoric layers (Trantalidou 2005). A minute sample of mollusk remains (18 specimens) from the site was also described in a preliminary report (Karali 2005b). Karali's species list consists mainly of *Cerastoderma glaucum* remains.

In 2004, the analysis of archaeozoological material from Ulucak was integrated into the same regional archaeozoological program already mentioned in Section 2.7.2. The faunal assemblages recovered from the Chalcolithic and Early Bronze Age contexts of the mound suggest that animal husbandry was of primary importance for the animal sector of the subsistence economy. [15] Cattle were the most important meat supplier, followed by pig, sheep, and goats in the Chalcolithic Age; whereas in the Early Bronze Age, pigs lost their significance. Hunted animals have a high representation in the faunal assemblages of both periods. Fallow deer appears to have supplied a significant portion of the protein diet in the Early Bronze Age, apparently more than domesticated pigs. Only three fish bones have been found in the studied portion of the assemblage, one of them belonging to a Sparid.

[14] Details of animal husbandry practices and hunting at Yenibademli are included in C.Y. Gündem's forthcoming doctoral dissertation.

[15] A comprehensive analysis of the mammal assemblages dating to the Bronze Age layers of the mound will be included in C.Y. Gündem's doctoral dissertation.

2.8. Earlier work on the mollusk remains of Troia

The first excavator of Troia, Heinrich Schliemann was one of the first archaeologists to give importance to the relationship between human settlements and their environment. This approach, like his other undertakings, was stimulated by Homer's detailed descriptions of the Trojan landscape. Not much space is spared to activities related to foraging in Homer's depictions of life and environment in the Troad. Throughout the Iliad and the Odyssey, Homer refers to invertebrates only once. In the Iliad (XVI, 745-748), Patroklos praises Hektor's chariot driver:

> See now, what a light man this is, how agile an acrobat
> If only he were somewhere on the sea, where the fish swarm,
> he could fill the hunger of many men, by diving for oysters;
> he could go overboard from a boat even in rough weather.

The Greek word translated here as "oyster" is "τήζεά". The first scientist to comment on the mollusk remains from Troia, von Martens (1879) questioned this transilation, pointing that although Homer's "τήζεά" is generally associated with oysters, a very similar word, "τήζυον", is used in the context of sea-squirts (or anemone of Ascidiacea) by other Greek authors such as Aristotle. He thereby opened a discussion regarding the identity of Homer's "τήζεά". Ascidiacea are benthic invertebrates living on hard surfaces. Certain species are used as a tasty appetizer in modern Mediterranean cuisine (Davidson 1972: 218). While Ascidiacea are not mollusks, they can be regarded as shellfish in terms of the techniques used for their acquisition, preparation and culinary meaning. Therefore, this excerpt from the Iliad indicates that Homer regarded –whether he meant oyster or sea-squirt– at least some shellfish gathering as a strength-requiring but calorifically rewarding task and acknowledged its social implications.

There is no hint in Schliemann's publications to even remotely imply that Schliemann himself was in a anyway inspired by this section of Homer's work to become particularly interested in the nature of shellfishing activity at Troia, but the data his team collected already made it manifest that Homer's description of shellfishing is not a sufficient representation of everyday activities involving shellfish procurement and consumption in Troia, but, like the rest of his discourse, a poetic depiction of heroic acts (Virchow 1881: 135).

Von Martens, who aided Virchow in identifying the mollusk remains from Troia, commented on a number of molluscan species represented in the earliest excavations in Troia (von Martens 1879, later quoted in Virchow 1881: 134).

> "In den Ausgrabungen von Troja fanden sich:
>
> *Murex trunculus* (L). *Purpura haemastoma* (L.). *Columbella rustica*. *Cerithium vulgatum* var. *spinosum* (Philippi). *Cypraca lurida* (L.). *Trochus articulatus* (Lam.). *Patella caerulea* (L.). *Ostrea lamellose* (Brocchi). *Spondylus gaederopus* (L.), *Pecten glaber* (L.). *Pecten glaber* var. *sulcatus* (Born). *Pectunculus pilosus* (L.). *Pectunculus violascens* (Lam.). *Mytilus edulis* (L.). var. *galloprovincialis* (Lam.); sehr zahlreich. *Cardium edule* (L.), var. *rusticum* (Lam.), sehr zahlreich.

Venus verrucosa (L.). *Tapes decussatus* (L.). *Solen marginatus* (Pulteney; *vagina*, auct.)."

Von Martens' record helps to bridge the gaps in the species spectrum produced by the archaeozoological data of the New Excavations, by listing species such as *Columbella rustica* and *Purpura haemastoma*, which were either not present in the faunal assemblages of the New Excavations or were registered as "unidentified mollusks," "unidentified marine gastropods," or included within the Muricidae by the archaeozoologists who worked on the site's faunal assemblage.

Von Martens (1879) also states that *Murex* (*=Hexaplex*) *trunculus* remains were found in great numbers in accumulations which did not include the remains of other mollusks. This is certainly information he gathered from Virchow who brought him the mollusk remains from Troia. Von Martens dedicated a paragraph to the discussion of the dye industry in Troia, based on his observations on the *H. trunculus* and *P. haemastoma* remains. He then concluded his discussion about the mollusk remains by making the correct judgment that most of the mollusk remains found in the deposits of Troia represent food refuse. He additionally made a few taphonomic comments on the mollusk shells that were presented to him in Berlin. For example, he noted that the aperture lip of the *C. vulgatum* specimens were broken, a breakage he found analogous to the way *C. vulgatum* is served in Venice of the time before it is served. He also noted that the margins of some *P. caerulea* specimens were broken. To von Martens, these breaks appeared to have occurred when the specimens were gathered off the rocks they lived on. Von Martens adds that *M. galloprovincialis* remains were mostly fragmented. He also reports that the largest *Cardium edule* (*=Cerastoderma glaucum*) valve was 40 mm in height and 41 mm in length. He commented on the curved shell morphology of *C. glaucum* valves, which he exemplified with an "especially oblique" specimen, measuring 32 mm in height and 43 mm in length. Von Martens interpreted the gray-black shells of *Ostrea* and *Cerastoderma* remains as having stained by anorganic sedimentary mud, whereas (he reports) Virchow was in the opinion that these specimens were exposed to fire.

In 1881, Virchow quoted much of von Martens' work in his contribution to Schliemann's book on his excavations in Troia. Although Virchow did not record the quantity of represented species in figures, he mentioned that *Mytilus galloprovincialis* and *Cardium edule* (*=Cerastoderma glaucum*) remains were "numerous." This information is, to a certain extent, in agreement with the quantity and the quality of the mollusk remains yielded from and around the Schliemann's Trench during the New Excavations (cf. Section 4.2.3). On the other hand, Schliemann reported that oysters and mussels are so numerous that whole layers consist only of their remains and they are especially abundant in the layers of "the oldest city". He added that *Cerastoderma* shells comprise the largest portion of the remains of Troia III, but the oysters were generally more common in all layers of the "prehistoric city" (Schliemann 1881: 359). According to Schliemann (1881: 579), masses of shells continued to be an important component of archaeological strata in his fourth city.

Probably upon von Martens' initial comments on the *H. trunculus* remains from Troia, Schliemann (1881: 359) paid particularly close attention to the presence of fragmented *Murex trunculus* and *Purpura haemastoma* remains, both of which are known to have been used in the dye industry in the ancient world. Schliemann reports that he found a layer consisting exclusively of crushed *H. trunculus* shells below the highest stratum of "the wall of Lysimachus" (1881: 359). According to R. Becks' stratigraphic reconstructions, Schliemann undoubtedly refers to the area of his southeastern trench, which cuts through House VI G and the Troia VI citadel wall (cf. Çakırlar and Becks in print). Like wise, the heaps of 'murex' shells Dörpfeld found in his 1894 excavations were also located inside the Troia VI citadel (Dörpfeld 1902: 9, also cf. Çakırlar and Becks in print).

Notes on the mollusk remains from Blegen's excavations have been added to the excavation monographs. Similar to the rest of the faunal remains, this data was not quantified (Blegen et al. 1950, 1951, 1953, 1958). Gejvall, who also studied the vertebrate remains from these excavations (1938 and 1939), summarized his observations on the mollusk remains in two preliminary archaeozoological reports. He (Gejvall 1938: 52-55) reported that,

> "the animal remains of this period [Troia I] show a distinct uniformity which is easily accounted for, since here we find mainly marine animals such as cockles, ark shells (*Arca Noae*), blue-mussels, knife-mussels, two species of purple shells, sepias (the rudimentary shells of these animals are very frail, but quite recognizable fragments have been found). Careful investigations have shown that no visible changes have occurred in the composition of the marine fauna since that period....The prehistoric settlement of Kumtepe, excavated in 1934, which begins at a somewhat earlier date than Troy I abounds in mussels. Here as in Troy I the fragments of bones take second place being only a small percentage of the finds compared with the shells. Nearly ninety per cent of the cockles have been opened with some kind of pointed tool and have distinct breaks in the same place close to the hind part of the shell. The ark shells, however, which are larger have no marks of any damage. Consequently it is very likely that the former were at that period eaten fresh and the latter were boiled. The ark shell is not so common in these strata as the cockle.
>
> ....
>
> Mussels are still found in great numbers [in Troia II].
>
> ....
>
> In Troy VI a great many purple shells were found in a small area and this suggests a purple industry.
>
> ....
>
> Among other animals not mentioned above I should like to name...the bivalves (*Patella*), *Ostrea*, *Venus*, *Tapes*, *Pecten*, and *Pinna nobilis*."

Part of the mollusk material from Blegen's excavations is currently stored in Duluth, Minnesota, U.S., part of it is in Stockholm, Sweden, and most of it has been discarded (per. com. D. S. Reese). Gejvall's account and interpretation of the mollusk remains from Blegen's excavations are discussed in the relevant sections of this work (cf. Sections 4.3.1, 4.3.2, 4.6.1, 5.4, 5.7, 6.1, 6.3).

Like Schliemann, the only mollusk remains Blegen discussed in some length were the *H. trunculus* deposits. These were discovered between House VIG and the fortification wall, corresponding to Quadrates J7-8 in association with wells, stone grinders, and fragments of worn millstones (Blegen 1937). Blegen reported that "these strata can be traced continuously some 20 or 30 m northward into Quadrate J6." and proposed that the dye production establishment may have occupied this place (Blegen 1937: 582, for a more detailed discussion of Blegen's 'murex' related findings, cf. Çakırlar and Becks in print). D.S. Reese, who analyzed the shell material in Minnesota also reported on samples of crushed *H. trunculus* remains from House VIF located in the southeastern sector of the Troia VI citadel (1987a).

Additional evidence about the mollusk remains from other sites in the Troad is present in the scientific literature of the late 19th century. Hanaytepe, a settlement mound and tumulus four km southeast of Troia (cf. Section 2.6.1), was initially investigated during the second half of 19th century (Calvert 1881). The excavator, Frank Calvert, reported that he consulted Virchow about the mollusk remains that were recovered from this site: "Virchow beobachtet, dass sich in Hanai Tepeh [=Hanaytepe] nur die *Ostrea lamellose* findet, während *Ostrea cristata* auf Hissarlik beschränkt ist.... (Calvert 1881: 787)". It was in fact von Martens (1880) who reported on the mollusk remains from Hanai Tepeh and made this observation. In modern malacology, *O. cristata* and *O. lamellose* are no longer considered to represent separate species, but only variations of *O. edulis* (Demir 2003).

Descriptive accounts of the mollusk remains from the 1980s excavations at Yassıtepe and Sivritepe at Beşik Bay have been published (Boessneck 1986, von den Driesch 1999). The implications of these accounts for the reconstruction of environment and economy in the Troad will be discussed in detail in the following chapters.

The following chapter, Chapter 3, will explain the methods used in this study.

Chapter 3. Methods

This chapter is devoted to the description of the methods utilized in this study. The rationale involved in choosing these methods is also explained.

3.1. Excavation techniques and sampling

It is very well known that abundant archaeobiological remains constitute the remnants of the biosphere surrounding a site. They are the logical source to consult in reconstructing the human relationship with formerly existing landscapes (Renfrew and Bahn 1996: 211). But reconstructing past human-environment relationships is not always among the priorities of archaeological projects. The research objectives of an archaeological project determine the attitudes and techniques that are adopted during excavation. Although one of the research concepts of New Excavations is "landscape archaeology" (Korfmann 1991), systematic recovery and careful recording of archaeobiological remains from homogenous stratigraphic units of the site was not made a priority during the recovery of archaeological deposits in Troia. The priorities of the archaeological projects at Yenibademli and Ulucak have been likewise oriented towards aims other than palaeoenvironmental and palaeoeconomic reconstruction.

The mollusk remains in Troia, Yenibademli and Ulucak have been retrieved through hand-collecting methods. Sampling was done in all excavation units. Sampling archaeozoological material by hand-collection is not very sensitive to the recovery of small species and specimens. As is always the case with hand-collecting methods, an indeterminate quantity of shells –depending on the care and interest of individual excavators and their teams– will have remained in the earth matrix. In many cases, this situation hampers archaeozoological interpretation (Clason and Prummel 1977, van Neer et al. 2005). With the exception of excavation units ("Behälter") 174, 286, 632, and 699 from Quadrate KL16/17 in Troia, none of the shell-bearing archaeological contexts that are dealt within this work have been wet-sieved.

In Troia and Yenibademli, mollusk remains have been retrieved and stored along with other animal remains, whereas in Ulucak, they have been registered and stored separately. Mollusk remains belonging to the same archaeological deposit were labeled with successive "Knochenbehälter" (i.e., bone container) numbers unique to specific excavation areas in Troia and Yenibademli (e.g., Beh. 530 in Quadrate A5/6). In Ulucak, mollusk remains that form stratigrafically meaningful groups bear the label-code of the context where they were recovered. The coding system is continuous through the site and each code is unique (e.g., COF, ADC, BTG, and so on). All archaeomalacological material from all three sites has been hand-washed with water immediately after recovery. This process removes foreign debris off the specimens and reveals taphonomic and morphological features of the bone and shell material.

Although the excavators in Troia were trained to collect all visible mollusk finds at the beginning of the New Excavations by the archaeobiologists working in the project, decisions related to collecting mollusk remains have been progressively left to the individual excavator's personal choice over the course of 19 years of excavations (per. com. H.-P. Uerpmann). This situation has created inconsistency in the sampling techniques used across the site's stratigraphy, hampering the accuracy of the analytical results obtained by quantitative archaeobiological methods, particularly, those presented in this study. It is highly probable that the picture depicted by the temporal distribution of the mollusk remains has been negatively affected by this practice, since individual excavators tend to focus on certain chronological periods that are of interest to them. An attempt was made to compensate for this inconsistency by deciphering the field notes in the excavation journals.[16] A survey of these hand-written notebooks demonstrated that the excavators chose only occasionally to make notes of the shell deposits they recovered from the archaeological matrix. Not infrequently, such deposits appear to be devoid of mollusk remains in the archaeozoological database (See Section 4.3.4). As a result, surveying yet-to-be digitalized excavation journals could help no more than recognizing certain conjunctures in the practice of archaeology at the site. Mention of ghost mollusk remains does not improve the accuracy of meaningful interpretations of the structure of shellfishing in Troia, which should be based on quantitative results.

For Yenibademli, it is presumed that all mollusk remains that were observed in the soil matrix were recovered without selective arbitrary decisions left to individual field coordinators. Again, presumably, mollusk remains received more attention from the excavators working at Ulucak. Since Ulucak is situated inland, mollusk shells of marine origin represent tangible evidence for contact with coastal areas. Evidence for inter-regional contact is a highly-rated in Aegean archaeology.

In summary, mollusk remains from different archaeological contexts were sampled to varying and unknown levels of completeness. It is impossible to assess how much of the material is actually missing. The intra- and inter-site inconsistency and the lack of standardization in recovery techniques constitute incalculable hindrances to the comparability of mollusk assemblages within and among sites.

[16] Excavation journals are property of the Troia Projekt. They are archived at the project's headquarters at Tübingen University.

3.2. Identification, description, and recording of specimens

Most of the palaeoenvironmental and palaeoeconomic interpretations laid out in this study are based on the quantitative record of archaeozoological assemblages including specimens of shell and bone. Their taxonomic identity and absolute quantity allow calculations of relative abundance of represented taxa, which provides a measure to determine what species were utilized and in which relative amounts. Essential to the understanding of the manner in which specimens were utilized are also descriptions of the mollusk specimens at the time of recovery and recognition of the formative agents that affected their state, and how their taphonomic state influences archaeozoological quantification. The comparability of the properties of the mollusk assemblages from the three primary sites that are dealt with in this study and the published archaeomalacological data from other Bronze Age sites in the Aegean is negatively affected by the diversity of the methods that were applied during their preliminary archaeozoological identification, description, and recording.

All mollusk remains that made their way into the hands of archaeozoologists at work in Troia were identified, counted, weighed, and digitally registered since 1989 along with other faunal remains. Examined as part of the faunal material of Troia, information about mollusk remains have been entered into a KNOCOD© database, which is a computer program designed to record and quantify animal bones from archaeological sites (Uerpmann 1974). With the start of archaeobiological research at Troia, certain common Mediterranean mollusk taxa were added to the 'type-list' available in the program. This amendment was expanded with the addition of new species in 2004 and was later modified and corrected using Microsoft EXCEL. However, due to the fact that the mollusk remains have not been studied by archaeozoologists with experience in Mediterranean mollusk taxonomy or researching specific questions relating to archaeomalacology, a good number of specimens belonging to different species were identified only at the family or genus level. Examples that were encountered only rarely in the assemblage were included in categories such as "unidentified bivalves," "unidentified land snails" et cetera. Mollusk specimens have not been examined systematically with regard to their taphonomic state.

At least 80 % of the mollusk remains that were recorded in this manner were subsequently discarded. They now form an anonymous post-excavation deposit on the site. For that reason, analytical techniques such as morphometric measurements and incremental growth examinations could not be employed on specimens selected through a systematic elimination of the entire assemblage, but simply on the bulk of the remaining ca. 20 % that were stilled kept in the storage rooms near the site. Taphonomic observations could also only be made on these remains.

Mollusk specimens in the faunal assemblage of Yenibademli were identified and registered in the KNOCOD© database by C. Y. Gündem and the author. Taphonomic descriptions of the specimens were noted in the "remarks" column of the KNOCOD© database. Taxonomic identifications were subsequently revised in Microsoft EXCEL. Animal bones from the Chalcolithic and Early Bronze Age layers of Ulucak were studied by C.Y. Gündem and the author, and computed by KNOCOD©. All mollusk specimens in Ulucak were examined and registered by the author using Microsoft Access. The archaeozoological databases are linked to Microsoft Access files that contain contextual information. This allows for the archaeozoological evaluation of faunal remains studied from different excavation units separately or merged, forming chronologically and stratigraphically meaningful combinations.

Taxonomic research of mollusks with calcareous body parts relies mostly on the morphological properties of shell for species determination. The same method is applied to ascribing subspecies and forms of species. Although the principles of malacological and archaeomalacological identification are essentially the same, a few differences, which occasionally pose problems for archaeomalacological studies, exist between the two. In malacology, it is preferable to work on specimen lots (i.e., sample populations) obtained from known locations for taxonomic research (per. com. B. Öztürk). Self-evidently, this is not possible in archaeomalacology. Additionally, not all archaeological mollusk specimens can be readily identified to the species level based on shell morphology, due to the differential taphonomic conditions of single specimens. Moreover, there is a good deal of disagreement between individual researchers about the taxonomic identification of certain aquatic mollusks. For instance, some species names tend to vary between publications. The present study does not intend to alter or contribute to molluscan taxonomy. Herein, the latest, most relevant taxonomic listings and their descriptions were used to define archaeological and modern specimens, according to CLEMAM –Checklist of European Marine Mollusca (www.mnhn.fr/biotaxis/clemam). The systematic list of mollusks provided in Table 15 follows Demir (2003). Specimens are assigned to species using illustrations in Fischer et al. 1987, Gianuzzi-Savelli et al. (1994, 1996, 1999, 2001, 2003), Nordsieck (1969, 1982), and modern specimens collected by the author. In cases of uncertainty about the specific identity of archaeological specimens, generic names with "sp." or "spp.", or family names are assigned to taxonomic groups. These sorts of cases usually occur when the preservation state of specimens did not allow specific identifications, or in the case of Troia, where the bulk of the archaeological specimens were no longer accessible and the interpretations were based on data compiled by archaeozoologists who previously worked at the site.

3.3. Quantification

In archaeology, a broad range of quantitative methods are used to retrieve dietary, ecological, demographic, and economic evidence from mollusk remains (Shackley 1981). Factors such as the differential preservation of organic remains and the distinctive properties of different mollusk species that occur in the archaeological record do not permit to consider any of the methods as presenting a truer picture of past activities and environmental conditions relating to mollusks. As Claassen (2000) argues, since each specimen represents a different human behavior and has a unique environmental and taphonomic history, "there can be no uniform methodological treatment of different species collected in the past."

Quantitative analysis of archaeozoological remains is performed to assess the relative dietary and economic importance of various animal species and their abundance in the environment. This process becomes complicated when faunal assemblages include multiple phyla of the animal kingdom. While mollusk shells generally preserve better in the archaeological soil matrix than the remains of other common biological consumption refuse such as vertebrates and plants (Meehan 1977, 1983, Meighan 1969, Moss 1993, Thomas 2002), and that the depletion of shell color patterning, which leaves the white coloring of the $CaCO_3$, in most shells, result in the "high visibility" of mollusk remains in archaeological stratigraphy (Shackleton 1983), it is impossible to express the consequences of these processes in mathematical terms with any accuracy. Nevertheless, the amount of mollusk remains relative to the amount of vertebrate remains in faunal assemblages is an important component of the archaeomalacological data, which should be evaluated in terms of the temporal and geographical differences between the relative contributions of mollusks to the diet of human societies (Glassow and Wilcoxon 1988).

In order to transcend the limitations imposed by taphonomic factors, researchers developed complicated calculations of energy and protein yields represented by shell and bone remains. Some have gone as far as estimating that the meat weight of one gram of bird bone "probably" represents the same amount of meat weight represented by 100 g of clam shells (Treganza and Cook 1950). These calculation methods are commonly applied in regions like the Americas and South Africa (e.g., Buchanan 1988, Erlandson 1988, Erlandson et al. 1999, McKillop 1984, Parmalee and Klippel 1974, Raab 1992). In these regions, the necessity of such quantitative applications was born from the need to decipher the subsistence systems of hunter-gatherer-fisher societies represented by coastal shell middens, where mollusk remains outnumber any other archaeological material or organic remains, thereby constituting the primary tool for understanding the past. Such quantitative reconstructions of calorific and nutritional yields of food taxa are useful to stimulate thoughts on food preferences, procurement strategies, and the relative importance of certain species in the diet of hunter-gatherer-fisher societies. Yet, they are not more accurate than the applications of other quantification methods, because the immeasurable taphonomic and cultural factors, as well as the presumed presence of archaeologically invisible foods (e.g., mushrooms, seaweeds, and herbs), will always distort the percentage values provided by these calculations and hence the informed guesses inferred from them.

In Europe and Southwest Asia, minimum number of individuals (MNI), number of identified specimens (NIS), and weight of identified specimens (WIS) are commonly and alternately used by archaeozoologists and archaeomalacologists alike, in order to assess the relative abundance of species in faunal assemblages (Grayson 1984, Davis 1987: 35-36, Claassen 1998: 106-107).

For archaeomalacological assemblages, calculations of MNI are derived from non-recurrent body part counts (Claassen 1998: 106). It is claimed that MNI estimates of mollusks are closer to the actual number of individuals represented in archaeological units (Mason et al. 1998). In the cases of assemblages including both vertebrate and invertebrate remains, MNI, just like calculations of flesh weights and calorific contents, is an estimate derived from the actual data that consists of NIS and element counts. An estimate can only multiply the default shortcomings intrinsic to the archaeological material (Uerpmann 1973). Therefore, problems associated with MNI estimates of both invertebrate and vertebrate assemblages are more than those arising from using NIS or WIS (Glassow 2000).

NIS and WIS are more accurate ways of quantification when mollusk remains are dealt with in their relationship to other faunal remains. Exclusive use of WIS per taxon results in an inaccurate understanding of the relative abundance of taxa, due to high variability between shell/meat ratios and differential preservation rates between different taxa (Mason et al. 1998). Recording and evaluating the present body parts of mollusk remains, in a similar manner to the methods applied when recording vertebrate remains, can bring valuable insights into the information obtained from NIS and WIS calculations. In the present study, NIS and WIS are concurrently used to describe the frequency of occurrence of specimens belonging to different taxonomic groups. In this way, (1) the basis for quantification is not an estimate, but the materially existing specimen in the archaeological matrix both for the invertebrates and the vertebrates; (2) calculation methods are consistent between the vertebrates and invertebrates; and, (3) the results of the two applications can complement each other in addressing research questions.

Counts and weights of mollusk specimens have been recorded in Troia since the beginning of archaeozoological study of the faunal material yielded from the New Excavations. In order to allow the compatibility of Yenibademli and Ulucak data with the studied shell assemblages from Troia, this method of recording was retained when studying mollusk assemblages from those two sites. Compatibility of the assemblages in quantitative terms was essential to optimize the accuracy of observations of inter-site and chronological patterns and thereby to partially compensate for the disparity caused by differential recovery techniques. Present body parts of mollusk specimens were not systematically recorded in Troia. For Yenibademli and Ulucak, care was taken to record this information into the database, but it was not quantified. Such quantified information has also not been published for any of the sites that provide comparative material for this study. In Aegean archaeozoology, NIS is commonly the sole information published for mollusk assemblages. Older excavation reports often provide only an estimation of the abundance of mollusks. In this study, NIS is utilized as the basis for inter-site comparisons.

NIS and WIS of the total amount of mollusk specimens in each site are analyzed in relation to NIS and WIS of the total amount of other animal remains, and in relationship to each other. Because relative expressions of species abundances too have their problems and there is no uniform definition of what constitutes a sufficient sample size for research questions (Claassen 2000), in this study, figures of relative abundances are always accompanied with absolute values. NIS and WIS are expressed in percentages of total number of specimens from temporally or spatially meaningful archaeological units and in absolute numbers. It is essential to recognize that neither of these quantitative methods will produce concrete and unchallengeable solutions to questions regarding the role and nature of shellfishing and shellfish taxa in subsistence economies that depend on diversified resources.

Relative and absolute quantities of specimens, as well as patterns in their morphometric properties and growth rates were calculated using KNOCOD©, Microsoft EXCEL, Microsoft Access, and SAS JMP-IN. Statistically significant differences between the morphological attributes of *C. glaucum* and *P. caerulea* specimens and the breadth of growth increments in *C. glaucum* populations were tested using analysis of variance (ANOVA) and Student's t-test assuming unequal variances.

In statistical analyses, an attempt was made to use as many individual specimens as the material at hand and time allowed. This decision was based on two reasons: (1) mollusk species that are evaluated here have high reproductive rates and are gregarious; and, (2) large quantities of mollusks are usually required to prepare a meal for a typical household. To exemplify, Meehan reports that the Anabarra of Australia would collect an average of 1200 individual shellfish on a typical gathering day (1983). Large amounts of mollusk remains are more likely to represent a larger variety of ecological relationships and human behaviors pertaining to mollusks.

3.4. Morphometry

Morphometry is the "assignment of quantities to biologic shapes" (Bookstein 1982: 451). In shelled mollusks, morphological features such as size, robustness, sculpture, convexity, concavity, and number of repetitive three-dimensional elements display striking variety at the inter-specific level and between geographically defined populations of the same species (Graus 1974, Tevesz and Carter 1980, Vermeij 1978, 1980, 1993). One of the most important functions of the mollusk shell is to form a protective exoskeleton for the soft body parts (with the exception of *Sepia*) (Vermeij 1993:4). Variety in form is related to shell function evolved through adaptive mechanisms. Size and other features of shell morphology are conditioned by endogenous factors such as population genetics and by exogenous factors such as the characteristics of the substrate, water temperature and salinity, food availability, and predation intensity.

In modern malacology, shell morphology is used to define the size range of geographically defined mollusk populations (e.g.; Gontikaki et al. 2003), to assess the relationship between growth, body size, and environmental parameters (e.g.; Ivell 1979a, 1979b, Sheldon 1967), and, to determine causalities and correlations between shell form and habitat (e.g., Graus 1974, Mariani et al. 2002, Tevesz and Carter 1980).

One of the most obvious morphological traits of an individual shelled mollusk is its shell size. Since mollusk shell grows incrementally to envelop the growing soft body parts, an allometric relationship develops between the animal's age and its shell size. A mollusk shell grows continuously until the death of the animal. There is also an exponential relationship between the animal's shell size and body weight (Sheldon 1967). At an intra-specific level, shell size of a mollusk is a good representation of the amount of flesh the animal contains. Therefore, shell size variability of a mollusk population is an indication of its age structure and a useful proxy to deduce the total amount of flesh weight the population represents. Additionally, in mollusk populations that are subject to human-exploitation, size variability can be further conditioned by the intensity of human consumption (Thomas 2002).

In archaeology, size profiles of mollusk shells have been used as a proxy measure of the intensity of the human exploitation of mollusk resources. Diminution in shell size throughout occupational periods of single sites or a group of sites, and reduced size of archaeological shells in relation to extant populations have been the most frequently utilized morphological characteristics for arguments suggesting the intense human exploitation of shellfishing grounds in the past (e.g., Jerardino 1997, Prummel 2005, Yesner 1984). Some researchers have been critical about relating shell size with human pressure on mollusk populations, arguing that environmental factors including predation by other animals can impact mean shell size of local populations (Claassen 1998: 49). Claassen (1986) suggested that the differences in the mean shell size of archaeological populations can be used as a measure of human mode of exploitation only when there are no notable differences in the mean amount of shell growth per age year between the study group and the control group. This was put into practice in various studies. For example, Mannino and Thomas (2001, 2002) quantified shell size together with age estimates using growth rings visible on the shell surface and successfully attested the intensification of shellfishing activity in Britain during the Mesolithic period.

Other morphological features of mollusk shells have been used to reconstruct the environmental conditions that were at play during the ontogenetic life of archaeological mollusk specimens. Cabral and da Silva (2003) found out that the shells of *Patella* spp. remains from the Iron Age fortress of Terroso in Portugal were significantly taller than modern specimens. They suggested that the morphometric attributes could be related to changes that occurred in the coastal zone (i.e., changes in wave action and intensity, alterations in coastal morphology, and changes in sea-levels), but also did not exclude the possibility that the morphological changes could have developed as a consequence of gathering strategies (Cabral and da Silva 2003). In another study, the differences between the morphology of archaeological *Argopecten irradians* assemblages and extant populations living in the Gulf of Mexico were utilized to interpret the effects of water quality, habitat alteration, and pollution on the decline of modern populations (Marelli and Arnold 2001). Elsewhere on the eastern U.S. coast, the valves of archaeological *Crassostrea virginica* populations have been used to reconstruct past aquatic habitats based on the observation that modern *Crassostrea virginica* shells become thicker with increasing average water temperature, salinity, and turbidity (Kent 1988, cited in Claassen 1998).

Figure 11. Descriptions of the morphological measurements on *C. glaucum* valves (Modified after Mariani et al. 2002, Figure 2).

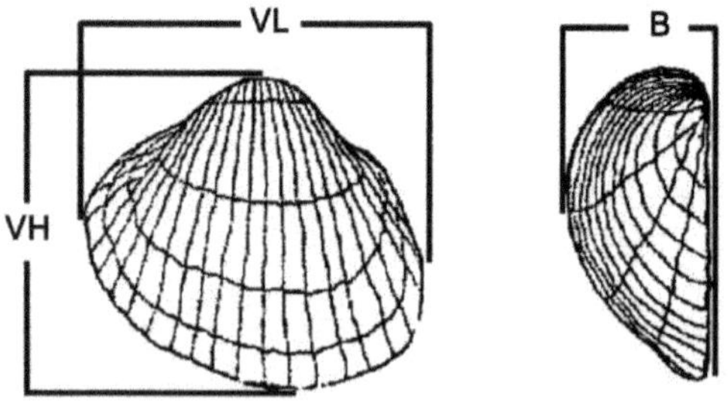

In this study, specimens of the most commonly represented and best preserved mollusk species were chosen for morphometric investigation. Morphological attributes of *P. caerulea* and *C. glaucum* specimens have been analyzed in order to disclose the size choice of prey, intensity of shellfishing activity, and the stable and changing characteristics of the water body where gathering occurred. The morphological traits of the shells of both of these species are considered to show variation in accordance with the environmental conditions of the coastal habitats they populate (Barnes 1880: 54-59, Cabral and da Silva 2003, Mariani et al. 2002). Recent malacological studies indicate that there is a positive relationship between the solidity (i.e., thickness/robustness) and curve (i.e., convexity/ obliqueness) of modern *C. glaucum* shells, and average water temperature and salinity (Mariani et al. 2002). It has also been shown that the thickness of the mollusk shell is generally related to shell calcification and to the availability of $CaCO_3$ in the external medium (Graus 1974). Graus (1974) demonstrated the positive relationship between shell thickness and latitudinal increase in temperature and precipitation of $CaCO_3$ in the North Atlantic.

Valve length (VL), height (VH), breadth (B), and weight (W) were measured for all accessible and complete right valves of *C. glaucum* from the Bronze Age contexts of Troia, Yenibademli, and Ulucak (Figure 11). Measurements applied on *C. glaucum* are adopted from Mariani et al. (2002). Shell length (L), height (H), and width (W) were measured for all complete examples of *Patella caerulea* from Yenibademli. Decarbonized (i.e., leached), encrusted, or otherwise modified archaeological shells were not measured. Measurements taken on *P. caerulea* are those used in Öztürk and Ergen (1999) and Cabral and da Silva (2003) (Figure 12). Samples were weighed singly using a precision scale to the nearest 0.1 g. Measurements were taken using a digital vernier caliper to a precision of 0.1 mm. The morphological characteristics of *C. glaucum* samples collected live from the delta area of the Karamenderes were recorded in the same manner. An individual number is given to each morphologically-investigated specimen. Throughout this study, these specimens are recalled with the number of the unit they belong to, followed by its individual number (e.g., 2345-5, 310-4 et cetera).

This raw data was then used to calculate the solidity and curve indices, as well as the relationship between the valve heights and age in archaeological and modern *C. glaucum* populations. Solidity and curve indices were determined following Mariani et al (2002):

$$\text{Solidity} = [(\text{Weight}*2)/\text{Valve height}^2]*1000$$

$$\text{Curve} = (\text{Breadth}*2)/\text{Valve length}$$

In *Patella caerulea* shells, height/length and length/width ratios were calculated.

Figure 12. Description of the morphological measurements on *P. caerulea* shells (Modified after Küçük et al. 1996, Figure 3).

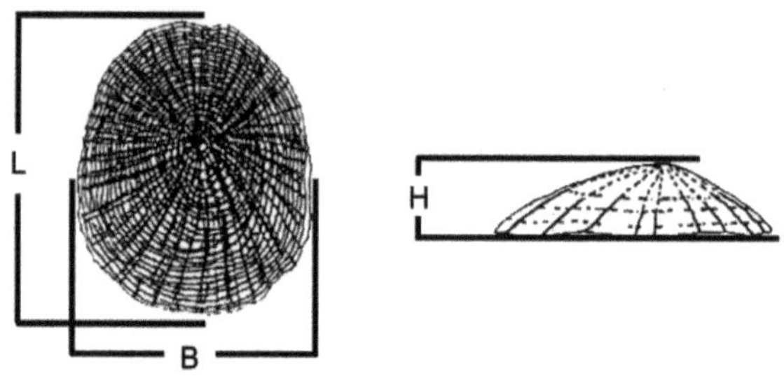

The indices of *C. glaucum* valve heights and the size of *P. caerulea* indicate the size range of the mollusk populations exploited by the inhabitants of settlements that are in question. This data was evaluated in the light of relative abundance of species in the faunal assemblages, age estimations in the case of *C. glaucum* in Troia, and archaeological and modern malacological indices available in the literature.

3.5. Incremental growth analysis

Calcified body parts of both aquatic and terrestrial mammals, fish, and invertebrates exhibit growth increments. An increment is a repetitive structural and compositional unit in the skeleton (Rosenberg 1975). In archaeology, biological increments are used to assess age and season at death of ancient populations represented by faunal assemblages

(Gordon 1993).

In those mollusk species where shell functions as an exoskeleton, the mantle (the organ which encloses the entire soft tissue including the head and the foot) secretes the shell by adding $CaCO_3$ to the shell margin (Sabelli 1979: 39-40, Barnes 1987: 342). In bivalves, shell secreting occurs each time the valves gape. When valves are closed, shell secreting ceases (Barnes 1987: 405, Claassen 1998: 22). Modern research has shown that in most bivalve populations, gaping (i.e., shell secreting) occurs periodically in accordance with cosmic rhythms affecting the earth, such as tidal or circadian rhythms, depending on the species and its environment (Pannella and MacClintock 1968, Farrow 1971, 1972, Deith 1983a, Richardson 1988, Richardson et al. 1995, Clark 2005). The rhythm of gaping, and thus the growth rate of shell increments are affected by a number of endogenous and exogenous factors. Endogenous factors include individual differences, age, and reproduction (Farrow 1971, Seed 1980). Exogenous factors can be related to the characteristics of the substrate, population density, tidal cover, temperature, extreme climatic events, salinity, and food availability (Farrow 1971, Seed 1980). One could add human impact (pollution, radioactivity, disturbance of substrate, et cetera) to these exogenous factors (Richardson 1988).

In temperate regions, the most important exogenous factors such as temperature, salinity and food availability have an annual pace which changes seasonally. Consequently, bivalve species living in temperate climates grow at predictable annual rates and at a seasonally determined pace (Richardson 2000, Schöne and Surge 2005). Changes in the incremental growth of bivalves are recorded in the shell microstructure and chemical composition. Increments, which are formed when conditions are favorable, are rich in $CaCO_3$ and conchiolin (the protein matrix of the shell). These appear white to the naked eye and opaque under transmitted light (Figure 30; Claassen 1998: 153, Quitmyer et al.1997: 826). Increments, which are formed when conditions are not favorable, represent slow growth. They are not rich in $CaCO_3$ content in relation to the conchiolin deposited (Claassen 1998: 153). These increments are gray (or yellow or brown) in normal light and translucent in transmitted light. When the growth rate is regulated in accordance with seasonal changes in water conditions and when the change in growth rate only occurs once a year, a set consisting of one white incremental zone and one gray incremental zone forms the annual growth (Quitmyer et al. 1997).

In many shelled mollusks (e.g., *Hexaplex trunculus*, *Arca* spp., *C. glaucum*), the borders of incremental zones are visible on the external surface of the shell (Figure 13). These, commonly called growth "lines" or "rings," can take a variety of forms on the same individual, such as depressions, notches, or discolored zones (Clark II 1979). Although external lines are almost always related to changes (periodical or abrupt) in the growth pattern of the shell, not all changes in shell growth is represented as a clear external line. In addition, external lines do not contain information regarding the extent of incremental zones, but usually only the beginning or the end (i.e., border) of the incremental zones. The extents of the incremental zones are more clearly visible in the valve cross-section (Figure 30).

Figure 13. External lines on modern C. glaucum valves and size variation in the modern populations of the Troad.

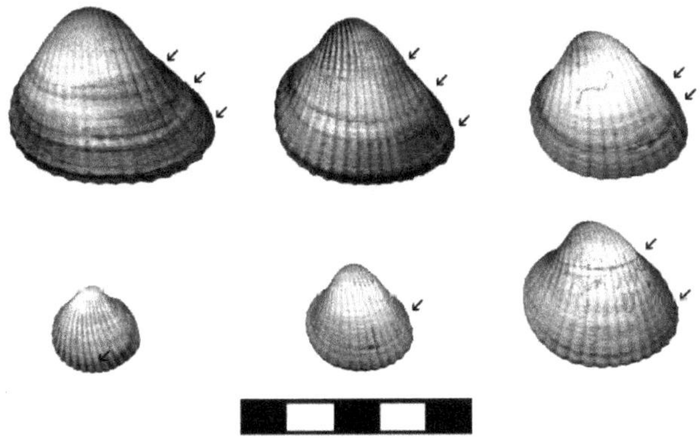

For these reasons, sciences interested in the growth pattern of shelled mollusks (geology, malacology, palaeontology, palaeoclimatology, et cetera) have developed a variety of analytical techniques that allow for more accurate analysis of the increments than that restricted to observing the shell surface (Richardson 2001, Schöne and Dunca 2005). Although the details of techniques vary depending on the level of resolution suited to answering questions particular to individual research fields, the basic approach is the microscopic observations of sections (thin sections, replicated sections, or thick sections) made perpendicular to the growth axis (e.g., Deith 1983a, 1983b, Ivell 1979, Milner 2001, Richardson 2001, Schöne et al. 2005).

In the present study, the incremental growth analysis of *Cerastoderma glaucum*, the most abundant species in Bronze Age Troia, was utilized to estimate the demographic structure and harvest time of the archaeological populations. Live collected samples of extant *C. glaucum* populations in the Karamenderes Delta (See Section 3.6 for details of the methods used in establishing modern collections) were used as a modern analogue and a control group to interpret the incremental growth pattern of the archaeomalacological material.

The most suitable technique for investigating the seasonality of shellfish collection at a large and complex site is to work on the thick sections of large amounts of archaeological bivalves (Claassen 1983). This technique is relatively less time consuming and suitable for most archaeological questions, such as seasonality and intensity of the gathering activity. In studies that depend only on the microscopic readings of incremental growth from acetate peel replicas or thin sections, the studied sample is always limited to small amounts because of the large consumption of time these methods require (e.g., Clark II 1979). The thick section technique has been applied on the estuarine Atlantic bivalve *Mercenaria mercenaria* and freshwater bivalves as a means of determining the season of death of archaeological assemblages by Claassen (1983, 1986). Quitmyer et al.

(1996, 1997) used the same technique on the *Mercenaria* spp. from shell middens in Florida along with the isotopic analysis of molluscan carbonate (cf. Section 3.7) in order to substantiate the results obtained through incremental analyses.

Maximizing concordance between data structures and empirical phenomena that are under investigation constitutes the backbone of a scientific study. In the present study, the incremental growth examination technique was applied to archaeological and modern shell thick sections. Growth patterns of both archaeological and modern shells were established by observing and measuring the breadths of sequential white (opaque) and gray (translucent) increments in the cross-sections of their valves. It is important to note that the measured breadths of incremental zones cannot be converted to time units.

To produce the thick sections, the existing laboratory machinery at the Archaeozoological Laboratory of Tübingen University was utilized. Prior to cross-sectioning, all right valves that were intended to be sectioned were individually weighed and necessary measurements were taken as explained earlier in Section 3.4. The valves were cross-sectioned along the growth-axis across the umbo and through one of the middle ribs, using a hand-operated Struers© Labotom speed saw (Figure 14a). The valves of *C. glaucum* proved to be strong enough, that embedding them in epoxy resin before cutting was not necessary. After sawing, the cut surface was polished with a Struers© LaboPol-1 polishing machine (Figure 14b), using 220 and 1200 grid carborundum paper. Polishing with the 220 grid paper produces a regular surface and applying 1200 grid paper permits better visualization of finer details. The thick sections were examined using an Olympos© light microscope at differing magnifications. The breadths of the increments were recorded using a digital vernier caliper to a precision of 0.1 mm. Each incremental zone is labeled with a number with respect to its accretion sequence and with characters "G" and "W", designating gray and white increments. For example, the nearest gray incremental zone to the umbo is labeled 1G, the white incremental zone that formed after 1G is called 1W, and so on. The characteristic (white or gray, and breadth) of the last incremental zone at the margin of the seasonally collected *C. glaucum* shells were used to assign the season of collection of archaeological *C. glaucum* populations.

The pattern of incremental growth was additionally used to estimate the age profiles of modern and archaeological *C. glaucum* populations from the Troad. Age profiles of populations are estimated by counting the gray and white zones. But the prerequisite to age estimation is the determination of the periodicity of alteration between white and gray zones. Although in most bivalve populations one-time alteration between white and gray incremental zones per year is the pattern of growth rate, this is not the rule for all populations. For instance, *M. mercenaria* populations from temperate latitudes of the North Atlantic form a gray incremental zone in the winter and a narrower gray zone in the summer (Quitmyer et al. 1997). In the present study, once the periodicity of alternation between white and gray zones was established based on the incremental growth pattern of modern populations and the isotopic analyses of archaeological specimens, the results could be translated into ontogenic stages, by counting the number of incremental zones present on the thick sections.

Figure 14. Laboratory equipment used for valve sectioning.

a. Speed saw

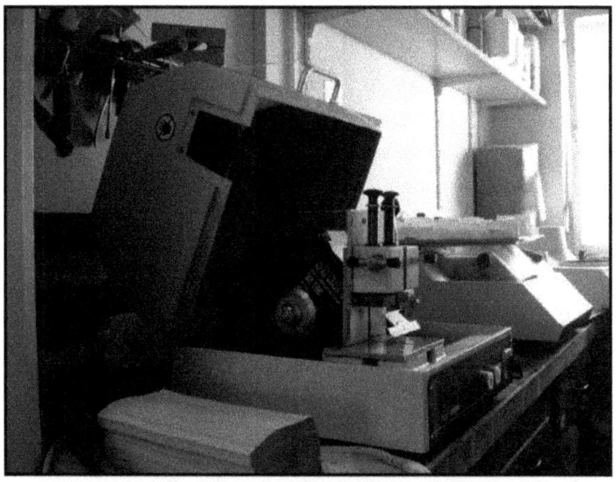

b. Polishing maschine

3.6. Sampling extant mollusk populations

Archaeomalacological interpretations largely depend on existing modern malacological work. Information retrieved from the literature on mollusk taxonomy, biology, and habitats is essential in making accurate inferences about archaeological mollusk assemblages.

Just as in archaeomalacology, growth patterns, age and size structures of mollusk populations are topics in modern malacology (e.g., Arneri et al. 1998, Ivell 1979a, 1979b, Keller et al. 2002, Leontarakis and Richardson 2005, Peharda et al. 2002, Richardson et al. 1995, 1999). Research in this area has proven that there are significant differences between the growth patterns, size-frequency distributions, age structures, and morphology of populations of the same species that grow in different environments. For example, growth rate and the season of incremental shell formation vary across the latitudinal range of *Mercenaria mercenaria*

(Ansell 1968, cited in Quitmyer et al. 1997). Arneri et al. (1998) demonstrated that while the populations of the bivalve *Venus verrucosa* grows slowly and poorly during the summer in the Aegean and the Adriatic, populations of the same species living off the coast in Normandy and Brittany grow faster during the summer. Various studies indicated the diversity in morphology, growth, reproduction, maximum age and size of geographically divided populations of *C. glaucum* (Ivell 1979a, 1979b, Zaouali 1979, 1980, Mariani et al. 2002, Gontikaki 2003). The results of the research conducted on modern *C. glaucum* populations for this study adds to this diversity (cf. Section 4.2).

Figure 15. Sampling of live *C. glaucum* at the deltaic lagoons of the Karamenderes.

a. Hand-sampling

b. Sieving

For the reasons described above, when it is necessary to build analogies with modern populations in order to interpret archaeological mollusk assemblages, it is essential that the sampled populations come from geographically coherent areas (e.g., Deith 1983a, 1983b, Mannino and Thomas 2001, Marelli and Arnold 2001, Milner 2001, Cabral and da Silva 2003). For the purposes of the present study, the lagoons and the deltaic distributary channels formed at the mouths of the Karamenderes and Dümrek Rivers were inspected for extant *C. glaucum* populations. Populations were located at three different stations at the Karamenderes River Delta (Figure 16). Sampling took place five times, covering a time span of 15 months: in August 2005, February 2006, June 2006, August 2006, and November 2006. Collection was done by hand and using a shovel and a sieve with a mesh size of less than 10 mm (Figure 15). The SST was recorded using a C&F© fishing thermometer. Salinity was measured with an ABBE© refractometer. Water samples were taken additionally. Location of the sampling areas, depth, characteristics of the substrate, and estimated population density were described for each station.

Other areas of the coastline in the Troad were surveyed in order to obtain an overview of the present state and location of the mollusk populations that are of concern for this study. The results of this fieldwork are reported in Section 4.2.

3.7. Isotopy

Isotopes are different forms of the same atom that have the same number of protons but a different number of neutrons (Hoefs 2004: 1). The two types of isotopes are the stable and instable isotopes. Instable isotopes are those that are subject to radioactive decay in detectable time ranges, such as the radioactive carbon isotope ^{14}C. Stable isotopes do not exhibit measurable changes through time and therefore they have the potential to explain various processes that occurred when chemical compounds were formed. Because the ratio of heavier stable isotopes to lighter stable isotopes in nature (isotope fractionation) is influenced by a set of interactions between different processes, measures of isotopic ratios are used as proxy data for explaining the phenomena that produced them (Hoefs 2004, Shackleton 1973). Of interest to the purposes of the present study are the oxygen and carbon isotopes in mollusk shells.

As mollusks grow (i.e., during the calcification process of the skeleton), calcium carbonate ($CaCO_3$) is produced through the biogenic processing of sources available in the host environment. These sources are those available in the meteoric water (mostly rain water) solution in the case of land snails, and freshwater or seawater solutions in the case of aquatic mollusks. Biogeochemical variability across the growth axis of mollusk shell is an ontogenetic record of variations in environment and climate (Arthur et al. 1983, Dettmann et al. 1999, Jones and Quitmyer 1996, Richardson 2001). For this reason, mollusk shells are considered to be high-resolution recorders of the chemical properties in their environment (Augierre et al. 2005, Balakrishnan et al. 2005, Emiliani and Mayeda 1964, Richardson 2001).

Calcium, carbon and oxygen atoms constitute the $CaCO_3$ compound of the mollusk shell. Stable calcium isotopes have a relatively large atomic mass and show little fractionation in nature, resulting in difficulties in determining calcium isotopic ratios using physical and chemical methods.

Figure 16. Location of sampling stations in the Troad.

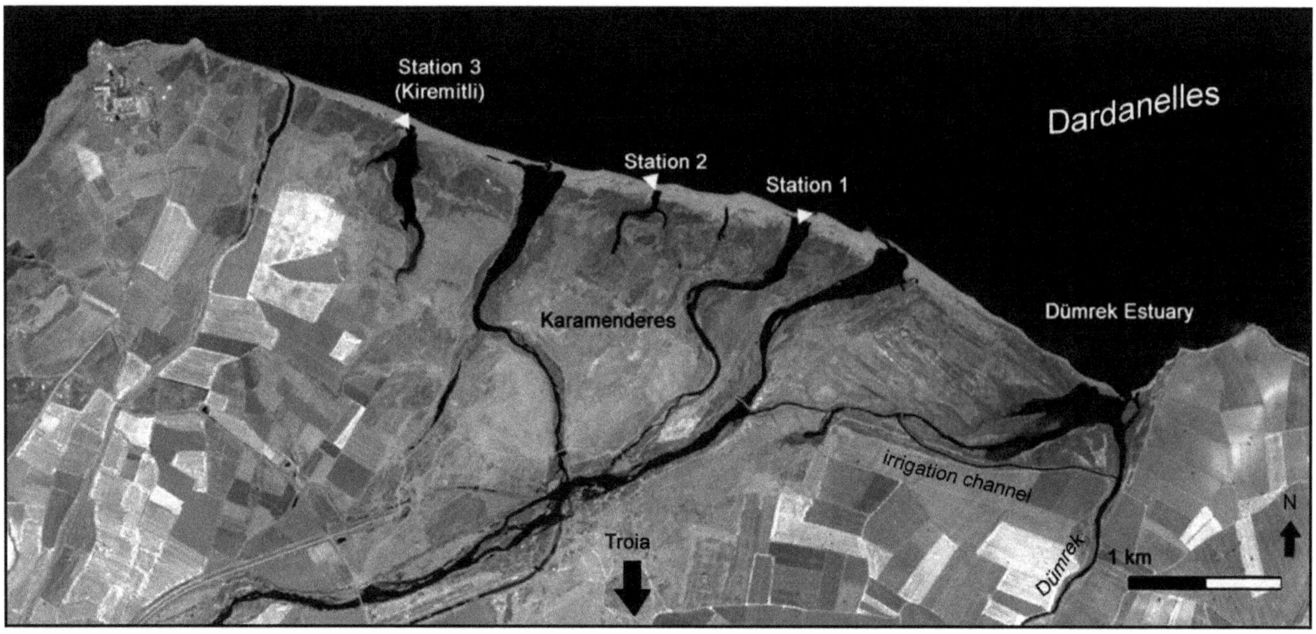

Moreover, factors leading to natural calcium isotopic fractionations are not well understood. For these reasons, while the geochemical investigations of the calcium content of mollusk shells are generally eliminated, the ratios of oxygen and –yet to a lesser extent– carbon isotopes are used to understand the short-term natural phenomena that lead to the isotopic fractionation of mollusk shell.

Three stable isotopes of oxygen (^{16}O, ^{17}O, ^{18}O) and two stable isotopes of carbon (^{12}C and ^{13}C) occur in nature. The ratios of ^{16}O to ^{18}O (expressed as $\delta\ ^{18}$O) and of ^{12}C to ^{13}C (expressed as $\delta\ ^{13}$C) can be measured using mass spectrometers (Shackleton 1973). Their ratios in various natural spheres (e.g., atmosphere, biosphere, lithosphere, et cetera) are largely predictable by certain equations after decades of empirical data compilation. In the literature, all isotopic contents are expressed in terms of their difference from international standard samples. Standards are used because the accuracy with which absolute isotope abundances can be measured is poorer than the precision with which relative differences in isotope abundances between two samples can be determined (Hoefs 2004: 23). The international standard used for carbon is the isotopic content of a *Belemnitella americana* specimen (abbreviated as PDB). For oxygen, it is standard mean ocean water measured in Vienna (abbreviated as V-SMOW or SMOW). For purposes of comparability, SMOW values are often converted into PDB values and expressed as such in the literature.[17]

Oxygen and carbon isotopes are incorporated into molluscan $CaCO_3$ following different paths. The mollusk shell extracts oxygen isotopes in equilibrium with the isotopic ratio of the ambient H_2O (e.g., Richardson 2001, Shackleton 1973). Thus, the isotopic composition of the aquatic mollusk shell mirrors the isotopic composition of the ambient water effective during ontogenetic life. Different forms of water vary considerably in their oxygen isotopic content. The mean $\delta\ ^{18}$O value of the surface ocean waters is 0 ‰, whereas the mean $\delta\ ^{18}$O value of meteoric water (e.g., rivers, lakes, rain, vapor, et cetera) is around -4 ‰ (Hoefs 2004: 104). The deviation from this mean is great: the $\delta\ ^{18}$O composition of meteoric water ranges between 9 and -48 ‰ depending on latitude and continentality (Hoefs 2004: 62). Meteoric water precipitated at sea-level has ca. -3 ‰ $\delta\ ^{18}$O PDB (Hoefs 2004: 116). Salinity, temperature, evaporation, precipitation, and continentality are among the factors that influence the oxygen isotopic composition of water. Temperature is considered to be the most important factor affecting the oxygen isotope composition of surface seawater. There is an inverse relationship between SST and the $\delta\ ^{18}$O content of surface waters. Thus, the δ^{18}O compositions of marine mollusks can be used to calculate SST values using a simple equation, when it is known that the isotopic composition of the surface water is influenced by temperature changes (For a review of the method: see Grossman and Ku 1986, Richardson 2001: 130-131, Shackleton 1973: 140-141, and Carré et al. 2005, for an improved equation). As the SST rises, the ^{18}O abundance in the mollusk shell is reduced by 0.2 ‰ for each degree rise in temperature. But as exemplified in Section 2.4.1, in the case of the Marmara and the Black Seas, there is a positive linear relationship between surface salinity and the proportion of $\delta\ ^{18}$O in the surface seawater (Hoefs 2004: 123). Therefore, the influence of evaporation and continental run-off of meteoric water (i.e., mixing of water bodies with different salinity ranges) complicates and hinders calculations of SST from the $\delta\ ^{18}$O content of mollusk shells, particularly those which grew in the littorals (Shackleton 1970).

The $\delta\ ^{13}$C ratios in mollusk shell are much more difficult to

[17] The formulas used for convertion are: $\delta\ ^{18}O_{(SMOW)} = 1.03091\ \delta\ ^{18}O_{(PDB)} + 30.91$ and $\delta\ ^{18}O_{(PDB)} = 0.97002\ \delta\ ^{18}O_{(SMOW)} - 29.98$ (Hoefs 2004: 55).

interpret, because, the precipitation of carbon during the formation of the shell $CaCO_3$ of calcareous organisms, contrary to the situation with oxygen isotopes, occurs in disequilibrium with the carbon isotopic composition in the ambient environment (Hoefs 2004: 150, McConnaughey 1989a, 1989b). Moreover, while the main causes of variation in the fractionation of naturally occurring oxygen isotopes in oceanic and meteoric water are known, factors controlling the $\delta\ ^{13}C$ contents in different water bodies are much more complicated and less well understood (Richardson 2001). The most important reasons for this are the variety of the sources producing the carbon reservoirs in the world and the high variability of the level of interaction between these reservoirs (Hoefs 2004: 48-49). Aquatic mollusks incorporate the dissolved inorganic carbon (DIC), the sum of inorganic carbon species (= types) in the ambient water into their shells (Dettmann et al. 1999, Geist et al. 2005, Schöne et al. 2006). DIC consists of CO_2, HCO_3^-, and CO_3^- , each of which has its own $\delta\ ^{13}C$ value (Schöne et al. 2006). The most important factors affecting the variability of $\delta\ ^{13}C$ composition of DIC in oceans is the level of primary production.[18] An increase in the $\delta\ ^{13}C$ value of oceanic DIC can be associated with an increase in the aquatic primary production, because photosynthesis of algae floating on the surface preferentially removes the ^{12}C from the surface DIC in the water (Hoefs 2004: 118, McConnaughey 1989a). The level of primary production varies depending on the amount of light, temperature and nutrients. While the amount of light and nutrients are largely regulated by seasonal cycles in mid and high latitudes, their ratios in surface waters are highly affected by upwelling events and influx of meteoric waters (Richardson 2001, Richardson et al. 2001). Upwelling events generally (but not always; cf. Sections 2.4.1 and 4.3.5.5) bring DIC whose $\delta\ ^{13}C$ values are more negative than those in surface waters (Richardson et al. 2001). The DIC content of meteoric waters, on the other hand, may exhibit an extremely variable isotopic composition, influenced by factors such as weathering of carbonates, freshwater plankton, CO_2 levels and types obtained from bacterial oxidation of organic matter, and exchange with atmospheric CO_2 (Hoefs 2004: 136). The $\delta\ ^{13}C$ of meteoric water DIC can be as low as -14 ‰ and as positive as +1 ‰ if the region it originates is carbonate rich (Hoefs 2004: 48 and 62, Schöne et al. 2006). While $\delta\ ^{13}C$ of marine carbonates fluctuate between 2 ‰ and -5 ‰, $\delta\ ^{13}C$ of freshwater carbonates can vary from 2 to -32 ‰, with a mean of -25 ‰ (Hoefs 2004: 48). Carbon isotopic composition of rivers is heavier near the mouth of the river than areas closer to the source (Hoefs 2004: 136). But the differences in carbon compositions between different river systems are highly variable (Hoefs 2004: 136).

The difficulties arising from predicting the phenomena causing variation in the $\delta\ ^{13}C$ values of DIC in water reservoirs constitute only one level of the difficulties involved in interpreting the $\delta\ ^{13}C$ record of aquatic mollusk shells. McConnaughey (1989a) calls these kinetic effects.

[18] Primary production is the production of organic compounds with reduced carbon from atmospheric CO_2 and oceanic carbonates, mainly through photosynthesis. The main sites for this production are the ocean surface, continental shelves, and continental biosphere. The organisms responsible for primary production in terrestrial ecosystems are plants, whereas in aquatic ecosystems, they are algae. (Albarède 2003: 205).

Other than kinetic effects, the isotopic fractionation of the mollusk $\delta\ ^{13}C$ is altered by metabolic effects (McConnaughey 1989a). It was mentioned above that disequilibrium between the $\delta\ ^{13}C$ of aquatic DIC and the $\delta\ ^{13}C$ of mollusk shell is the rule. That is because the extrapallial fluid (responsible for shell production) of the shelled mollusks modifies the $\delta\ ^{13}C$ composition of the shell $CaCO_3$, by incorporating lighter carbon isotopes into the shell through the respiration of CO_2. The $\delta\ ^{13}C$ content of aquatic mollusk shell is lighter than that of the DIC in the ambient water (Kennedy et al. 2001). But there is no clear equation between the $\delta\ ^{13}C$ of the aquatic mollusk shell and the $\delta\ ^{13}C$ of the DIC in the ambient water, because the metabolic rate of mollusk growth has no clear formula. However, since the metabolism of shelled organisms has an effect on $\delta\ ^{13}C$, growth rates, which are also metabolic events, should be related to the $\delta\ ^{13}C$ variations in the mollusk shell (Schöne et al. 2006).

Despite all the complications involved in predicting the natural phenomena affecting the isotopic compositions of aquatic mollusk shells, because shell growth is largely in accord with the changes in the host environment (Shackleton 1970 and 1973) and the steps in the life of a shelled mollusk can be investigated with high precision using the incremental growth techniques mentioned in Section 3.5, the isotopic composition of the $CaCO_3$ within structurally different incremental zones can be investigated separately and used to characterize the changes in the environment during sequential phases of growth (Arthur et al. 1983, Kennedy et al. 2001, Richardson 2001: 130-131, Richardson et al. 1999).

The $\delta\ ^{18}O$ and $\delta\ ^{13}C$ record of fully marine, estuarine, and non-marine mollusk shells have implications pertaining to the annual rhythm of pluvial regimes, precipitation rates, SST, salinity, of freshwater influx, and upwelling events (Shackleton 1970, Goodfriend 1991, Kennett and Voorhies 1995, Dettmann 1999, Ricken et al. 2003). In archaeology, $\delta\ ^{18}O$ and $\delta\ ^{13}C$ compositions of bivalves measured along the growth axis have been used to determine the season of shellfish collection and to reconstruct the environmental factors that affected the isotopic compositions of archaeological mollusk shells (e.g., Bailey et al. 1983, Deith 1988a, 1988b, Deith and Shackleton 1988, Kennett and Voorhies 1995, Killingley 1981, Mellars and Payne 1971, Shackleton 1970, 1973 and 1974, Quitmyer et al. 1997, Richards and Mellars 1998).

In this study, isotopic analysis was conducted on *Cerastoderma glaucum* shells from Troia. *C. glaucum* is not a fully marine species (cf. Section 4.1.2.9). It inhabits estuarine and lagoonal environments, where the isotopic compositions of both the host water and of the DIC in its solution are complicated by several factors besides seasonal changes in SST. The exchange between fresh and marine waters and the added influence of rainwater obscure the predictability of δ ^{18}O values in lagoonal environments connected to freshwater bodies. Fluctuations in marine surface waters and evaporation rates further influence the isotopic content of estuarine and lagoonal waters; hence the molluscan isotopy. For these reasons, the δ ^{18}O ratio of *C. glaucum* $CaCO_3$ from the Troad is not a reliable proxy in reconstructing actual SST values. The δ ^{13}C composition of *C. glaucum* specimens from the Troad is similarly difficult to interpret as a proxy for the exchange ratio between freshwater and seawater, as well as pluvial regimes because of the reasons described above. Therefore, in the present study, isotopic analysis is used, (1) to describe the general character of the growth environment of archaeological *C. glaucum*; and, (2) to test the seasonal pattern of visually differentiated incremental zone formation in archaeological *C. glaucum* shells, thereby serving as a control mechanism for the reliability of the incremental growth method in determining the schedule of shellfish gathering. Additionally, time constraints and financial shortages were also influential in assigning the geochemical investigations of the stable isotopic contents of *C. glaucum* shells such a role in this research.

Isotopic measurements were carried out on carbonate samples collected from the outermost layer of two archaeological shells. Shells that demonstrate easily distinguishable alternating growth zones were chosen for analysis. Prior to sampling, shells were weighed, measured and cross-sectioned. The shells were scrubbed using a toothbrush to remove the remaining foreign debris on the surface. They were dipped into a 0.01 % HCl solution for approximately 10 seconds and thoroughly rinsed with water immediately after. Some of the samples for isotope analyses were taken from the clean sections following Shackleton's description of sampling techniques (1973). Carbonate samples (each ca. 0.15 mg) from distinct white and gray incremental zones were obtained by grinding into the outer shell layer of each specimen, using a small dental burr (<1 mm) mounted in a hand-held drill. The limitation of this sampling technique is the time-averaging of the carbonate samples, especially of those taken from gray incremental zones. Isotopic variations that occurred within this time period are conflated in the end result. A Finnigan Mat 252 mass spectrometer at the Department of Geochemistry at Tübingen University was used for isotopic measurements. The reproducibility of the measurements is ± 0.1 ‰ PDB for δ ^{13}C and ± 0.2 ‰ PDB for δ ^{18}O.

3.8. Ethnographic analogy

Ethnographic documentations of shellfishing practices constitute an important guide to interpret variability in archaeomalacological record. Both ethnographic and ethnoarchaeological research (i.e., ethnographic research specifically designed to produce analogies to explain archaeomalacological finds) have greatly aided the interpretation of archaeomalacological record and led to previously disregarded considerations about past shellfishing practices (e.g., Meehan 1977, 1982, 1983).

It is an unfortunate situation that the ethnographic and ethnohistorical accounts concerning shellfishing are very scarce both in Turkey and Greece. In Turkey, it is plainly not possible to set up local ethnographic models that can aid the interpretation of archaeomalacogical record. The only commonly consumed aquatic mollusk species in Turkey is *Mytilus galloprovincialis* (the common Mediterranean mussel). The species is highly commercialized; often procured from aquaculture for the mass-market. The reason behind the avoidance of other mollusk species is hidden in complex religious and cultural practices, rather than environmental factors: The demographic composition of the Turkish Aegean coast has been fully transformed during the events that followed the First World War. Populations in coastal villages were replaced by peoples of Muslim/Turkish decent living in former European parts of the Ottoman Empire or by people of inland Anatolia. This new population, partially originating from pastoral villages in the Bulgarian hinterland and partially from the Anatolian Plain, is traditionally unfamiliar with marine resources. Today, shellfishing is no longer common practice on the Turkish coasts.

To compensate for the scarcity of ethnographic works from the study region under study and the low feasability of conducting ethnoarchaeological work in the vicinity of the sites discussed in this study, references to ethnographic records from other areas of the world, the scarce ethnohistorical and ethnoarchaeological accounts from the Aegean and the Mediterranean, interviews with recreational shellfishers, and even cookbooks with international recognition have been utilized to exemplify the possible reflections of the archaeomalacological record. In the absence of descriptions concerning local modern shellfishing, such information brings valuable insights to the possible explanations contained in the archaeomalacological record.

The following chapter; Chapter 4 presents the analytical results attained using the methods described above.

Chapter 4. Analytical results

This chapter reports on the results of the archaeomalacological analyses of three primary sites: Troia, Yenibademli, and Ulucak. The methods used in the analyses were provided in the previous chapter (Chapter 3). The analytical results are evaluated in Chapters 5, 6 and 7.

4.1. Aquatic mollusks

Mollusks are invertebrate animals that make up the phylum Mollusca. Despite their variable –yet widely common– contribution to human diet and use as ornaments for human dwellings and bodies, molluscan biology is not common knowledge. While the general biology of these invertebrates is not the focus of this study, the palaeoeconomic and palaeoenvironmental interpretations of the mollusk assemblages that are dealt with here are largely based on the biological properties of the species represented in the archaeomalacological assemblages. Because the ecology of each species has had a fundamental effect on the relationship humans developed with them, current malacological data such as habitat, population density, growth and reproductive rates, and mobility of mollusks is relevant to the analysis of archaeomalacological assemblages. Therefore, it is necessary to precede the presentation and discussion of the analytical results by an introduction to aquatic mollusks. This section reviews the taxonomy, life habits, biology, and habitats of the mollusks present in the archaeological assemblages under discussion.

There are eight classes of shelled mollusks living in aquatic environments: Caudofoveata, Aplacophora, Polyplacophora, Monoplacophora, Bivalvia, Scaphopoda, Gastropoda, and Cephalopoda (Figure 17). The archaeological deposits that are dealt within the present study only yielded the calcareous shells of gastropods and bivalves. A systematic list of the mollusks represented in the archaeological assemblages discussed in this study is given in Table 15. Scientific and common English and German names of the taxa represented at the sites under question are provided in Table 16.

Figure 17. Shells of different mollusk classes that frequently occur in archaeological contexts. (1) Scaphopoda; a Dentalium shell from the Pacific Ocean, (2) Gastropoda; a land snail from southern Germany, (3) Gastropoda; a Cypraeidae shell from the Arabian Gulf, (4) Gastropoda; a *Patella rustica* from the North Aegean, (5) Cephalopoda; *Sepia* spp. from the Arabian Gulf, (6) Bivalvia; a *Chamelea gallina* valve from the North Aegean.

Class Gastropoda is the largest class of mollusks with more than 35000 existing species (Barnes 1987: 347). Characteristically, they have a single shell and move by means of a wide muscular foot.[19] Gastropods have invaded a variety of marine, freshwater, and terrestrial habitats and have developed a great range of feeding, defensive, and respiratory mechanisms (Barnes 1987: 348). Bivalvia possess a shell with two valves that enclose the body. They only dwell in aquatic environments and, with the exception of some deep sea species; generally survive by filtering organic particles in the water. Many species of bivalves and gastropods have a periostracum, an elastic layer made of chitin that covers the shell and protects it during their life time. This outermost layer degrades rapidly after the animal dies (Vermeij 1993: 46).

The occurrence solely of gastropods and bivalves in archaeological deposits is not the rule. Scaphopods are represented in various archaeological contexts, including the human burials of the Natufian Levant, with shells of Dentaliidae (commonly called *Dentalium* shells in the

[19] Slugs that have no shell or a reduced one are also included in the class Gastropoda.

archaeological literature) (Bar-Yosef 1991). The calcareous parts of Cephalopods, most frequently the internal shell of *Sepia* (cuttlefish) are also encountered in the archaeological sites of the Mediterranean, along with other food refuse (Çakırlar 2003, Lamb 1936: 217. Also see Gejvall 1938 for a *Sepia* record from the early excavations in Troia). Accumulations of archaeological polyplacophoran (chiton) remains have been found on the coast of Pakistan and in pre-Columbian settlements in the West Indies (Desse and Desse-Berset 2004, Steininger and Mauser 2002). While *Dentalium* shells are commonly found as culturally modified forms, most *Sepia* and polyplacophoran remains have been found unmodified and interpreted as dietary refuse. Classes Caudofoveata and Aplacophora are worm-like mollusks. These classes have not been recorded from archaeological sites.

Figure 18. Definition of bivalve (left) and gastropod (right) shell part terms used in this study (Modified after Claassen 1998: 20-21, Figures 5 and 6).

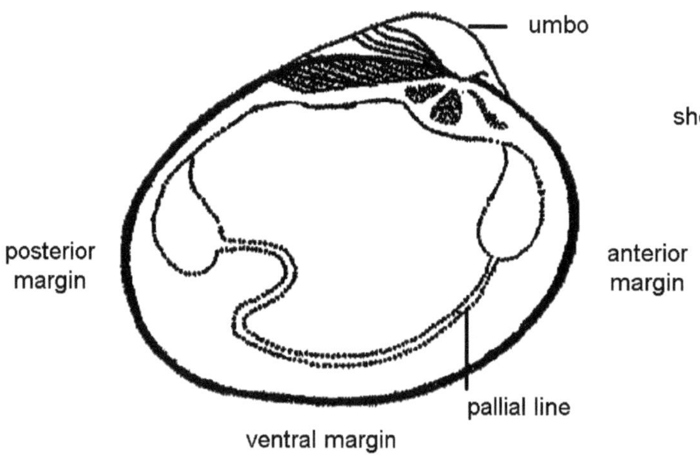

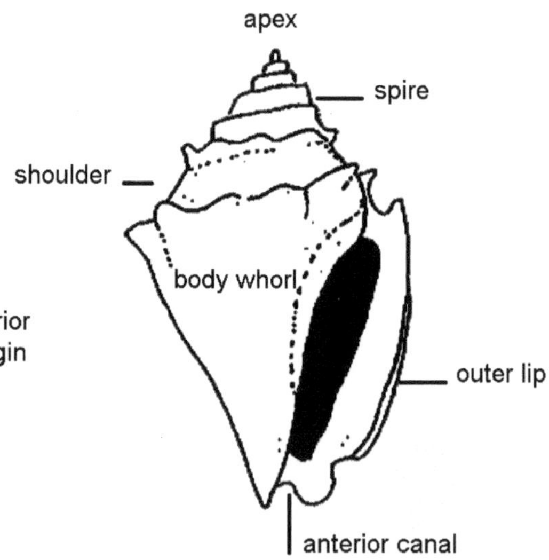

In broadest terms, the species represented in the archaeological assemblages of Troia, Yenibademli and Ulucak inhabit marine waters, with the exception of freshwater clams; *Unio* spp. Terrestrial and freshwater gastropods were rarely and randomly sampled during the excavations at these sites. Their archaeozoological registration was also done only partially. This is unfortunate since land snails are important tools in reconstructing various aspects of palaeoenvironment and palaeoeconomy (Evans 1972, Lubell 2004, Shackley 1981).

Figure 19. Definition of aquatic ecological zones.

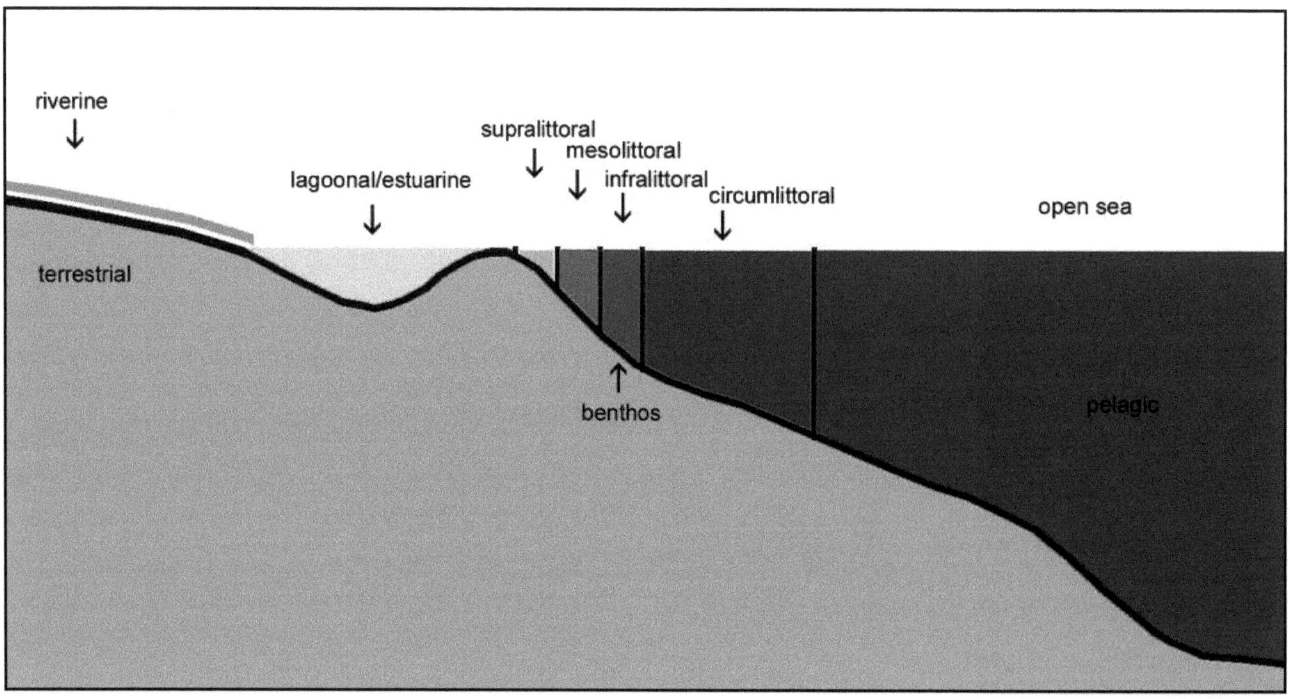

The following habitat and habitual descriptions are based on modern malacological literature. As in the case of the tables and figures presented in this paper, they follow the taxonomical systematic used in Demir 2003. Terminology used in describing mollusk shell parts is illustrated in Figure 18. Terms used to describe aquatic ecological zones are depicted in Figure 19.

4.1.1. Gastropods

4.1.1.1 *Patella* spp.

Patellidae, or limpets, are herbivorous gastropods that graze on algae living on hard surfaces. They have a conical shell varying in color from brown to bluish and sculptured with radial furrows in some forms (Sabelli 1979). *Patella* spp. here refers to any three of the species of the genus *Patella* reported from the Aegean: *P. caerulea, P. rustica,* and *P. ulyssiponensis* (Öztürk and Ergen 1999, Demir 2003). *Patella* species show high phenotypic plasticity, i.e., their shells assume a variety of forms within each species. This situation may cause taxonomic confusion when identification is based on shells alone. Archaeozoologists working in Troia did not distinguish between the different species and recorded all *Patella* specimens as *Patella* spp. It was also not possible to identify each archaeological *Patella* specimen to the species level in other sites. The two definitively represented species of the Patellidae family in Troia and Yenibademli are *P. caerulea* and *P. rustica* (Figure 20a). *Patella caerulea* can reach a maximum shell length of ca. 90 mm, whereas *P. rustica* are generally smaller (maximum ca. 70 mm), but higher than *P. caerulea* shells (Gaillard 1987).

Table 4. Morphological properties of the modern *Patella* populations in Saros Bay (After Öztürk and Egemen 1999, 517: Table 1).

Species	No. of specimens	Length (Mean)	Length (Maximum)	Width (Mean)	Height (Mean)
P. caerulea	270	31.1	50.2	25.9	8.3
P. ulyssiponensis	33	36.4	48.6	28.9	9.5
P. rustica	37	27.5	35.5	21.9	10.0

All measurements are given in mm.

All three of the above mentioned *Patella* species colonize supralittoral and mesolittoral zones of high-energy, fully marine environments. Littoral zones are populated vertically in a species-specific order. *P. rustica* is distributed at the upper mesolittoral zone, extending towards the supralittoral zone (generally 40-90 cm. above water level at the wave splash zone), on more or less upright surfaces. *P. caerulea* is found in mesolittoral and upper infralittoral zones, generally below *P. rustica* colonies. This species rarely colonizes areas below 1 m in the infralittoral zone. *P. ulyssiponensis*, on the other hand, colonizes exclusively the infralittoral zone, where its habitat range overlaps with *P. caerulea*, but does not touch the habitat range of with *P. rustica*. *P. caerulea* is the most common Patellidae inhabiting the coasts of Saros Bay (northeast of Gökçeada) today. (Öztürk and Ergen 1999).

Patella spp. are generally confined to habitats with salinities above 25 ‰. For example, in Güllük Lagoon where water salinity ranges between 7 ‰ and 14 ‰ *Patella* spp. populations have not been encountered (Egemen et al. 1999). Koutsoubas et al. (2000) also did not record any *Patella* specimens in the Gialova Lagoon in the Ionian Sea, where salinity levels fluctuate between 13 ‰ and 60 ‰. *Patella* species avoid habitats with salinity fluctuations. Low salinity might be one reason why only *P. caerulea* was able to invade the Black Sea (Gaillard 1987). Moreover, *Patella* spp. avoids sheltered locations lacking wave action (Hayward et al. 1995).

The relationship between the age and shell dimension of *Patella* populations in the Aegean is unknown. In western Mediterranean (Tuscany, Italy) a *P. caerulea* population whose individuals reached 30 mm in diameter or more during the first year of growth has been found (Bianucci and Longinelli 1982). It has also been found that the oxygen isotope intake of the same population is in equilibrium with the isotopic content of the host water. The individuals of this population secrete their shells throughout the year, with increased rates during the summer (Biannucci and Longinelli 1982). It is known that *Patella* spp. size and shape vary greatly due to ecological factors (Bianucci and Longinelli 1982, Cabral and da Silva 2003). The above mentioned survey in the Saros Bay gives a rough idea about the expected shell size structure of present populations in the North Aegean (Table 4).

4.1.1.2. *Haliotis tuberculata*

Haliotiidae, or abalones, are ear-shaped gastropods that live on hard substrates. Their shells bear perforations that serve as ventilation holes (Barnes 1987: 352). Only one species of abalone, *Haliotis tuberculata*, have been reported from the Turkish Mediterranean (Demir 2003).[20] The *H. tuberculata* shell can be green, brown or reddish; the inner shell is nacreous (i.e., it has a layer of mother-of-pearl) and can grow up to 15 cm in length (Gaillard 1987). *H. tuberculata* inhabits infralittoral and circumlittoral zones of the sea and, similar to Patellidae, is an herbivorous grazer.

[20] Some malacologists (e.g., Sabelli 1979, de Smit and Bába 2002) consider *Haliotis lamellosa* as a second *Haliotis* species that occur in the Aegean. Here, Gaillard's (1987) and Demir's (2003) taxonomy is used, where *Haliotis tuberculata lamellosa* is considered a subspecies.

Figure 20. Shells of *P. caerulea, P. rustica,* Monodontinae, and *H. trunculus.*

a. Modern specimens of *P. caerulea* (left) and *P. rustica* (right). Inner shell surface (top) and outer shell surface (bottom)

b. Modern specimens of *M. articulata* (left?) and *M. turbinata* (right?)

c. A well-preserved archaeological specimen of *H. trunculus* from Troia (Unit 993).

4.1.1.3. Monodontinae

Seven species belonging to this subfamily of Trochidae, the Monodontinae (topshells), are recorded on the Turkish coasts (Demir 2003). Mollusk remains of this subfamily were not identified to species level in Troia. In Yenibademli and Ulucak, specimens of both *Monodonta articulata* and *Monodonta turbinata* have been identified, but they were not recorded separately. The Monodontinae shell consists of a cone ornamented with spiral belts. Genus *Monodonta* has a nacreous inner layer. The background color is white, yellow, or gray, with a series of reddish-brown irregular zigzags in *Monodonta articulata* and regular patches in *Monodonta turbinata* (Figure 20b; Sabelli 1979, Gaillard 1987). Both species reach a maximum height of ca. 35 mm (Gaillard 1987). Occasionally sharing the same rocky surfaces with *Patella* spp., Monodontinae also colonize supralittoral, mesolittoral, and infralittoral zones and graze on algae. Monodontinae are known to practice vertical migration seasonally, settling in upper sections of rocky surfaces in the supralittoral zones in the summer months and moving to the mesolittoral zone in the winter (Little 1996: 22).

4.1.1.4. *Cerithium* spp.

Seven species of the genus *Cerithium,* the hornshells, exist in the Mediterranean (Gianuzzi-Savelli et al. 1996: 28-42, Demir 2003). Two of these are lessepsian[21] migrants (Gianuzzi-Savelli et al. 1996). In Troia, *Cerithium* spp. specimens were not designated to species, but it is unlikely that the archaeozoologists have encountered one of the three *Cerithium* species that only reach a maximal height of ca. 25 mm. The *Cerithium* specimens from Yenibademli and

[21] Lessepsian migrants are species which have entered the Mediterranean, from the Red Sea through the Suez Canal.

Ulucak are all larger than 30 mm. They can either belong to *C. vulgatum* or *C. alucastra*, but most probably, the majority belongs to *C. vulgatum*. *C. vulgatum* reaches a height of ca. 65 mm, while *C. alucastra* can grow slightly higher up to ca. 75 mm (Gaillard 1987). Demir (2003) did not record *C. alucastra* in his surveys in the Aegean. Gaillard (1987) reports that *C. vulgatum* are very common on rocky and sandy bottoms of the mesolittoral and infralittoral zones, whereas *C. alucastra* are much less frequent. *Cerithium* spp. shells form a very high conical spire, with a sculpture consisting of thin grooves. Coloring is highly variable (Giannuzzi-Savelli et al. 1996). *C. vulgatum* has adapted to both sandy and rocky bottoms as mentioned above, but it is also found among sea grass (*Posidonia*) meadows. It is a vagile epifaunal herbivore, i.e., it remains on the sediment surface and moves freely, feeding on plant matter. Although it is commonly regarded as a fully marine species, it is also known to colonize estuarine and lagoonal waters. The fact that it was present in the Gialova Lagoon in northwestern Greece, where salinity and temperature vary respectively between 13-60 ‰ and 14-24 ° C, confirms that this species is tolerant of such fluctuations (Koutsoubas et al. 2000). It was absent in the Güllük Lagoon however, in southwestern Turkey, where water salinity stays below 10 ‰ most of the year (Egemen et al. 1999).

4.1.1.5. Cypraeidae

Cypraeidae, commonly known as cowrie shells, have porcelain-smooth spherical shells, in various colors. They rarely have any sculpture. They inhabit tropical and warm-temperate seas of the world. Among the primary mollusk assemblages that are studied here, only the assemblage from Yenibademli yielded examples of Cypraeidae. One specimen of *Erosaria spurca*, one *Luria lurida*, and an unidentified cowrie shell come from the archaeological contexts of Yenibademli. Cypraeidae may have occurred in the studied faunal assemblages from the New Excavations at Troia, and categorized among "unidentified marine gastropods". Four species have been recorded from the Aegean: *Erosaria spurca*, *Luria lurida*, *Purpuradusta gracilis*, and *Zonaria pyrum* (Demir 2003). *E. spurca* is one of the few Cypraeidae with shell sculpture; it bears vague pits (Sabellli 1979). Adult individuals reach a maximum height of ca. 40 mm. *L. lurida* and *Z. pyrum* can grow as high as 55 mm (Giannuzzi-Savelli et al. 1996). The shell of *Z. pyrum* is pear-shaped (Sabelli 1979). The smallest of the Aegean Cypraeidae is *P. gracilis*, reaching only a maximum height of 17 mm (Giannuzzi-Savelli et al. 1996). Aegean Cypraeidae inhabits the infralittoral zone, living underneath or between stones (Sabelli 1979, de Haas and Knorr 1965). *Z. pyrum* is also known to live among *Posidonia* beds (Sabelli 1979). Both *E. spurca* and *L. lurida* are enlisted as endangered species in the Mediterranean (European Environmental Agency 1999).

4.1.1.6 *Phalium granulatum*

Phalium granulatum, the lipped helmet shell, is a carnivorous mollusk feeding especially on echinoderms by injecting a salivary juice that paralyzes its prey (Gaillard 1987). Adult individuals possess a big robust shell (maximum length: 100 mm) with an elongated spire and a convex outer lip. Shell sculpture consists of spiral ribs. They occupy the sandy-muddy bottoms of the circumlittoral zone and are fairly common (Gaillard 1987).

4.1.1.7. *Charonia tritonis*

Charonia tritonis, commonly known as the triton shell, is the largest mollusk of the Mediterranean Sea with individuals reaching over heights of 400 mm (Gaillard 1987). It bears a very high spire, smooth sculpture with wide ribs, and colorful zigzagging patches on a yellowish white background. The outer lip is rather sharp. Like *P. granulatum* (cf. Section 4.1.1.6), *C. tritonis* is also an active predator feeding on echinoderms, especially starfish (Kang and Kim 2004). It is fairly common on rocky surfaces in the circumlittoral zones of the eastern Mediterranean (Sabelli 1979, Gaillard 1987). *C. tritonis* is also considered among the endangered species of the Mediterranean Sea (European Environmental Agency 1999).

4.1.1.8. *Bolinus brandaris*

Bolinus brandaris[22], the spiny murex, is a fair-sized gastropod belonging to the Muricidae family, reaching a height of 92 mm (Gaillard 1987). Its shell is uniformly cream-white in color. The body whorl and siphonal notch are elongated; the spire is round with raised spiral ribs. *B. brandaris* is a carnivorous species that lives on a variety of substrates, including coarse, sandy, or muddy areas and sometimes on rocky bottoms in the circumlittoral zone (Gaillard 1987). The hypobranchial gland, present also in other gastropods, secretes a purple substance (Sabelli 1979).

4.1.1.9. *Hexaplex trunculus*

Hexaplex trunculus[23], the so-called common 'murex' shell, (Figure 20c) is another member of the Muricidae family represented in the archaeological assemblages. This carnivorous gastropod dwells on sandy and rocky bottoms in shallow bays and lagoons (also brackish) or deeper in wave-exposed coasts of the infralittoral zone (Rilov et al. 2004). Colonies prefer quiescent waters. Muricidae prey on a variety of gastropods and bivalves by means of drilling into their shell (Rilov et al. 2002). Because they also scavenge, they are common at the bottoms of fishing harbors (Ruscillo 2005). *H. trunculus*' solid shell consists of a triangular spire and a short body whorl, both with rough spiny sculpture. Individuals reach a maximum height of 83 mm (Gaillard 1987). Shell color is a grayish green, striped with purple, especially visible on the inner side. Similar to *B. brandaris*, this species secretes a purple substance.

[22] The generic name used for this species in both malacological and archaeomalacological literature has been *Murex* until recently (Sabelli et al. 1990).

[23] The generic name of this species, used both in malacological and archaeomalacological papers, has been *Murex* until recently (Sabelli et al. 1990).

Figure 21. Shells of *A. noae*, *M. galloprovincialis*, *F. glaber*, *O. edulis*, *S. marginatus*, and *T. decussatus*.

a. Inner (top) and outer (bottom) views of an *A. noae* valve from Ulucak (Unit 513).

b. Outer (left) and inner (right) views of a modern *M. galloprovincialis* valve.

c. Inner (left) and outer (right) views of a *F. glaber* valve from Troia (Unit 2045).

d. Left and right valves of *O. edulis* from Troia (Unit 2214).

e. Outer (top) and inner (bottom) views of modern *S. marginatus* valves.

f. Outer (top) and inner (bottom) views of left *T. decussatus* valves from Troia (Unit 1243).

4.1.1.10. *Fasciolaria lignaria*

Fasciolaria lignaria, the tulip shell, is another carnivorous snail, living on the shelly and detrital bottoms of the mesolittoral and infralittoral zones (Gaillard 1987). Its shell is thick and solid with a very high spire and a long siphonal canal. The color is variable. The shell can reach a maximum height of 70 mm (Gaillard 1987).

4.1.1.11. *Columbella rustica*

Columbella rustica, the rustic dove shell, is a smooth shell with a cone-shaped body whorl and spire. The lip is weakly toothed. The shell reaches a maximum height of 30 mm only. The color pattern generally is brown spots on white background. *C. rustica* is a predator that lives among sea grass like *Posidonia* and inhabits sponges such as *Sarcotragus* (Chemello and Milazzo 20002, Çınar et al. 2002).

4.1.1.12. *Conus mediterraneus*

Conus, or cone shells, are spread over the tropical and warm temperate seas of the world. They prey on worms, other gastropods, and fish by stabbing and poisoning them (Barnes 1987: 372). The sting of some South Pacific *Conus* species is highly toxic to humans (Barnes 1987: 373). The genus is represented with only one species in the Mediterranean (de Haas and Knorr 1965, Demir 2003). *C. mediterraneus* has a robust, conical yellow-brown shell with no sculpture. Individuals grow up to 55 mm. They dwell on sandy bottoms among seaweeds along with other mollusks such as *Pinna* spp. and *Columbella rustica* (Russo 1997).

4.1.2. Bivalves

4.1.2.1 *Arca noae*

The valves of *Arca noae* (Noah's Ark shell) are roughly trapezoidal in shape and have ribs (Figure 21a). Specimens in southern Turkish waters reportedly reach 68 mm (Bodur et al. 2004), whereas specimens whose valve lengths reach 90 mm have been reported from elsewhere in the Mediterranean (Poppe and Goto 2000). *Arca noae* lives attached to solid substrates such as rocks or shells in the infralittoral and circumlittoral zones (Poppe and Goto 2000). Although they may occur down to depths of over 100 m (Poppe and Goto 2000), population densities are highest between 3-5 m (Bodur et al. 2004). Water circulation and low levels of turbidity are important factors for this species (Poutiers1987). Populations are known to share marine habitats with *Ostrea edulis* and *Mytilus galloprovincialis* (Peharda et al. 2002). *Arca noae*'s salinity tolerance is limited to seawater. The species avoids littoral areas with freshwater influence (Bodur et al. 2004). This species' natural lifetime in the Adriatic Sea is estimated to be 16 years or more (Peharda et al. 2002). Specimens that form blister pearls have been reported from the eastern Mediterranean (Mienis 2000).

4.1.2.2. Glycymerididae

Four species of Glycymerididae (dog cockles) are recorded from the Turkish seas, all of which occur in the North Aegean (Zenetos 1996, Poutiers 1987, Demir 2003): *G. bimaculata*, *G. glycymeris*, *G. violacescens*, and *G. pilosa*. Shell morphologies of Mediterranean Glycymerididae are very similar, causing taxonomic confusion (Poutiers 1987). Their valves are large (up to 70 mm in length), rounded, and very solid, with no or very vague sculpture. Their color is brown and, sometimes, purplish. They are shallow burrowers in fine gravel or sandy/muddy bottoms and can be found offshore to approximately 100 m depth in the infralittoral and circumlittoral zones (Poutiers 1987).

4.1.2.3. *Mytilus galloprovincialis*

The valves of *Mytilus galloprovincialis* (Figure 21b), the common Mediterranean mussel, are equal, relatively thin, and smooth. Color ranges from dark brown to dark blue and black. A thick periostracum covers its thin shells (Sabelli 1979). *M. galloprovincialis* lives attached on substrates that can range from gravelly to rocky bottoms using their bysuss (silky filaments). Communities occupy mesolittoral and infralittoral zones and are known to penetrate the saline waters of coastal lagoons and estuaries (Poutiers 1987). There, they particularly colonize the sea-water influenced zone and with the entrance channel to the lagoon or estuary, where water velocity is relatively high (Barnes 1980: 44). They can survive in depths down to 6-9 m, forming very dense single-species communities. Individuals can reach 150 mm, but in most communities valve lengths between 50-80 mm are more common. (Poutiers 1987). Optimal growth occurs between 15 and 20 ° C and in ca. 27-40 ‰ in the Mediterranean (Kumlu 2001). Cultivated *M. galloprovincialis* in the Dardanelles showed growth rates from 10 to 40 mm a year, growing at a minimal rate in January and February, and rather fast between May and November (Yıldız and Lök 2005).

4.1.2.4. *Pinna* spp.

Three species of *Pinna* (fan mussels) live in the Mediterranean: *P. fragilis*, *P. nobilis*, and *P. rudis*. These large bivalves (maximum lengths in respective order: 350 mm, 1100 mm, and 300 mm) have equal, thin, and roughly triangular valves. The inner shell is nacreous; the outer shell is white to light brown. The animals settle among meadows of sea grass on loose substrates, usually on coarse sands, in the infralittoral and circumlittoral zones. *P. nobilis* lives up to 20 years in the Western Mediterranean (Richardson et al. 1999). Some individuals of *P. nobilis* secrete small pearls. Both *P. nobilis* and *P. rudis* are listed among the threatened species of the Mediterranean (European Environmental Agency 1999).

4.1.2.5. Pectenidae

Pectenidae, the scallops, are represented with at least 25 species in the Mediterranean (Poutiers 1987). Archaeological specimens in Troia were not identified to the species level and most of them were discarded after the preliminary archaeozoological examination. Two pairs of almost complete valves that were accidentally kept and subsequently examined by the author belong to *Flexopecten glaber* (Figure 21c; Units 1100 and 2075; dating to Troia III and Troia VI respectively). Pectenidae have fairly flat, unequal valves, armed with low ribs. Color varies between species and among individuals. The family consists of vagile epifaunal bivalves that attach themselves to the sea bottom with the help of a byssus at the juvenile stage and move freely when adult on sandy and detrital bottoms in the littoral and circumlittoral zones, most frequently resting on their right valves (Sabelli 1979, Poutiers 1987). They can swim to escape predators (Poutiers 1987). Some species of Pectenidae are known to migrate seasonally from shallow to deeper waters (Manuel et al. 1996). In the seas of Çanakkale, their upward migration seems to take place during the winter months (per. com. recreational fishermen from Çanakkale). *F. glaber* is essentially infralittoral and is known to penetrate coastal lagoons (Poutiers 1987).

4.1.2.6. *Spondylus gaederopus*

Spondylus, the spiny oysters, are distributed in tropical and warm temperate seas of the world. *Spondylus gaederopus* has a very thick shell of unequal bivalves sculptured with radial ribs and long spines. The top valve is nearly always purple; while the lower is usually white (Lamprell 1987). Individuals can grow as big as 150 mm long (Poutiers 1987). The right valve is cemented on rocky and detrital surfaces in the infralittoral and circumlittoral zones (Poutiers 1987).

4.1.2.7. *Ostrea edulis*

The valves of *Ostrea edulis*, the European oyster, (Figure 21d) are unequal and the overall shape is quite variable. Its color ranges from white to bluish gray. Individuals can reach a maximum height of 200 mm, but the common size is between 60 and 90 mm (Poutiers 1987). *O. edulis* create banks on sandy, gravelly, and rocky bottoms in the mesolittoral and infralittoral zone (Poutiers 1987). The left valve is either attached to a hard surface, or the animal lives freely on the substratum (Atay 1984: 111). Although *O. edulis* penetrates into lagoons and estuaries, it does not tolerate water turbidity or salinities below 23 ‰ (Milner 2001). Like *M. galloprovincialis* it only settles in areas with sea-water influence near the entrance of the lagoon or estuary (Barnes 1980: 44).

4.1.2.8. *Acanthocardia tuberculatum*

Acanthocardia tuberculatum, the red cockle, has equal, solid, and convex valves with well-developed ribs and under-developed spines towards the posterior margins. Its color varies from reddish-brown and yellow to white and yellow. Individuals can grow up to 90 mm in height, but commonly do not exceed 60 mm. They burrow in sandy bottoms just below the surface in the mesolittoral and infralittoral zones. Intolerance to wide salinity fluctuations confines these animals to areas with little or no freshwater influence. (Poutiers 1987).

4.1.2.9. *Cerastoderma glaucum*

Populations of *C. glaucum* (the lagoon cockle) (Figure 13) are distributed along the eastern coasts of the North Atlantic, in the Mediterranean Sea, and in the Black Sea (Poutiers 1987).[24] *Cerastoderma glaucum* is the sole species of the genus *Cerastoderma* living in central and eastern Mediterranean.[25] Its morphology is very similar to the edible cockle, *Cerastoderma edule.* The two species have slight morphological differences, but they can be positively distinguished by the differential relationship between their shell solidity and curve (Brock 1991: 21, Figure 11). The difficulty in identification has caused a great deal of confusion between the two species, and consequently in defining their geographical distribution and biological attributes. Specimens from Mediterranean sites (whether modern, archaeological, or geological) have often been incorrectly designated as *Cardium edule* or *Cerastoderma edule* (e.g., Karali 1999 and Kayan 2003). Phylogenetic research has shown that *C. edule* is decisively not present in the eastern Mediterranean (Brock 1991, Nikula and Väinölä 2003). Its geographical distribution might penetrate into the western Mediterranean (Poutiers 1987).

C. glaucum has equal valves (Sabelli 1979). The shell is sculptured with ribs, the number of which range between 21 and 28 (Zaouali 1979, Sabelli 1979). There are considerable morphological differences between populations. Mariani et al. (2002) reports a positive correlation between the number of the ribs in *C. glaucum* and temperature and salinity values of the host water. This data, however, does not seem to correlate well with Zaouali's (1979: 406) values from her study in the Bou Grara Sea in Tunisia, where *C. glaucum* bear a mean rib number of 21 (± 1,6). Temperature and salinity fluctuate seasonally in the Bou Grara Sea and reach 45 ° C and 30 ‰ respectively during the summer (Zaouali 1979). Mariani et al. (2002) found a positive relationship between curve (the relationship between breadth and length) and solidity (the relationship between shell weight and height), and temperature and salinity. Both Brock (1991) and Mariani et al. (2002) agree that the relationship between curve and solidity is not latitudinally defined. Zaouali describes her samples from the Bou Grara Sea as "...très dissymétriques. Leur partie postèrieure est, en general, très élargie et aplatie. Le crochet est presque toujours élevé. Les valves sont fortement convexes..." Its color also varies, but most adult individuals are yellowish with the posterior portion of the valves an olive green, whereas juveniles are mostly white or yellow (Sabelli 1979). The soft body is bright orange.

C. glaucum shell size is generally considered to reach about 50 mm in height, with most individuals commonly reaching 30-40 mm (Poutiers 1987). The relationship between age and valve height seems to become slightly more positive from northern to southern latitudes (Table 5).

C. glaucum inhabit muddy-sandy and coarse-bottoms in large communities (Sabelli 1979, Poutiers 1987). Juveniles with valve heights between ca. 2 mm and 14 mm live suspended by means of a bysuss, dwelling over floating

[24] *Cardium (=Cerastoderma) lamarcki* is the synonym of *Cerastoderma glaucum*, often used for populations in the Atlantic and the northern seas (Boyden and Russel 1972, Brock 1991).

[25] The former name of this genus is *Cardium.*

macrovegetation in the lagoons (Brock 1991: 15, Ivell 1979b). Adults usually live buried just beneath the surface of the bottom (Boyden and Russel 1972). *C. glaucum* is tolerant of various environmental factors such as temperature and salinity: around the British Isles, populations were found in waters with temperature fluctuations between 0-25 ° C and salinity ranges between 5 and 38 ‰ (Boyden and Russel 1972). In the Mediterranean, *C. glaucum* populations were located in lagoons where salinity conditions are more severe, e.g., in the Gialova Lagoon where highest salinity record was 60 ‰ in autumn (Koutsoubas 2000). It was found that exposure to air, exposure to wave action and tidal amplitude are the predominant factors restricting the occurrence of *C. glaucum* in the Atlantic (Boyden 1972a, Boyden and Russel 1972). Since tidal amplitude and consequent exposure to air are factors that are irrelevant for populations in the Mediterranean, it is reasonable to state that a preference for sheltered areas with very limited wave action is the sole limiting factor in *C. glaucum* colonization: the species inhabits predominantly enclosed or semi-enclosed lagoons.

Table 5. Relationship between valve height and age in modern *C. glaucum* populations in North Atlantic and the Mediterranean.

Location	Latitude (ca.)	Valve height reached (mm)	In the first ...months	Maximum VH (mm)	Life span (years)	Source
Essex, North Sea	52° N	11.0	12	31.0	6	Boyden 1972b
Sussex, English Channel	51°N	11.7	6	24.3	5	Ivell 1979b
Lago Lungo, West Italy	41°N	27.0	9	30.0	Not known	Ivell 1979a
Vouliagmeni Lagoon, Aegean	38°N	30.8	12	30.8	1	Gontikaki et al. 2003
Bou Grara, Tunisia	36°N	19.3	12	32.0	2	Zauali 1979

Although the species is tolerant of the above mentioned physical changes in the host water, the metabolic activities of *C. glaucum* are regulated in accordance with environmental changes and follow a seasonal pattern. To conserve energy, the animals reduce oxygen consumption at elevated temperatures during the summer (Wilson and Elkaim 1997). Atlantic populations spawn in May (Boyden 1972, Ivell 1979b), whereas in the Mediterranean, spawning takes on a polycyclic rhythm. Among the studied populations, those in Italy observe three spawnings a year, with a break during summer, and the populations in the Bou Grara Sea observe three spawning periods a year with a rest period from January to March (Ivell 1979a, Zaouali 1980, Gontikaki et al. 2003). The age of sexual maturity is also habitat-dependent. In laboratory conditions, seven-week-old individuals are already mature (Yankson 1986). Natural populations, on the other hand, do not become sexually mature until one-year-old in a lagoon in England, while in western Italy populations could already spawn in under six months (Yankson 1986, Ivell 1979b).

The incremental growth rate of the *C. glaucum* shell changes throughout the year. In Sussex, 50 % of the observed specimens produced a "winter ring", i.e., they stopped secreting their shell during the winter (Ivell 1979b). Ivell (1979a) found the increments of *C. glaucum* populations in Lago Lungo in western Italy too "ambiguous" to be used as evidence for a growth pattern. Contrariwise, he reports that the water temperature was too high (34.5° C in July) between July to October for the adult community to even maintain existence (Ivell 1979a). Based on this information, it could be argued that this community ceased or slowed growth during this period. Populations in Bou Grara in Tunisia form a "ring" during summer (Zaouali 1979).

4.1.2.10. *Solen marginatus*

Solen marginatus, the razorclam, has equal, fragile, narrow, and elongated valves that can grow up to 170 mm (Figure 21e). The common valve size is 100-110 mm. Shell color is a yellowish white. The species is a fast and deep burrower, sometimes burrowing up to 50 cm deep in sandy-muddy bottoms in low-lying beaches of the infralittoral zones. The species does not colonize turbid waters. (Poutiers 1987).

4.1.2.11. *Donax* spp.

Demir (2003) found three *Donax* (wedge shells) species in the Aegean: *D. semistriatus, D. venustus, and D. variegatus*. Zenetos (1996), however, reports two additional species from the Aegean: *D. vittatus* and *D. trunculus*. Donax spp. are the smallest bivalves represented in the archaeological assemblages studied for the present research. They grow only up to 35 mm, commonly only up to 25 mm in valve length (Poutiers 1987). Valves are equal, smooth, roughly triangular, and elongated. Shell color is a variable gray and light brown patterning on a grayish white background. *Donax* spp. are shallow burrowers, dwelling in sandy substratum in the mesolittoral and infralittoral zone commonly only down to ca. 10 m. (Poutiers 1987).

4.1.2.12. *Tapes decussatus*

Tapes decussatus, the carpetshell, has moderately convex, robust, and smoothly triangular valves (Figure 21f). The color pattern varies from yellowish to a reddish brown background with broken reddish-brown broken zigzag lines. Shell sculpture consists of moderately impressed concentric growth lines (Sabelli 1979). The common length of adult individuals is 40-50 mm (Poutiers 1987). *T. decussatus* lives in sandy to gravelly-sandy substrates in the mesolittoral and infralittoral zone and can infiltrate lagoons. (Poutiers 1987). They burrow only to shallow depths, down to 15 cm (İlhan and Gülyavuz 2004).

4.1.2.13. *Unio* spp.

The taxonomy of freshwater clams in northwestern Anatolia has not been studied (per. com. T. Bodur). It may be presumed, however, that *Unio crassus* –distributed in Europe, Russia, Kazakhstan, and Anatolia (İlhan et al. 2004) – is the prevailing *Unio* species in Çanakkale. *Unio* valves are solid, equal, oval, and elongated. Shell morphology in Unionidae varies depending on substrate, stream size, velocity, water temperature, trophic degree, and alkalinity (Tevesz and Carter 1980). Valve lengths of *Unio crassus* can reach up to 110 mm, but commonly remain at 40-70 mm (İlhan et al. 2004). Shell color is whitish-yellow and transparent, covered with a thick green-brown periostracum. The innermost shell layer is nacreous. *Unio* spp. inhabit both small and large rivers, as well as freshwater lakes. Known to prefer sandy and gravelly bottoms (İlhan et al. 2004), they are very shallow burrowers, with the larger part of the shell remaining on the substratum. Vicenti (2005) observed that *Unio crassus* individuals creep up to the shoreline in the morning and return to the deeper benthos in the late afternoon, exclusively during the reproductive season.

4.2. Modern collections

Fieldwork conducted in the Troad provided two types of information about the habitats and habits of the mollusk taxa represented in the assemblages: (1) the present locations and ecology of the extant mollusk populations; and, (2) the morphological properties and incremental growth pattern of *C. glaucum* populations in the Troad. These data complement existing malacological knowledge and constitute the modern analogy for interpreting the archaeomalacological assemblages of commonly represented taxa.

4.2.1. Survey of modern mollusk populations in the Troad

Three areas were surveyed in the Troad: the deltaic area of the Karamenderes and the Dümrek, the coast of the Sigeon Ridge, and the Beşik Bay (Figure 2). The deltaic area consists of lagoons and estuaries. The lagoons are small areas of shallow water separated from the sea by low sand dunes, which are not connected to the rivers with surface waters but are part of the river delta system by being fed by subterrenian waters connected to the rivers. Estuaries are parts of the river mouths where the distributary channels of the rivers meet the sea (Figure 22). They, too, are partially separated from the sea by low sand dunes.

As described in Section 2.4.1, the lagoons and estuaries of the Karamenderes and the Dümrek are only diminished relicts of the lagoons and estuaries that dominated the Troad in the remote past and until recently (Figure 16). Present day lagoonal and estuarine biotopes are very small in area. The Karamenderes flows into the Dardanelles creating three estuaries (Figures 16 and 22). The estuary of the Dümrek is located to the east of these. It is connected to the Karamenderes by an irrigation channel. Another artificial channel is located to the west near the Sigeon Ridge. Three lagoons are found in the area; two very small lagoons are located between the distributary channels of the Karamenderes and one larger lagoon (the so-called Kiremitli Lake) is located to the west of the distributary channels.

The mouth of the Dümrek is a semi-enclosed estuary (Figure 16). The substrate here consisted of empty, but fairly intact *C. glaucum* shells and black mud between 2003 and 2006, indicating anoxic conditions. *Cyclope neritea*[26] colonizes the shallow waters and soft bottoms at the mouth of the Dümrek. Some detrital rocks (probably artificially) create shallow and narrow pools here, protected from the waves of the Dardanelles. A few adult individuals of *H. trunculus* were located in these pools. The estuary is relatively deep. No *C. glaucum* communities were located in this estuary.

In 2005 and 2006, three communities of *C. glaucum* were located at the semi-closed closed middle estuary and at the larger two of the lagoons of the deltaic area of the Karamenderes (Figure 16). The communities are densely populated but they are not frequent and do not form large beds. The two bigger estuaries and the smallest lagoon did not yield any *C. glaucum* communities. In general, areas that did not yield *C. glaucum* communities were the semi-open estuaries with less sheltered areas and greater freshwater influence. The substrate of the larger estuaries consists of gravelly sand in shallow areas and organically rich mud in deeper areas. Water at these locations were clear at the times visited. Samples were collected from these communities five times in 2005 and 2006. *C. glaucum* collection sites are called "stations" and are numbered from east to west (Figure 16). Each sampling has been given a unit number similar to archaeozoological unit numbers (Table 6). The amount of specimens collected at each site on different dates, surface temperature and water salinity at the time of collection, and features of the substrate are given in Table 6. The community in the so-called Kiremitli Lake (Station 3) was located at a depth of 0-50 cm. The substrate here was muddy and shelly, consisting of dense *Abra* spp[27] communities, empty *Unio*

[26] *Cyclope neritea* is a gregarious carnivorous gastropod able to tolerate conditions of relatively low salinity (Shackleton 1988a).

[27] *Abra* spp. are typical lagoon bivalves whose valve length does not exceed 10 mm (Gontikaki et al. 2003 and personal observations). The shell is smooth and transparent yellow. They can tolerate salinities that range between 3 and 60 ‰ (Barnes 1980).

spp. and *C. glaucum* shells, and even one adult specimen of *Mytilus galloprovincialis*. At the other two locations (Station 1 and 2) where *C. glaucum* beds were found, the substrate was muddy, with a few occasional empty shells. Communities were settled between 20-50 cm depth. The substrates colonized by the *C. glaucum* communities varied from muddy to sandy and shelly, while sandy substrates with small-sized gravel did not yield living *C. glaucum* specimens. All *C. Glaucum* communities dwell in sheltered areas close to the northwestern shores of the water bodies they were located.

Figure 22. A lagoon (Station 2) at the deltaic area of the Karamenderes in February 2006.

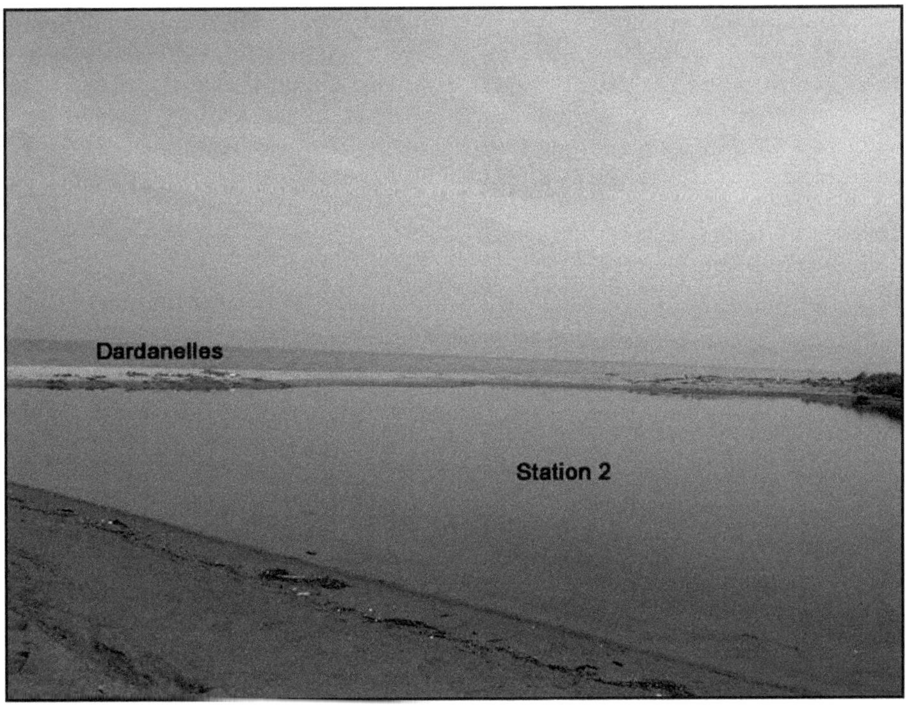

Table 6. Characteristics of *C. glaucum* collection stations in the Troad.

Date	Station	Unit	No. of specimens	SST	Salinity	Substrate
10.08.05	3	3	67	26 ° C	36 ‰	Muddy-shelly
04.02.06	1	6	19	8 ° C	19 ‰	Muddy
04.02.06	2	5	22	8 ° C	26 ‰	Muddy
04.02.06	3	4	16	6 ° C	20 ‰	Muddy-shelly
31.05.06	1	9	14	22 ° C	26 ‰	Muddy
31.05.06	2	10	25	25 ° C	33 ‰	Muddy
31.05.06	3	11	43	24 ° C	26 ‰	Muddy-shelly
02.06.06	3	12	44	24 ° C	22 ‰	Muddy-shelly
04.08.06	1	14	30	30 ° C	36 ‰	Muddy
04.08.06	2	15	30	29 ° C	34 ‰	Muddy
04.08.06	3	16	30	29 ° C	35 ‰	Muddy-shelly
05.11.06	1	17	25	11 ° C	18 ‰	Muddy
05.11.06	2	18	29	11 ° C	27 ‰	Muddy
05.11.06	3	19	45	12 ° C	30 ‰	Muddy

Young specimens were concentrated among the weeds near the shore at 0-25 cm, while adult sepecimens were found at slightly deeper locations. Juveniles and young specimens were buried somewhat deeper than the adult shells, between ca. 5-10 cm below the surface of the substrate, while the adults were found directly underneath the surface of the bottom and sometimes (ca. 30 %) semi-immersed on the surface.

The water conditions at the collection stations vary seasonally and between the stations. Seasonal variations are caused by seasonal changes in the air temperature, precipitation rates, evaporation rates, and the changing influences of seawater and freshwater. The most important factor affecting the differences observed between the environmental conditions of the collection stations is the differential influx of freshwater and seawater. Seawater from the Dardanelles and freshwater from the Karamenderes penetrate into Station 1 by channels above the surface, whereas at Stations 2 and 3 both freshwater and seawater intake occurs at lower rates: indirectly through winter splashes in the case of seawater and via groundwater movements in the case of freshwater. SST values were very low during the winter in February at all stations (6-8 ° C) (Figure 23). The upper parts of the distributary channels, ca. 700 m south of the stations, had patchy ice on the surface in February 2006. The collection in February took place one day after a four-day-long snowstorm, but on the day of the collection the air temperature rose up to 19 ° C at midday. SST values were already elevated at the end of May (22-25 ° C) and peaked to 26-31 ° C in August (Figure 23). Salinity rates at the collection sites fluctuated considerably; between 19 and 36 ‰ (Figure 23). Evaporation rates are high in these shallow waters, especially during the summer, a factor easily depleting the influence of and meteoric water on the lagoons and estuaries. Station 2, where salinity fluctuated only between 26 ‰ and 34 ‰ (Figure 23), has no direct connection to the Karamenderes river system. It is small and shallow.

Figure 23. SST (° C) and salinity (PPT) values at the time of collection.

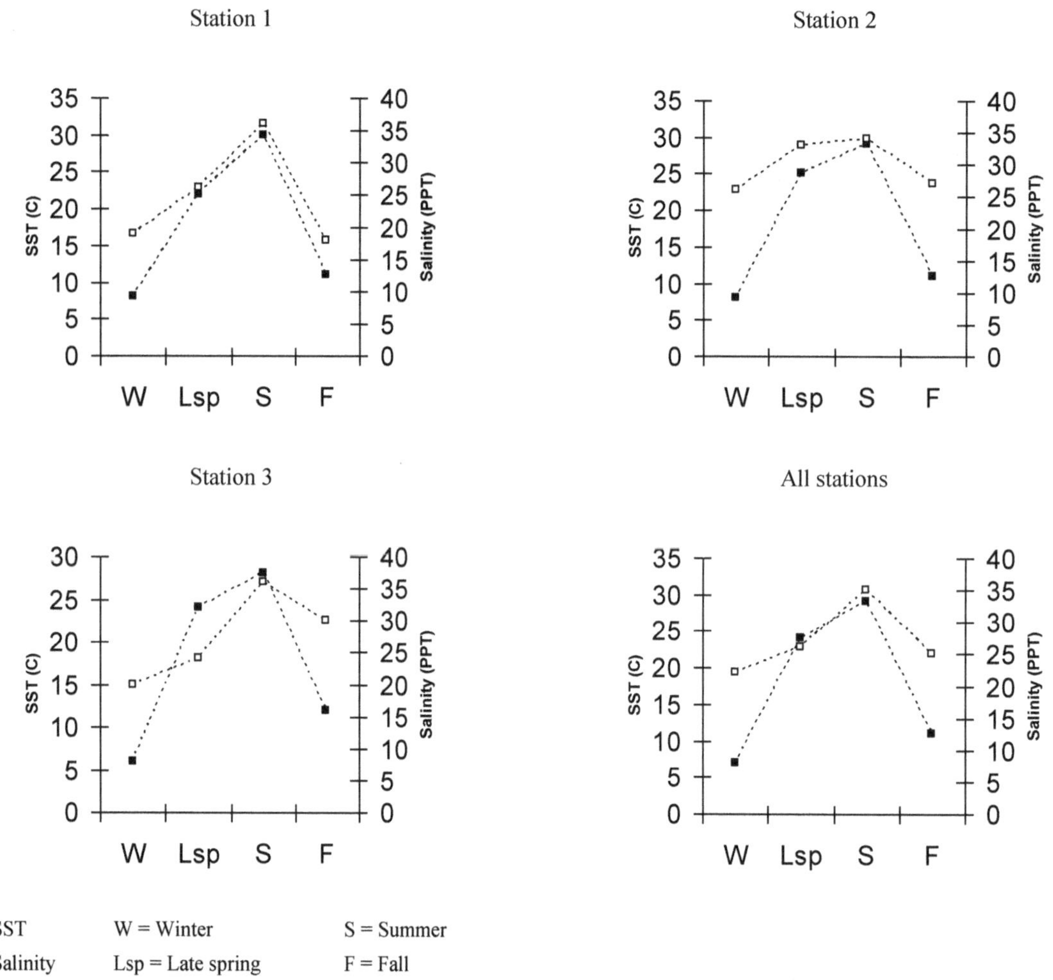

From the end of the Dardanelles in the north to the Beşik Bay in the south, the coastline along the western edge of the Sigeon Ridge is exposed to wind and wave effects. The coastline is rocky and steep, at parts aligned with narrow, gravelly beaches. The mesolittoral zone is narrow, quickly ascending to the fairly deep littoral zone. Here, a few isolated communities of *M. galloprovincialis* were located in the mesolittoral zone, and *P. caerulea* and Monodontinae were found in the supralittoral and mesolittoral zones. Although this coastline would theoretically provide an ideal habitat for *O. edulis*, no specimens of this species have been encountered. In Beşik Bay, between 30-60 m off the coast, already in the infralittoral zone, sparsely populated communities of *A. tuberculatum* and *Mactra glauca*[28] were found in depths from 75 to 200 cm.

[28] *Mactra glauca* is an infralittoral sand burrowing bivalve, whose shell is smooth, roughly oval, and solid (Poutiers 1987). Its color is gray and white.

4.2.2. Properties of modern *C. glaucum* populations in the Troad

4.2.2.1. Morphology

A total of 413 living *C. glaucum* specimens were collected from three different collection stations. They represent benthic populations.

Valve heights of modern specimens ranged between 11.4 mm and 45 mm with a mean value of 33.2 mm (Table 17, Figure 24). The variance of valve heights is slightly skewed towards larger specimens (Figure 24). This situation is due to the differential depth of the collection stations. Nearby sea grass beds, which probably host juvenile specimens, were not searched. Specimens from Station 1 represent the community with the largest valve heights (Figure 25). Specimens collected at Station 2 are slightly smaller than those from Station 2. The valve heights of the specimens from Station 3 demonstrate the largest variance among the stations (p = <0.0000-0.0027). This station yielded a relatively large amount of specimens and the communities were located between 0-50 cm deep. A comparative analysis of valve heights and growth patterns (Section 4.2.1.2) also demonstrates that the inter- and intra-station variance in valve heights is attributable to small differences in shell secretion rates between different communities.

Figure 24. Morphology of modern *C. glaucum* populations from the Troad (Distributions).

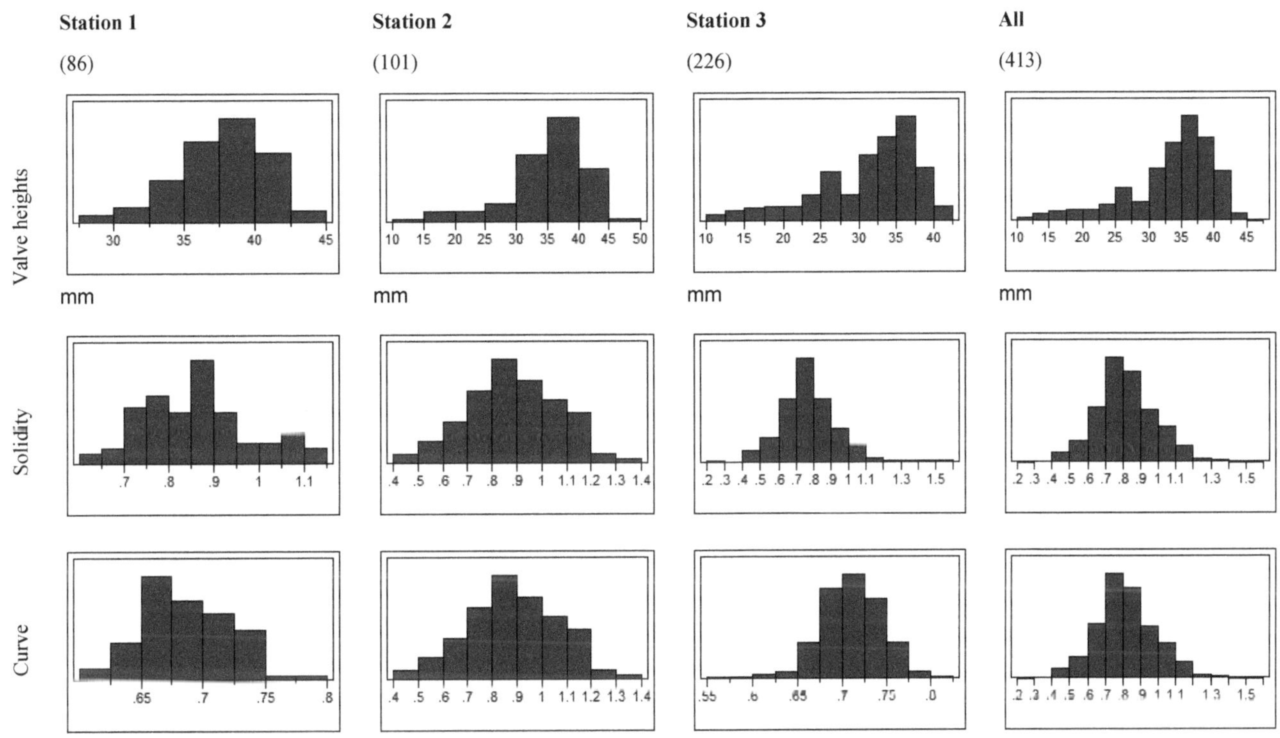

(86) = No. of specimens.

The solidity of *C. glaucum* shells collected from Stations 1 and 2 are not significantly different from each other (p = 0.3618), whereas communities living in Station 3 were significantly less solid than of those living in Stations 1 and 2 (Figure 25; p = <0.0000). The difference between the mean solidity of different populations is less than 0.09. Shells from Station 1 are less curved (p = <0.0000) than those from Stations 2 and 3, whose shell curvatures are comparable to each other (p = 0.6581). The difference between the mean curve values of different populations is less than 0.02 (Figure 25).

A meaningful relationship could not be constructed between the water conditions (i.e., SST and salinity values measured seasonally during the course of this study) and the morphological properties of the *C. glaucum* populations at the different stations. This situation can be attributed to methodological shortcomings, mainly to the infrequent monitoring of the water conditions and the uneven representation of age and size ranges at different stations.

4.2.2.2. Description of incremental shell growth

External lines on the surface of *C. glaucum* form a prominent feature of this species (Figure 13). Even a casual examination of the empty valves found on the beach confirms that variations in shell secretion rates are present. External lines are not uniform in appearance: some are grooves caused by a

Figure 25. Comparison of the morphology of modern *C. glaucum* populations from the Troad by station (ANOVA).

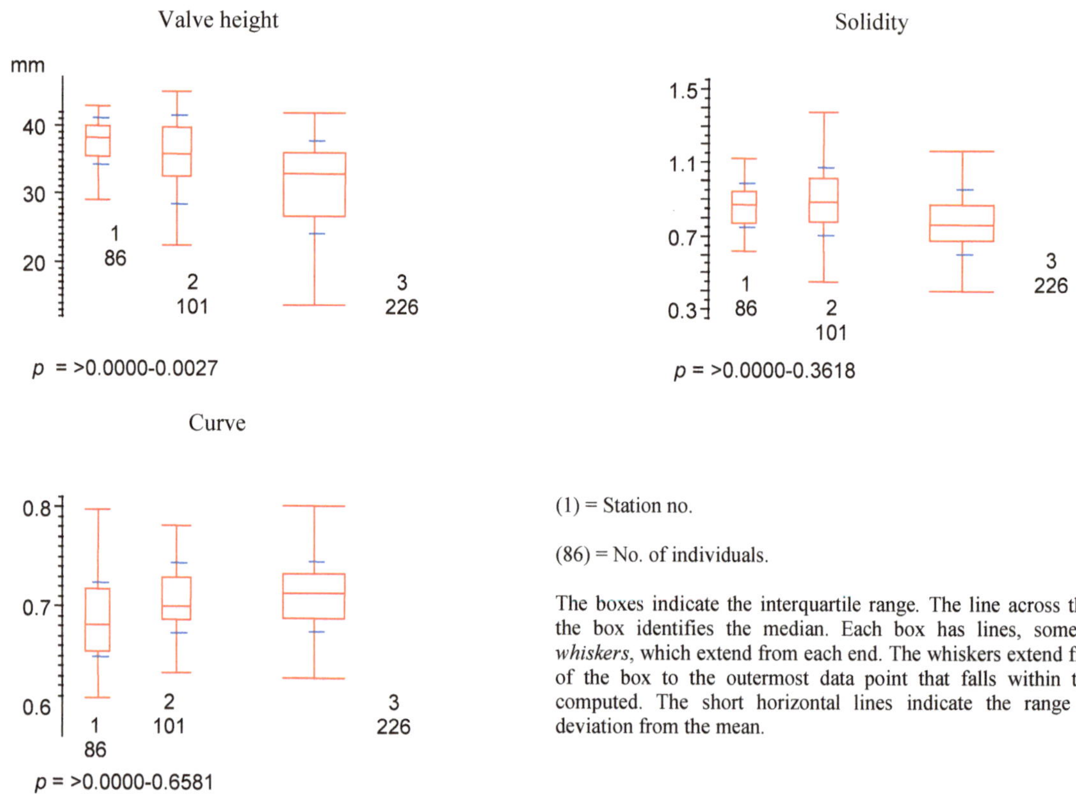

(1) = Station no.

(86) = No. of individuals.

The boxes indicate the interquartile range. The line across the middle of the box identifies the median. Each box has lines, sometimes called *whiskers*, which extend from each end. The whiskers extend from the ends of the box to the outermost data point that falls within the distances computed. The short horizontal lines indicate the range of standard deviation from the mean.

thinning of the shell, some are only slightly visible. They can be obscured by the occurrence of successive indentations located closely after one another or by the occurrence of a wide depression. They can also appear as very thin lines. While many of the external lines are highlighted by a change in shell color, this is also variable. Almost all external lines are related to internal features observed in the radial cross-sections of the shell. The ones that are not associated with a change in the color of internal features represent disturbances. These are rare. The comparison between external lines and internal incremental zones reveal that the external lines most often mark halting events in growth but they do not provide information about the amount of shell secreted during a growth period. Like in many other bivalves, in *C. glaucum* this information is best attained by observing the internal features of the shell, i.e., the increments (e.g., Lutz and Rhoads 1980, Schöne et al. 2005). Neither method is, however, helpful in explaining the duration of discontinuities (halting events) in shell secretion; but as stated, growth stops are rare in the *C. glaucum* specimens observed for this study. This can be explained by the species' high tolerance for a wide range of salinities and temperatures (Boyden and Russel 1972).

Incremental zones are differentiated optically as gray and white bands. Gray bands sometimes tone towards brown, due to the coloration of the shell. Under transmitted light, the gray bands appear as translucent zones and white bands appear as opaque zones. The degree of transparency decreases as the valve thickness increases. Although the transition between two incremental zones does not constitute a clear border, for the purposes of seasonal growth research, the observer needs to make a subjective decision as to where growth periods border each other.

The incremental zones are formed by microscopic increments. The formation of microscopic increments in other bivalves is influenced by rhythmic patterns in the environment such as day/night cycles or tides (e.g., Clark II 2005, Leontarakis and Richardson 2005, Richardson 2001, Richardson et al. 1979, 1980). The observations made for the present study are not precise enough to disclose the periodicity of single microscopic increments in *C. glaucum* populations in the Troad. It can be inferred, however, that they form daily. The rhythm of the marine life in the Mediterranean generally takes on a daily pattern, for tidal effects are insignificant on the coasts of this sea.

Successive growth increments tend to become progressively narrower in all species for which growth has been observed (Claassen 1983, Monks and Roberts 1983). The breadths of incremental zones in *C. glaucum* are related to body size and age. Growth rates in juveniles are higher than in adults. The distance between the umbo and the first gray increment, namely 0W, is not discussed in terms of its breadth or the relationship between time and breadth of incremental zones, because the time of recruitment (i.e., the beginning of shell formation) for the *C. glaucum* populations from the Troad is unknown. Based on other studies on *C. glaucum* populations elsewhere in the Mediterranean (Ivell 1979a, Zaouali 1980), it can be suggested that at least three spawning periods occur in the Karamenderes Delta populations. Therefore, although it is not rare that the breadth of the 0W is larger than that of 1W, it is considered that the fastest growth takes place with 1W. Mean breadth of the 1W zone is 7.5 mm. The mean breadth of the white incremental zones decreases as the shell grows (Figure 26). The decrease in the size of successive incremental zones continues steadily until the completion of incremental zone 4W. The mean breadth of 5W is only slightly smaller than 4W, but almost the same as the breadth of 6W. The mean breadths of 5W, 6W, and 7W are very small, making the determination of these incremental zones difficult. The trend of variation in the breadth of successive gray incremental zones is much less clear. The mean differences between successive gray incremental zones of the modern populations are small. Incremental zones 6G and 7G are difficult to distinguish from one another and from the white incremental zones between them.

Figure 26. Comparison of completed breadths of successive white (a) and gray (b) incremental zones in modern *C. glaucum* populations (ANOVA).

a. Breadths of white incremental zones.

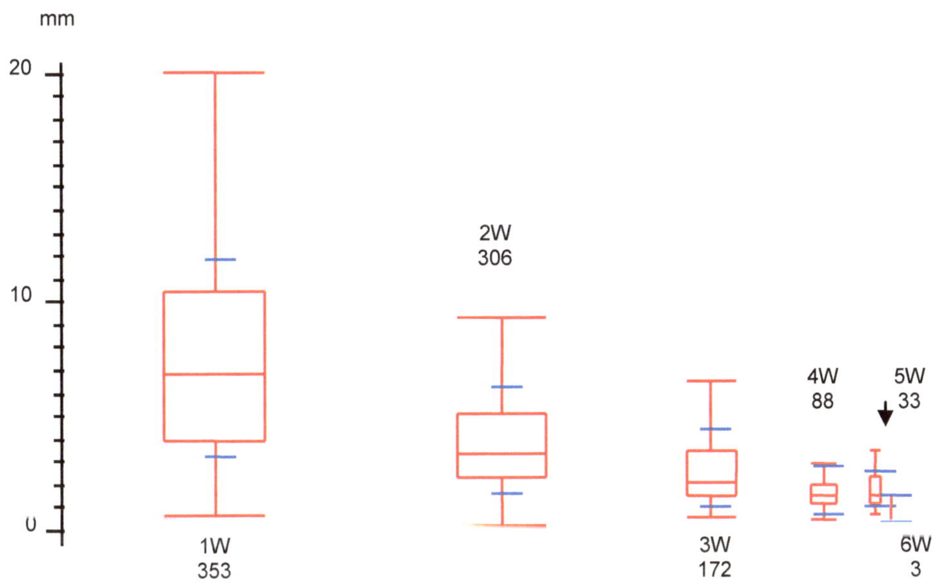

b. Breadths of gray incremental zones.

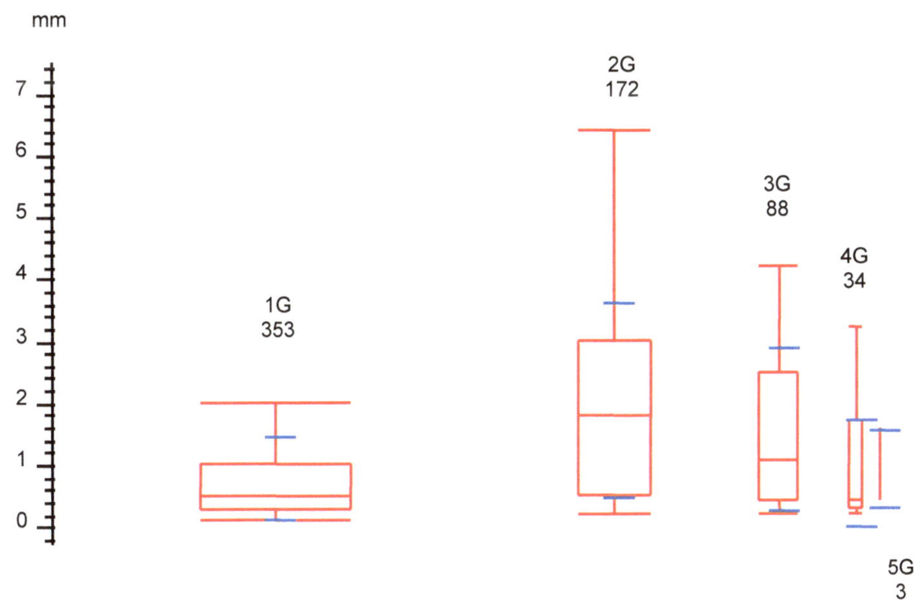

The variability of the breadths of incremental zones is great among individuals (Figure 26). One of the factors which determine variation in increment size is year-to-year environmental and climatic variations (Jones and Quitmyer 1996). The duration of field observations conducted for this study did not permit the determination of the effects of year-to-year changes on the size of successive increments. During the 15 months that *C. glaucum* populations were observed, the added breadth of completed incremental zones showed variability between the collection stations (Figure 27). Individuals in Station 1 could secrete slightly larger increments than the individuals in Stations 2 and 3 and individuals living in Station 3 secreted the narrowest increments among the modern populations. The variability between the growth rates of populations can be attributed to the differences between the environmental conditions prevailing at the different stations (cf. Section 4.2.1). But, it should be kept in mind that factors that were not measured or otherwise observed for this study, such as food supply, population density, and genetic differences are also known to affect growth rates in bivalve populations (Seed 1980).

Figure 27. Mean valve height (VH) achieved at each growth stage in modern C. glaucum populations.

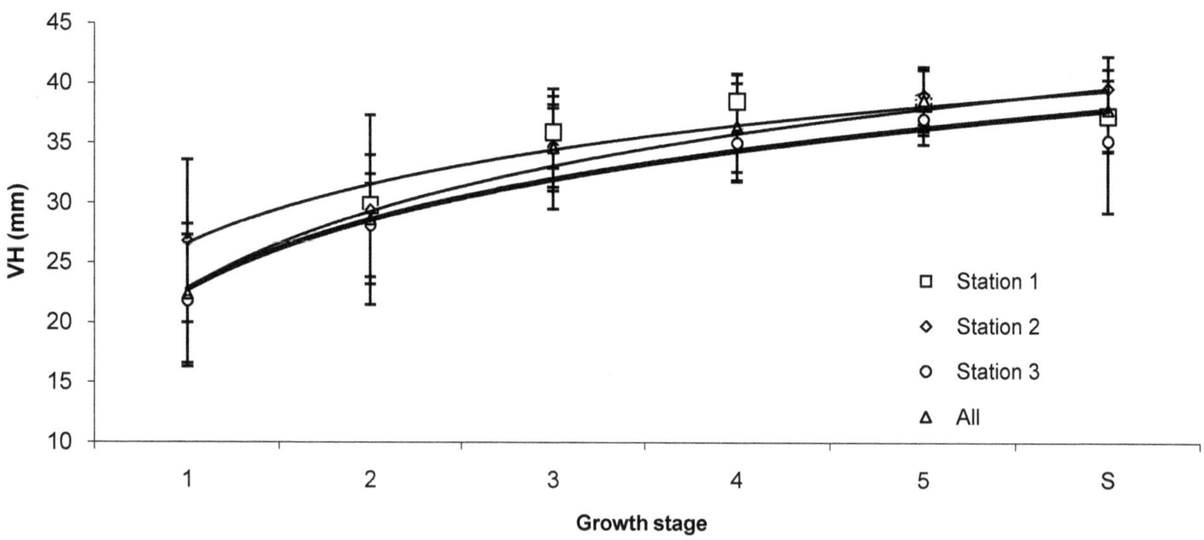

4.2.2.3. Age profiles

Observations on the living specimens collected periodically throughout 15 months allowed for the conclusion that annual growth consisted of one white and one gray zone (See Section 4.2.2.4 below). In this way, it could be established that one white incremental zone, together with a gray incremental zone form a growth stage more or less equivalent to one year. For example, the first white incremental zone, 0W, forms Stage 1 together with 1G. Zone 1W, together with zone 2G, forms Stage 2 and so on (Table 7). To overcome the difficulties posed by the variability in the appearance of the last incremental zones, the rarely occurring growth stages of 6, 7, and 8 are together categorized in the "senile" stage.

By following the methods laid out above, the age profiles of 413 modern individuals were plotted (Figure 28). All age stages between 1 and 7 are represented by the study samples taken from the modern populations. Senile stage is only very rarely encountered among the sampled individuals, indicating that the maximum age these populations reach is ca. 5 years. This time span is similar to the maximum age observed on *C. glaucum* populations in the North Atlantic (cf. Table 5).

The estimated age range of the population in Station 2 is evenly distributed; representing few young and senile animals, with the bulk of the sample belonging to adults. The age profile of the sample population from Station 1 is skewed towards older animals. Stage 1 is not represented by the sample collected from this station. At Station 3, the age profile of the sampled specimens is patchy, with a large emphasis on individuals collected while in Stage 3.

4.2.2.4. Seasonal incremental growth pattern

Seasonal collection of living specimens revealed the relationship of incremental zones to seasons. The timing of the change between white and gray shell secreting is largely related to the seasonally changing environmental conditions of the host water, but it is also largely dependant on the age of the individuals.

Specimens with similar valve heights and number of incremental zones, collected from the same station showed very similar growth patterning during the height of winter, in February, and during the height of the summer, in August (Figures 29, 30a, and 30b). In February 2006, only two out of 57 individuals (ca. 3.5%) were secreting gray increments at their margin during the time of collection (Figures 29, 30c, and 30d). In August, both in 2005 and 2006, the situation was reversed, with only 26 individuals out of 167 individuals (ca. 16%) secreting white increments at the time of collection (Figures 29, 30e, and 30f). Individuals secreting white increments at the time of collection consisted mainly of very

Table 7. Method of growth stage determination.

Growth stage		Individuals with incremental zone	Character of incremental zone at margin
1		0W	W
		1G	G
2		1W	W
		2G	G
3		2W	W
		3G	G
4		3W	W
		4G	G
5		4W	W
		5G	G
Senile	6	5W	W
		6G	G
	7	6W	W
		7G	G
	8	7W	W
		8G	G

Figure 28. Estimated age profiles of modern C. glaucum populations in the Troad.

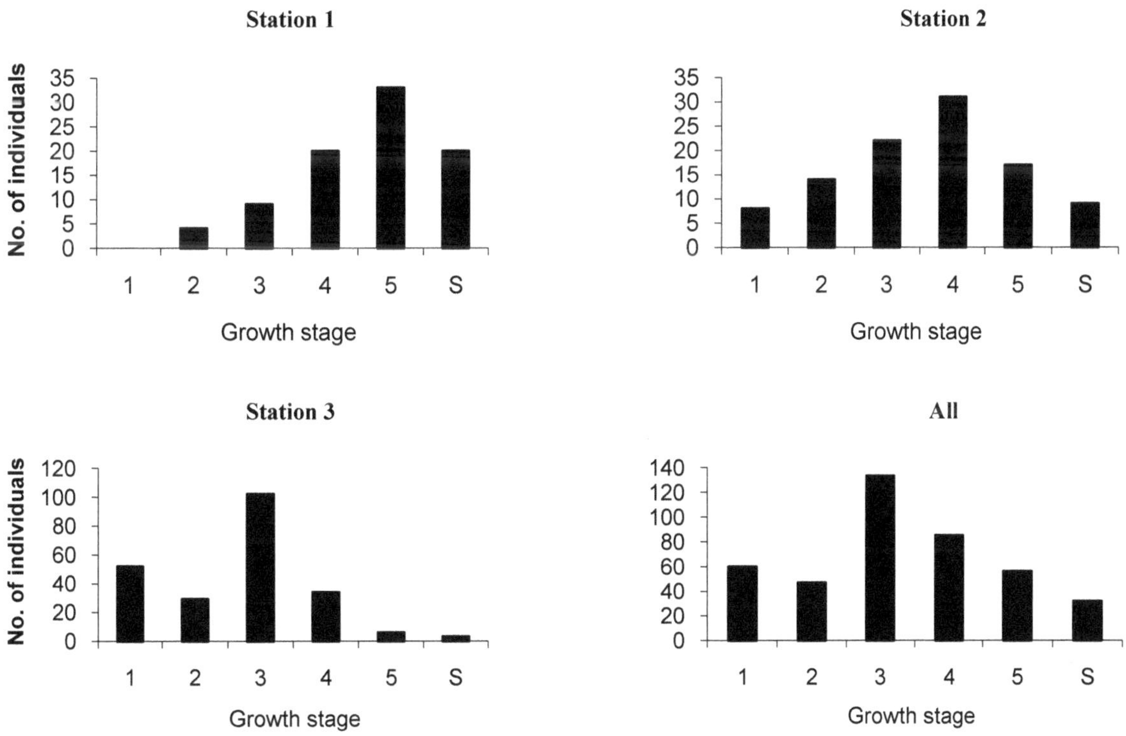

young individuals. In late May-early June, there is considerable variation between stations in terms of observed growth patterns (Figures 29 and 31a). At Station 1, only very few individuals were secreting gray increments, while at Station 2, almost all individuals were secreting gray increments. In Station 3, as some individuals of the same age group were secreting gray increments, others were secreting white increments. At Station 2, where most individuals were secreting gray increments already in June or perhaps even earlier, the SST and salinity were very close to August values. Similar to the situation observed during the late spring, there was variability in growth patterning among the

Figure 29. Seasonal growth pattern of modern *C. glaucum* populations.

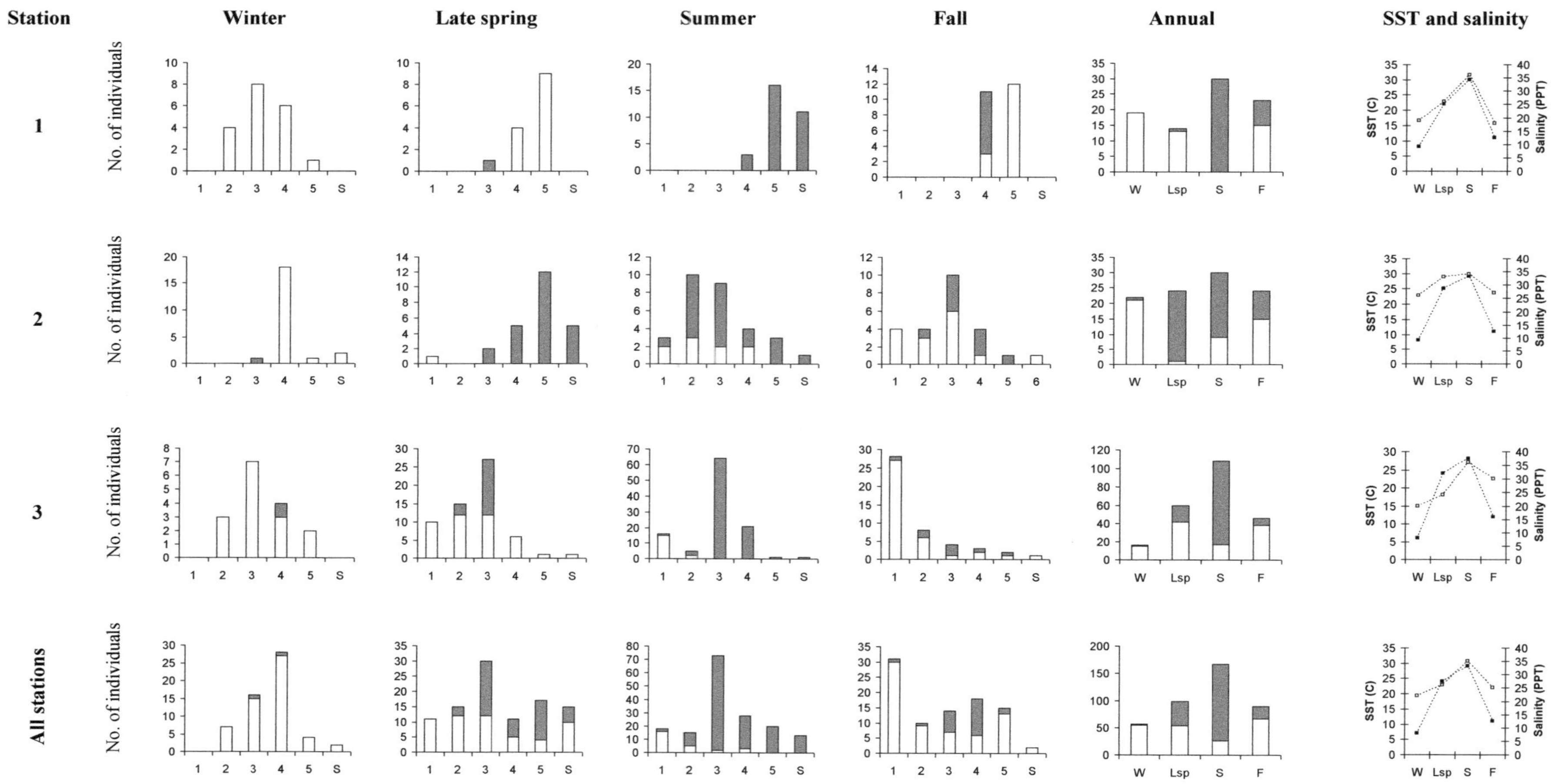

(■) = SST. (□) = Salinity. S = Senile. Gray fills represent individuals which were secreting gray increments, whereas white fills represent individuals which were secreting white increments at the time of collection.

stations and among the individuals from the same stations in November. While 23 out 90 individuals (ca. 25.5 %) were secreting white increments in November, a considerable number of the old individuals were secreting gray increments (Figures 29, 31b, and 31c).

When all stations are considered as a group, it appears that younger individuals, mainly those at growth stages one and two, secrete gray increments only for a short time period during summer months, whereas older individuals secreted gray increments significantly longer; some of them between June and November (Figure 29). Differential environmental conditions at the three stations may also have played a role in the timing of alteration between gray and white incremental zones. Most individuals, excluding very young ones, show gray incrementing at their margins during summer months. Older individuals are more sensitive to seasonal changes in their environment. The time span of gray increment secretion stretches from late spring to early fall in older individuals. Some individuals, with no clear relationship to their age or size, have been complacent to seasonal changes. During winter, almost all individuals secrete white increments.

Figure 30. Growth pattern of modern *C. glaucum* specimens (1).

a. Individual no. 3-57

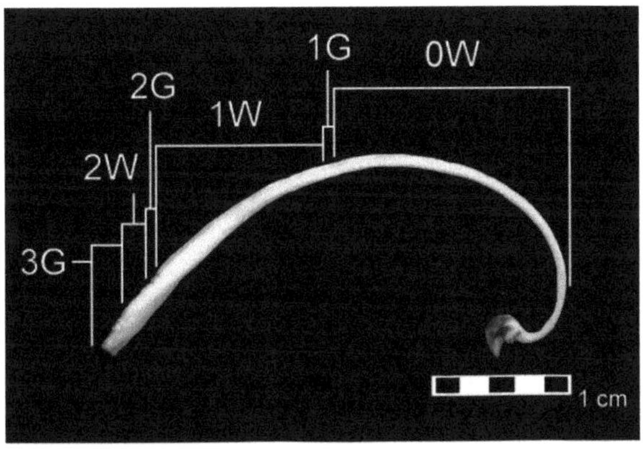

b. Individual no. 3-67

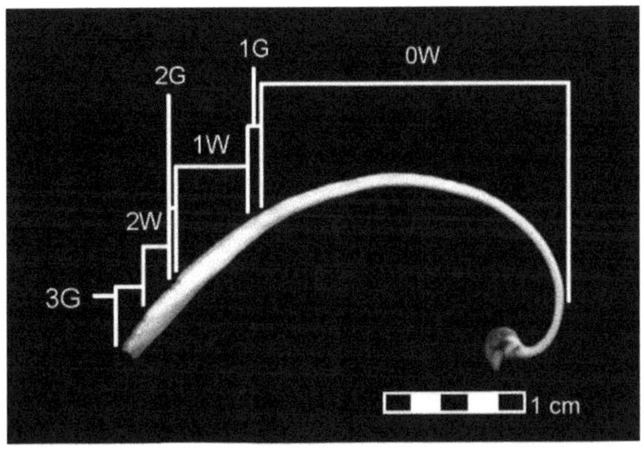

c. Individual no. 4-7

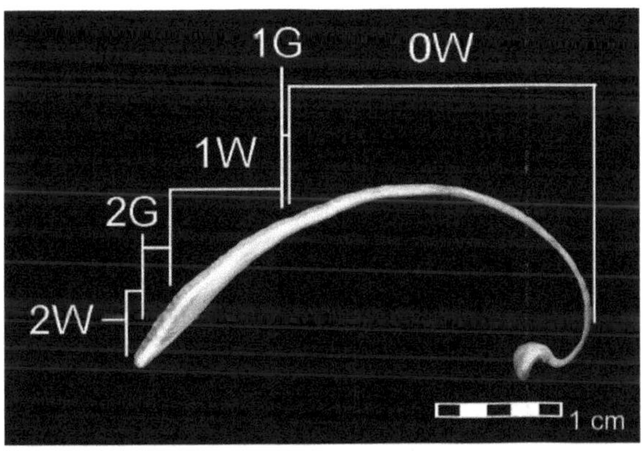

d. Individual no. 4-8

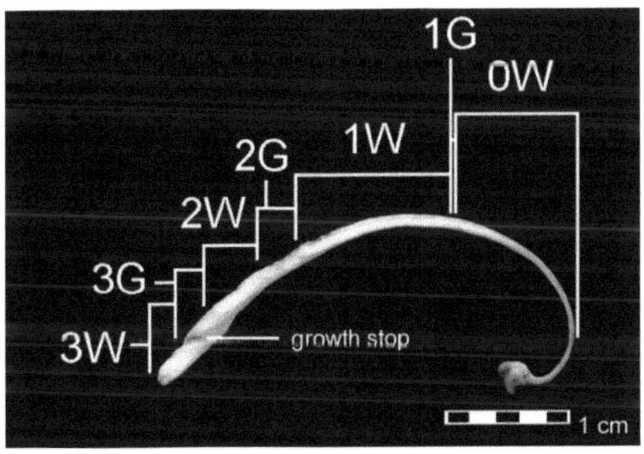

e. Individual no. 15-12

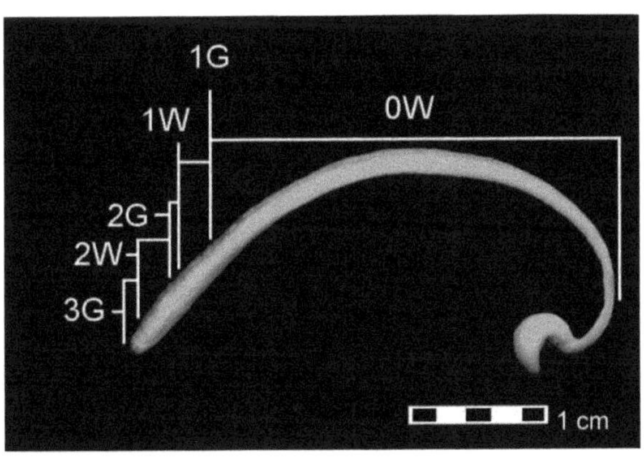

f. Individual no. 15-17

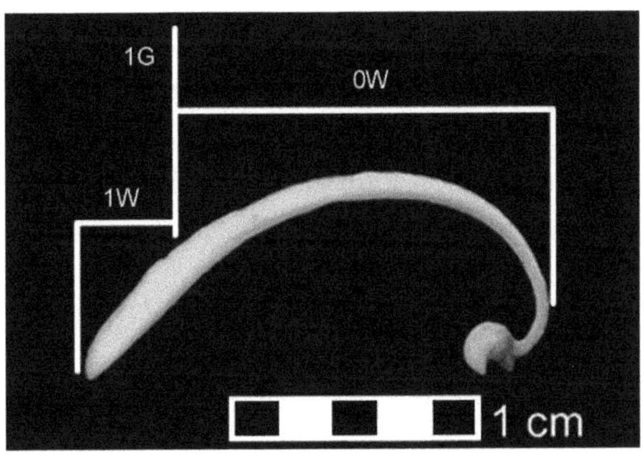

These quantified results are correlated with seasonal changes in SST and salinity values recorded at the collection stations (Figure 29). The majority of the specimens displayed a gray incremental zone at the shell margin, when SST and salinity values were elevated due to high levels of evaporation, increased air temperature, and low precipitation. The formation of gray incremental zones at the shell margin also coincides with elevated SST values reported from the Çanakkale region, as well as high air temperatures and low precipitation (cf. Sections 2.3 and 2.4). The secretion of white increments corresponds with low SST and salinity values. The secretion of white incremental zones also correlates well with high DO and pH values in the Karamenderes (cf. Section 2.4.1), indicating that summer dystrophy, resulting from low freshwater influx depleted in DO and pH, can be an additional factor causing the secretion of gray increments between May and November. Although it is commonly suggested that mid- to high-latitude bivalves grow in reduced rates in winter as a result of declining seawater temperatures and decreased food availability, and faster at warmer temperatures (Peharda et al. 2002), modern *C. glaucum* populations in the Troad do not constitute a single case where warmer temperatures hinder shell growth in bivalves. Bivalve populations, whose growth rates are negatively affected by elevated temperatures, were found both in the Mediterranean and elsewhere along the coasts of temperate regions (cf. Section 3.5 and 3.6; e.g., Ansell 1968, Arneri et al. 1998).

Figure 31. Growth pattern of modern *C. glaucum* specimens (2).

a. Individual no. 12-25

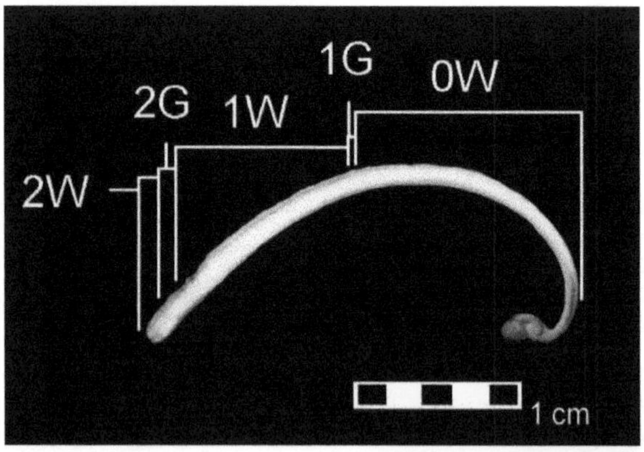

b. Individual no. 18-1

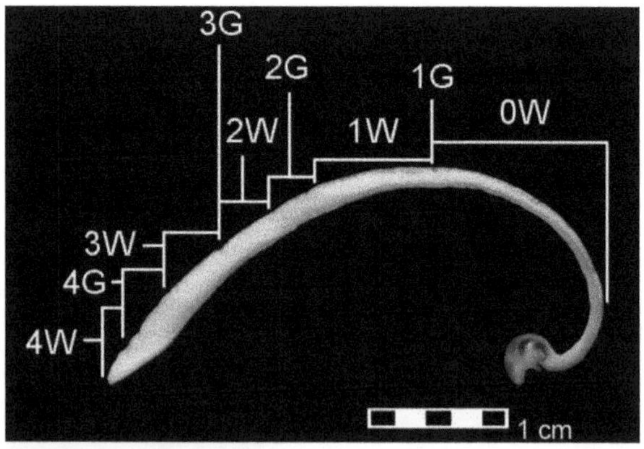

c. Individual no. 18-15

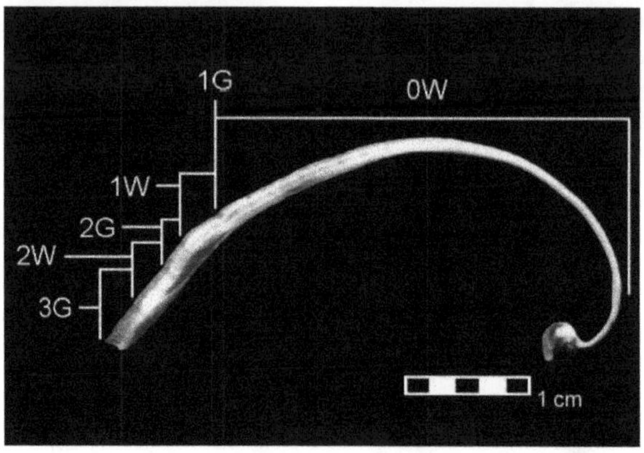

When the differential number of specimens collected at each season is evened out, it appears that through one seasonal cycle, ca. 39.7 % of the individuals were secreting gray increments at the time of collection (Figure 29). The fact that the age profile of the sampled portion of the extant populations is skewed towards older individuals (cf. Section 4.2.2.3) inflates the proportion of the individuals secreting gray increments at the time of collection.

4.2.3. Summary

The present coastal environments in the Troad constitute a rich biotope for small-sized mollusk species, such as *C. neritea* and *Abra* spp. Communities of archaeologically important species (such as *M. galloprovincialis*, *O. edulis*, et cetera), however, are not common features of the coastal biotopes in this region. Pollution, over-exploitation, but more importantly, the loss of former habitats are among the factors responsible for the limited species diversity represented by the mollusk populations in this environment.

Small but dense communities of *C. glaucum* were found in the shallow, sheltered areas of the Karamenderes Delta, in muddy to muddy and shelly substrates. The host water of *C. glaucum* can be defined by seasonal fluctuations in SST and salinity. Sampling stations differ from each other in terms of water conditions. The size and age range of the sampled specimens is large, yet skewed towards older and larger animals. While there are morphological differences among the communities, these differences could not be readily related to differential water conditions among sampling stations. In general, the valves of the extant *C. glaucum* populations in the Troad are neither very solid nor well-curved. The external growth lines of the *C. glaucum* valves are differential in appearance, whereas the internal incremental zones can be differentiated as gray and white bands with a high level of certainty. The mean breadth of white incremental zones decreases as the shell grows. Size variability in gray incremental zones does not demonstrate a clear trend. Examination of the incremental zones of individuals collected at seasonal (ca. 3-4 months) intervals indicate that alteration between gray and white zones occur once a year. In most individuals, the gray incremental zone forms during warm and drier months. Younger individuals secrete gray increments only for a short time during the year, whereas the duration of gray increment formation in older individuals can span from June to November. High DO and pH values in the Karamenderes, low SST, low salinity, and nutrient rich freshwater influx appear to be among the factors that affect *C. glaucum* growth positively.

4.3. Assemblages of Troia

4.3.1. Proportion of mollusk remains in faunal assemblages

More than 200,000 faunal specimens from the New Excavations in Troia have been studied. Out of that total, 161,920 specimens could be ascribed to meaningful chronological units (Table 18). Altogether they amount to 989,704.9 g, i.e., almost one metric ton. Included in this assemblage are 2265 excavation units distributed in diverse areas of the site and occupational phases, from the beginnings of the settlement in Troia, i.e., Troia I, until the Byzantine Period, i.e., Troia X. The assemblages from the Post-Bronze Age layers, although not the focus of the present study, will be discussed here briefly, in order to give a more complete picture of the structural changes observed in the archaeozoological and archaeomalacological assemblages.
The non-mollusk remains in Troia include domesticated mammals, wild mammals, birds, reptiles, amphibians, fish, and a minute amount of crustaceans (Table 18). About 21.1 % (NIS)/ 9 % (WIS) of the faunal remains –excluding unidentified mollusks– could not be attributed to any taxa due to their fragmented state. Mollusk remains account for 33.7 % of the NIS and 19.5 % of the WIS (Table 18 and Figure 32). The difference between NIS and WIS calculations is mainly due to the smaller size of mollusk specimens in comparison to most of the other common faunal specimens (e.g., large and medium mammals). Mollusks remains are present in all occupational phases and in 57 % of all studied Bronze Age units.

The amount of studied faunal remains is not proportional for each chronological unit. General quantitative differences between the chronologically categorized assemblages are as divergent as 3192 specimens (20.822.7 g) for Troia X and 32,699 specimens (166213.1 g) for Troia I (Tables 19 and 18). Large differences between the amounts studied from different chronological units in Troia may be deceptive. All faunal assemblages in Troia are large enough to infer about the diverse aspects of former animal economy and ecology at the site. While it is true that different levels of information can be extracted from archaeozoological assemblages with varying sizes (Davis 1987: 46), even those absolute sample sizes that are considered statistically large enough might not result in well-structured patterns in the archaeozoology of a site (Uerpmann 1973). Differential patterns caused by multiple variables, such as the divergent character of excavated areas and units, taphonomic conditions, and excavation techniques occur independently from sample size and their effect on archaeozoological results cannot be estimated through quantification methods (Uerpmann 1973). Differential sample size is only one of the factors that affect the archaeozoological structure of a faunal assemblage.

The contribution of mollusk remains to the faunal assemblages in Troia does not remain unchanged through time. Mollusks constitute 37.9 % of the total sum of all specimens in count and 24.7 % in weight of the faunal assemblage dating to Troia I (Table 19 and Figure 32). The amount of mollusk remains increases up to 62.1 % in NIS and 50.4 % in WIS in the Troia II assemblage, representing the largest contribution of mollusk remains to any chronologically categorized faunal assemblage in Troia (Table 19 and Figure 32). The predominating presence of mollusk remains in the Troia II deposits uncovered by the New Excavations is in good agreement with Gejvall's (1938) comparison of shell and bone remains from the Troia I layers excavated by Blegen. In great contrast with Troia I and II, the sum of mollusk remains drops abruptly in Troia III, making up only 2.1 % of the NIS and 2.2 % of the WIS of the total assemblage (Table 19 and Figure 32). Such extreme rarity of mollusk remains is unique to the faunal assemblages of this period. Despite the sudden decrease of mollusk remains in Troia III, their contribution to the total assemblage –by 46 % in NIS and 31.4 % in WIS– is higher in the Early Bronze Age than any successive occupational phase in the settlement (Table 19).

Mollusks are considerably rare in the faunal assemblages of Troia IV, by 9.8 % in NIS and 6.2 % WIS (Table 20 and Figure 32). The picture drawn from calculations of NIS for Troia V is one in which mollusk remains are becoming more frequent (28.4 %) than the previous period. WIS calculations however reflect a different sketch for the period, with 5.8 % (Table 20 and Figure 32). The deviation between NIS and WIS counts of the Troia V assemblage is caused by a change in the mollusk taxa represented and the taphonomic state of the specimens (cf. Sections 4.3.2 and 4.3.3). This situation is inevitably reflected in the share of mollusks in the Middle Bronze Age assemblage of Troia, as 24 % and 5.5 % in NIS and WIS respectively (Table 20).

The share of mollusk remains rises up to 47.7 % in NIS and 34.4 % in WIS in the faunal assemblage yielded from the earliest occupational phase of Troia VI (Table 21 and Figure 32). It becomes as low as 25.7 % in NIS and 19.5 % in WIS in middle Troia VI, and remains essentially at about the same level in the latest Troia VI assemblage with 31.6 % in NIS and 21.5 % in WIS (Table 21 and Figure 32). While mollusk remains are slightly less frequent in the Troia VIIa assemblage than they are in the late Troia VI assemblage, their occurrence is still in the range of Troia VI, with 27.3 % in NIS (Table 21, Figure 32). Their weight makes up 15.8 % of the Troia VIIa assemblage. On average, mollusks constitute 31.7 % of all Late Bronze Age assemblages in count, and 20 % in weight (Table 21).

Figure 32. Proportion of mollusk remains in the faunal assemblages of Troia.

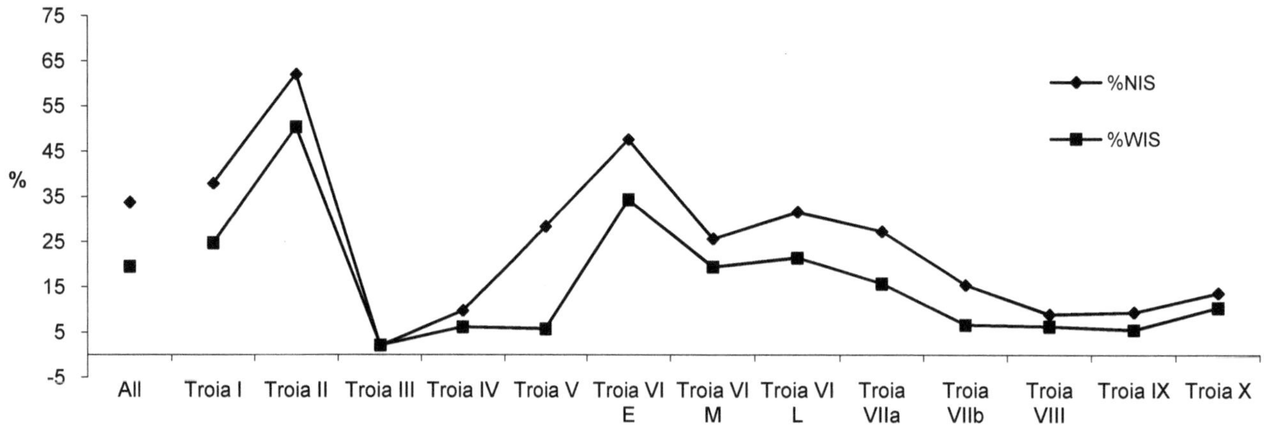

Occupational phases designated as Troia VIIb constitute a cultural entity different from the preceding Troia VI and VIIa phases (cf. Section 2.6.1). The contribution of the mollusk remains to the faunal assemblage is smaller during this period than the preceding Troia VI periods (Table 22 and Figure 32). Mollusks make up 15.5 % of the NIS of the faunal remains from excavation units designated to the Troia VIIb period. They constitute 6.7 % of the weight of the entire assemblage of the period. Several excavation units that lay above late Troia VI levels could not be designated to either of the two periods of Troia VII. The main quantitative characteristics of the fauna from those units are also given in Table 22.

The ratio of the mollusk remains in the faunal assemblages coming from the Post-Bronze Age levels in Troia is generally lower than it is in the Bronze Age assemblages. Mollusk remains comprise 12.1 % in NIS and 7.4 % in WIS of the Post-Bronze Age assemblage in average (Table 23). Only 9 % of the count of identified faunal specimens (6.4 % in WIS) dating to the Troia VIII period are mollusk remains (Table 23 and Figure 32). Their contribution is not greater in the subsequent Troia IX period: 9.5 % in NIS and 5.6 % in WIS. The percentage of mollusk remains increase slightly up to 13.7 % in NIS and 10.5 % in WIS in Troia X assemblages.

Altogether 54,548 mollusk specimens were found among the faunal assemblages studied in Troia, making up 33.7 % of the specimens' count (Table 18). Their total weight adds up to 192,735.3 g, about 19.5 % of the total weight of studied faunal specimens from the New Excavations in Troia.

4.3.2. Archaeomalacological assemblages

At least 15 families of aquatic mollusks are represented by the mollusk remains from the New Excavations in Troia (Table 24). The limitations of the species spectrum are partially due to methods used to identify archaeomalacological specimens at the site (cf. Section 3.2). Gastropod remains could be identified to one subfamily, two genera, and one species. Based on the author's observations on some limpet samples, *Patella caerulea* is thought to be the most common limpet at the site. Among the *Cerithium* spp., *Cerithium vulgatum* is the most likely candidate to be the most frequently represented species. While it is possible that a small number of *B. brandaris* and *P. haemastoma* may have been registered along with specimens of *H. trunculus*, at least in the assemblages examined by the author, *B. brandaris* and *P. haemastoma* remains were absent. Bivalves are represented with at least ten marine genera and one freshwater genus.

More than half of the taxa represented in Troia are only very rarely encountered. Monodontinae and *Cerithium* spp. among the gastropods, and *A. noae*, Glycymerididae, *Pinna* spp., Pectenidae, *S. gaederopus*, and *Unio* spp. each constitute less than 0.5 % of the total number and weight of identified mollusk specimens (Table 24, Figure 33). *C. glaucum* remains are conspicuously the most common mollusk remains in Bronze Age Troia. They make up 74.2 % of the NIS and 71.1 % of the WIS of the total assemblage (Table 24). Remains of *Patella* spp., *H. trunculus*, *M. galloprovincialis*, *O. edulis*, *S. marginatus*, and *T. decussatus* together constitute most of the remaining ca. 25 %. The relative abundance of represented mollusk taxa varies greatly among chronologically distinct assemblages.

Figure 33. Changes in the relative abundance of species represented in the mollusk assemblages of Troia.

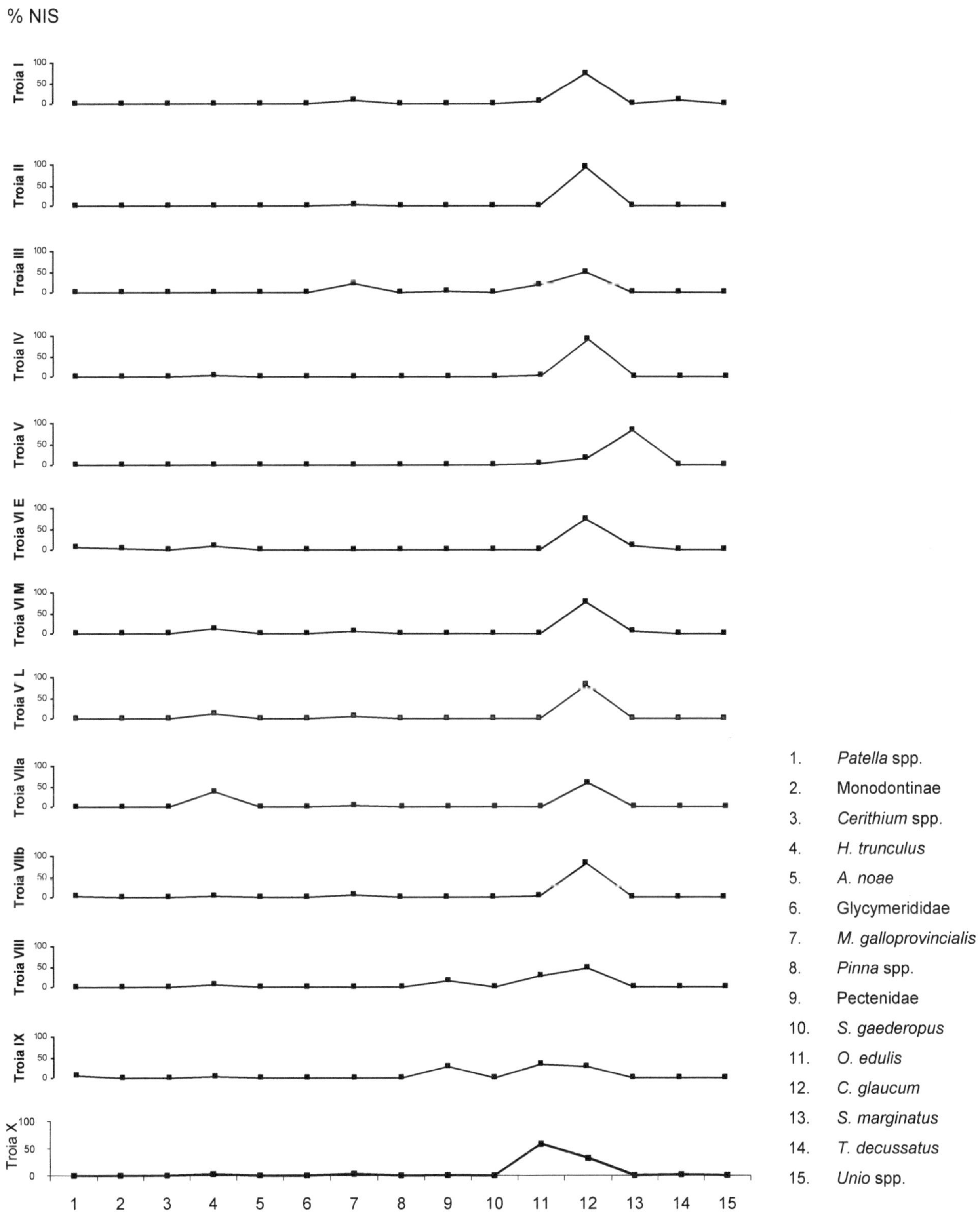

C. glaucum's prominent presence is already well-established in the archaeomalacological assemblages from Troia I (Table 25 and Figure 33). Remains of *M. galloprovincialis*, *O. edulis*, and *T. decussatus* are also not uncommon, especially when compared with the subsequent Troia II phase. They represent 8.9 %, 6.5 %, and 8.9 % of the NIS of the Troia I mollusk assemblage respectively. The amount of *T. decussatus* remains, however, is questionable (cf. Section 4.3.4 below). The relative abundance and absolute amount of *A. noae* remains from Troia I and II layers recovered during the New Excavations contrast with Gejvall's (1938) account of abounding *A. noae* remains from the Troia I layers excavated by Blegen (cf. Section 2.8). The ratio of *C. glaucum* remains increases up to 95.4 % in NIS and 93.5 % in WIS in the Troia II assemblages, while the contribution of other formerly abundant species like *M. galloprovincialis* and *O. edulis* is considerably diminished (Table 25, Figure 33). *Pinna* spp. is absent from all assemblages dating to the Early Bronze Age. Mollusk remains dating to the Troia III period are represented only by 68 specimens, altogether weighing only about 459 g (Table 25). This amount is not sufficient to be evaluated accurately in terms of species abundance. The dominance of *C. glaucum* prevails throughout all the Early Bronze Age assemblages, including those excavation units that could not be assigned to either one of the stages of the Early Bronze Age.

The relatively small size of the assemblages from Troia IV and V is reflected in the number of taxa absent from these assemblages (Table 26, Figure 33). Monodontinae, *A. noae*, Glycymerididae, *Pinna* spp., *S. gaederopus* are absent in both the Troia IV and V assemblages. Otherwise, the assemblage composition of Troia IV is similar to the preceding Early Bronze Age assemblages, in particular to that of Troia II. In both Troia II and Troia IV, the percentage of *C. glaucum* is above 90 %. The species composition of Troia V mollusk assemblage appears to be completely different from preceding and succeeding periods. *S. marginatus* remains make 81.5 % in NIS and 75.9 % in WIS of the mollusk assemblage (Table 26, Figure 33). *C. glaucum* remains comprise only 14.2 % of the number of the mollusk specimens. What this divergence could possibly mean, will be discussed below.

C. glaucum resumes its significant contribution to the mollusk assemblage in the earliest phase of Troia VI, with 72.7 % in NIS and 66.5 % in WIS (Table 27). *H. trunculus*, which has already started to appear in slightly higher amounts in the assemblages of preceding Troia IV and V periods (Table 27 and Figure 33), makes its first noteworthy appearance in the Early Troia VI assemblage. *H. trunculus* remains make 8.2 % of the NIS and 7.4 % of the WIS of this period's assemblage. The Troia VI assemblage includes remains of *Patella* spp. and Monodontinae in larger amounts also for the first time in Troia's occupational history. Contributions of *M. galloprovincialis* and *O. edulis* are minor, whereas *S. marginatus*, which had made its first noteworthy appearance in the archaeomalacological assemblages dating to Troia V, continues to be an important part of the assemblage, constituting 16.6 % of the NIS and 8.6 % of the WIS. The proportion of *Unio* spp. remains also becomes slightly larger than it was in the previous occupational periods (Table 27, Figure 33).

C. glaucum's significance is sustained in the subsequent phases of Troia VI. The share of *H. trunculus* remains continues to be high in Middle and Late Troia VI (Table 27, Figure 33). *Patella* spp. and Monodontinae virtually disappear, whereas the ratio of *M. galloprovincialis* becomes more visible in Middle and Late Troia VI assemblages, with ca. 6 % of the NIS and 3 % of the WIS (Table 27, Figure 33). *S. marginatus* appears in smaller amounts in the Middle Troia VI assemblage and loses its ca. 5 % share in Late Troia VI (Table 27, Figure 33).

A large amount of the excavation units belonging to the Troia VI period are yet to be dated to the different phases of Troia VI (i.e., Early, Middle or Late Troia VI). The composition of the mollusk assemblages yielded from these units is similar to those that are assigned to distinct phases of the Troia VI occupation: *C. glaucum* makes up 63.3 % of the NIS and 68.9 % of the WIS (Table 27). *H. trunculus* and *S. marginatus* constitute a larger portion of the mollusks; the former with 18.5 % of the NIS and 20.3 % of the WIS and the latter with 12.2 % of the NIS and 1.6 % of the WIS. *M. galloprovincialis* remains also appear, although in smaller amounts, making up 3.3 % of the NIS and WIS.

The mollusk assemblage dated to the Troia VIIa period, the last phase of the Late Bronze Age occupation at Troia, proportionally includes lesser amounts of *C. glaucum*. The species comprises 58.6 % of the NIS and 38.9 % of the WIS of this assemblage (Table 27, Figure 33). The share of *H. trunculus* remains, on the other hand, constitutes a much greater part of the mollusk assemblage, with 35.8 % in NIS and 53.3 % of the WIS (Table 27, Figure 33).

In short, the dominating proportion of *C. glaucum* and significant proportions of *H. trunculus* marks the Late Bronze Age mollusk assemblages of Troia. *S. marginatus* is the third most abundant species of the period.

The composition of the mollusk assemblage dating to Troia VIIb does not represent a large contrast with Troia VI and VIIa. *C. glaucum* remains the most frequent species, making up 80.3 % of the NIS and 72.8 % of the WIS of this relatively small assemblage (Table 28, Figure 33). The share of *H. trunculus,* however, decreases significantly down to 4.5 % of the NIS and 5.6 % of the WIS, while *M. galloprovincialis* reappears as an important part of the mollusk assemblage with 7.1 % of the NIS and 9.4 % of the WIS. *S. marginatus* disappear completely from the record. It must be noted, however, that this assemblage consists only of 966 specimens and the absence of some species is likely to be a consequence of limited sample size.

Troia VIIb is followed by the so-called Post-Bronze Age phases of Troia VIII, IX, and X, after a ca. 300 year hiatus in the stratigraphy of the settlement (Korfmann 2003). The relative abundance of the mollusk taxa in Post-Bronze Age assemblages are unlike those of the previous periods. The high proportion of *C. glaucum* is gradually replaced by *O.*

edulis. Already in Troia VIII, *C. glaucum* remains comprise only 47 % of the NIS and 20.2 % of the WIS of the mollusk assemblage, while *O. edulis* remains make up 28.3 % of the NIS and 67.2 % of the WIS (Table 29, Figure 33). In the mollusk assemblages of Troia IX and X, *O. edulis* remains become even more abundant, taking over *C. glaucum*'s place as the most common species. *O. edulis* remains constitute 32.1 % of the NIS and 55.5 % of the WIS of the Troia IX mollusk assemblage, and their percentage increase up to 58 % in NIS and 68 % in WIS of the Troia X assemblage (Table 29, Figure 33). Pectenidae also becomes an important component of the mollusk assemblages in the Post-Bronze Age period. They comprise 14.8 % in NIS and 9.2 % in WIS of the Troia VIII mollusk assemblage. Their contribution to the Troia IX assemblage is even higher; with 27.4 % in NIS and 19.4 % in WIS. Troia X assemblage, however, includes only 2 specimens of Pectenidae. (Table 29, Figure 33). Other noteworthy components (above 1 %) of the Post-Bronze Age assemblage are the remains of *H. trunculus* and *M. galloprovincialis*. Mollusk taxa, such as Monodontinae, *Cerithium* spp., *A. noae*, Glycymerididae, *Pinna* spp., *S. gaederopus*, which are minor elements in the assemblages of preceding periods, are completely absent from Post-Bronze Age assemblages. Specimens belonging to *Unio* spp. appear only very infrequently. (Table 29, Figure 33).

Figure 34. Taphonomy of *Patella* spp., *C. vulgatum*, *H. trunculus*, and *Pinna* spp. in Troia.

a. *Patella* spp. shells from Unit 3254

b. Crushed *H. trunculus* shells from Unit 1917

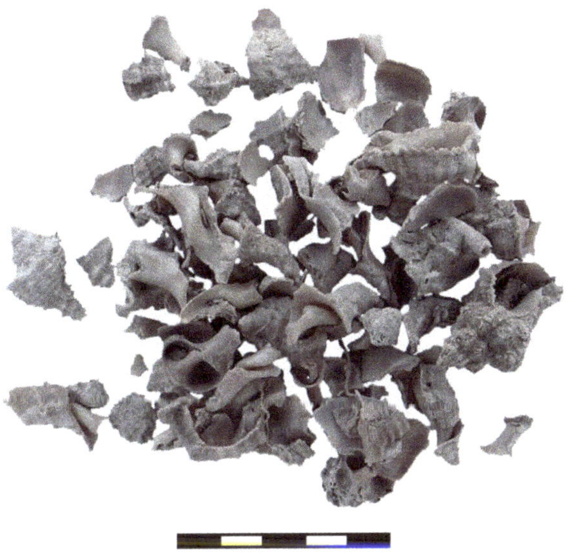

c. Naturally abraded and possibly modified *C. vulgatum* shell from Quadrate L5, Beh. 49. Unstratified.

d. Outer (left) and inner (right) surfaces of a *Pinna* spp. shell fragment from Unit 1918

4.3.3. Taphonomic description of mollusk specimens

About half of the *Patella* specimens bear –irregular to regular– round holes on and around their apices. Some specimens are broken at the margin or the margin is completely missing. The appearance of the *Patella* specimens resembles that of a large ring or a small bracelet. Originally larger individuals also appear smaller because of missing margins (Figure 34a). This type of fragmentation, occurring at structurally weak points where growth interruptions are located or where shell represents earlier stages of growth, is typical for limpet shells (Zuschin et al. 2003). It is caused by physical and chemical agents, and can occur before or after the burial of the shell. Wave action in the sea, alone, has such an effect on empty limpet shells. After deposition, the same effect is related to sediment compaction and to the properties of the soil (e.g., alkalinity). The irregularity of the holes and breaks on the *Patella* shells in Troia demonstrates that fragmentation mainly took place after the deposition of the specimens. Fresh breaks in the same form shows that the dissolution of the shell matrix at weak points is continued during the transportation and handling of the specimens after their excavation. Some specimens are encrusted, a process that can occur already during life in epifaunal shelled mollusks.

Nothing is known about the preservation state of the Monodontinae specimens included in the mollusk assemblage discussed above. However, it should be noted that two Monodontinae rings, similar to those found in other eastern Mediterranean sites (Reese 1984) have been recovered in the archaeological deposits dating to the Late Bronze Age layers of the site (per. com. P. Zidarov).
At least one *Cerithium* specimen (from the Late Troia VI assemblage) is a naturally abraded example (Figure 34c). Its sculpture is smoothed away; the outer lip is almost completely missing. Coloration is more or less intact.

Almost all *H. trunculus* specimens are exceedingly fragmented (Figure 34b). They are crushed. More intact specimens belong to younger (smaller) individuals, but they too are missing at least the aperture.

A. noae shells are mostly intact except some with broken margins. Some are encrusted. Encrustation may have occurred during the specimens' lifetime.

All *M. galloprovincialis* specimens are fragmented.

Not more than ca. 30 % of the shell, usually the sections closer to the umbo and the umbo itself, remain intact. The abrasion of mussel valves begins during ontogenic life through jostling by wave action (Light 2005). When the organic-rich nacreous inner layer of the *Mytilus* shell decays, shell strength decreases (Zuschin et al. 2003). As a result, the foliated shell is left vulnerable to the effects of compression in the sediment. Other archaeomalacological studies that report on *Mytilus* specimens elsewhere are affirmative about the natural processes that cause shell fragmentation in archaeological Mytilidae (e.g., Light 2005).

Only a single specimen of *Pinna* spp. could be observed by the author. In this specimen, only the portion of the shell close to the anterior margin remains (Figure 34d). The prismatic structure of *Pinna* shells exfoliate very easily in the sediment, leaving only the strongest shell parts. From the NIS/WIS relationship of the *Pinna* remains found in the assemblages in Troia it can be assessed that all *Pinna* shells in Troia were recovered in a similar state of fragmentation, posing an obstacle against identifying archaeological specimens of *Pinna* shells beyond the genus level.

Pectenidae valves are mainly intact, but at least a few are encrusted by the calcareous parts of other marine organisms on the inner shell side.

Shell layers of *O. edulis* are exfoliated. Folia flake and splinter off as the protein matrix between the calcium carbonate layers decay through time with the help of physical and chemical agents. All *O. edulis* remains are fragmented to varying degrees.

C. glaucum specimens were found mostly complete. The homogeneous and solid structure of *C. glaucum* shells is fairly resistant to breakage. A portion of the ventral-posterior margin of the valve, the thinnest part of the shell, is missing in several specimens (Figure 35a). This section is also the part where growth increments crowd after one another, as the shell growth in *C. glaucum* occurs faster along the valve height but slower lengthwise (Brock 1991, Mariani et al. 2002). Some valves of *C. glaucum* bear holes on the upper section of the valve near the umbo (Figure 35b). It can be easily seen that this section is also one of the thinnest parts of the shell when the valve is observed from the cross-section along the growth axis. In some valves, the outer shell layer is peeled off at the same part of the valve and the innermost layers remain (Figure 36a). These holes and peeled off parts occur as a consequence of gradual post-depositional shell degradation. Some *C. glaucum* valves have a gray and gray-brown color (Figure 35b). Contrary to von Martens' opinion (1879, cf. Section 2.8), this coloration does not stem from long contact with sedimentary mud; instead, the red-brown and gray color of the inner shell layers indicates that these specimens were completely burnt. Some Early Bronze Age contexts yielded shells that are partially encrusted with lime concretion. All valves have lost their original olive green and yellow color pattern and have assumed a pink and white pattern.

Not a single *S. marginatus* valve remains intact. All specimens are fragments, consisting of less than an estimated ca. 10 % of the valve. The thin and elongated valves of this species are easily subject to breakage as soon as the periostracum decays. The distance between the WIS and NIS ratios of this species is due to the high level of fragmentation of these specimens, with the exception of examples from a single context dated to Early Troia VI, because only the umbones of this sample were counted (Table 27, Figure 33).

Figure 35. *C. glaucum* valves from Troia : Burnt, partiallly broken, and (naturally ?) holed specimens.

a. Unit 1201

b. Unit 1103

The rather robust shells of *T. decussatus* are mainly intact. Some are broken at weak points of slow shell growth.

The oval shells of *Unio* spp. are largely complete with some specimens missing the ventral-posterior margin and some bearing a hole on the valve body occasionally close to the umbo (Figure 36b).

Figure 36. Examples of bivalve shell corrosion in Troia.

a. Early stage of shell fragmentation near the umbo of *C. glaucum* (Unit 1250).

b. Outer (top) and inner (bottom) shell surfaces of a *Unio* spp. valve with a (natural?) hole near the umbo (Unit 843).

4.3.4. Archaeological context of Bronze Age mollusk remains

The faunal remains that will be discussed in the sections below belong to diverse archaeological units recovered during the New Excavations in Troia. They represent 2056 excavation units ("Behälter" in Troia excavation nomenclature), coming from 61 different excavation areas (so-called Quadrates) of the site.

The majority of the studied faunal units come from fill deposits. There are two types of fill deposits: those associated with architectural remains; and, those from arbitrarily defined fills occasionally covering entire quadrates or representing the removal of steps that are artificially built to ease access to the excavation areas. Other faunal units are related to the removal of walls. The rest of the faunal units have been recovered from pits, floors, and hearths. A few units come from graves.

The studied faunal remains that belong to the Early Bronze Age have been recovered from Quadrates C3, D 2-3-4-5, and E 4-5 in and immediately around the Schliemann's Trench, from Quadrate D7/8 immediately outside of the citadel walls, from the Megaron in G6, and from the vicinity of the remains of the Troia II timber fortification in the lower city in KL16/17 (Figures 7, 8 and see Figure 37 for a typical example of an excavation unit with high shell content). Faunal remains that are dated to Troia IV and V have been studied from Quadrates x7, A5/6, A7, D8, E4/5, H6, FG5/6, and H17. The bulk of the studied faunal remains that were dated to the Late Bronze Age were yielded from Quadrates I8 and K8 immediately outside the East Tower of the Troia VI fortification walls, and KL16/17 in the southern section of the lower city. Smaller samples come from Quadrates x7, z7/8, A5/6, A7/8, D9, E9, E8, FG5/6, G9/10, H6, K7, K13, G28, and y29. The fauna of the Troia VIIb period is represented by excavation units recovered in Quadrates D9, E9, K4, KL16/17, and z7/8. Faunal samples from z6/7, A8/9, D9, H17, H6, K4, KL16/17, t14, and y29 dating to Troia VIII, C29, M18, w28 dating to Troia IX, and p28, tu14/15, x7, y7/8, dating to Troia X are included in the discussion of Post-Bronze Age mollusk remains.

Besides proportional changes in the abundance of mollusk species across occupational periods (cf. Section 4.3.3), an attempt to detect changes in the ubiquity of the mollusk remains across different chronological phases could have been made, but unfortunately neither the archaeozoological nor the archaeological databases of the Troia Project is suitable for such an analysis. Besides, some information provided in the excavation journals suggest that a quantified analysis of the proportional distribution of mollusk remains across the site can lead to erroneous interpretations. Such information also gives clues about the source of inconsistencies observed in other quantified archaeomalacological results. Excavators made note of the mollusk content of some archaeological units in their journals. According to the archaeozoological database, however, the majority of these units contains very few or no mollusk remains at all (Table 9). These accounts show that some mollusk remains, of unknown quality and quantity, are not represented by the quantified archaeomalacological analyses. The bulk of these contexts belong to the Middle Bronze Age phase of the site.

In general, the mollusk content of each faunal unit resembles the picture drawn from the cumulative quantified results of the occupational period to which they belong (cf. Section 4.3.3). This degree of homogeneity confirms the conclusion already reached by the analysis of animal bones that most excavation deposits in Troia do not represent consumption units (Uerpmann 2003). A few deposits yielded from the Bronze Age layers of the site disrupt the general homogeneity of the Bronze Age assemblages (Table 8). The species these units contain constitute a divergence from the quantified pattern of the mollusk spectrum.

Figure 37. Example of a typical excavation Unit with high mollusk content from Troia (Quadrate C3, Courtesy of the Troia Projekt).

Four fill deposits in Quadrate D5 (Beh. 363, 370, 374, and 380) yielded relatively larger amounts of *M. galloprovincialis* remains than reflected by the species abundance profile derived from the cumulative quantitative account of the archaeomalacological assemblages of Troia I. These four deposits together represent a wide "shell layer" (Figure 38). Another fill deposit from the same excavation (Beh. 377) area contained a very large amount of *T. decussatus* remains, altogether 999 specimens. This situation, however, might reflect a coding mistake that occurred during the registration of archaeozoological material on the site. Contexts with unusual mollusk contents dating to the early phase of Troia VI were found in Quadrates K8 and KL16/17. In K8, one deposit (Beh. 635) yielded a large quantity of crushed *H. trunculus* remains. Another deposit containing large amounts of crushed *H. trunculus* shells was found in a fill directly outside of Tower VIh of the citadel wall (Quadrate I8, Beh. 50). This fill is dated to the Troia VI Middle period. Three other deposits (Beh. 299, 448, and 752) containing large amounts of crushed *H. trunculus* shells were found in Quadrates I8 and IK8. These deposits are all related to Tower VIh and are dated to the Troia VI Late period. Another accumulation of *H. trunculus* remains belonging to this period was recovered from an oven (Beh. 632) in KL16/17, dating to Troia VI Early. A large accumulation of primarily *Patella* spp. shells in KL16/17 (Beh. 690) was denoted as a "shell midden" by the excavator of the Quadrate. A leveling deposit in the same excavation area (KL16/17; Beh. 699) consisted primarily of fragmented *S. marginatus* remains. The postholes of what is thought to be a sunken house dating to the Troia VI Middle period in Quadrate A8 (Beh. 1302) contained primarily *S. marginatus* shells. Another *H. trunculus* deposit immediately outside of the citadel was located outside the Northeast Bastion (Beh. 1040). A few archaeological units noteworthy due to their *H. trunculus* contents belong to Troia VIIa layers of Quadrate KL16/17 in the lower town. Two of these (Beh. 225 and 236) represent pits and one of them (Beh. 174) is a fill deposit (For further stratigraphical details on the contexts of *H. trunculus* accumulations in Troia, see Çakırlar and Becks in print).

Figure 38. "Shell layer" at Quadrate D5 in Troia (Courtesy of the Troia Projekt).

Table 8. Some special mollusk accumulations in archaeological deposits of Bronze Age Troia.

Period	Square	Behälter	Context	Species Composition
I	D5	363	Fill	57 % *M. galloprovincialis*, 35 % *C. glaucum*, 6 % *O. edulis*, and *T. decussatus* and *S. marginatus*.
I	D5	370	Arbitrary fill	59 % *M. galloprovincialis*, 35 % *C. glaucum*, 6 % *O. edulis*, and *S. marginatus*, *T. decussatus*, and Monodontinae.
I	D5	374	Arbitrary fill	47 % *C. glaucum*, 42 % *M. galloprovincialis*, 9 % *O. edulis*, and *T. decussatus* and Monodontinae.
I	D5	377	Fill	76 % *T. decussatus* (999 specimens), 20 % *C. glaucum*, 4 % *M. galloprovincialis*.
I	D5	380	Fill	46 % *M. galloprovincialis*, 31 % *C. glaucum*, and *O. edulis* and *T. decussatus*.
V	A5/6	267	Ashy layer above indoors floor	96 % *S. marginatus*.
VI Early	K8	635	Fill area with fallen stones	70 % *H. trunculus* and 30 % *C. glaucum*.
VI Early	KL16/17	632	Oven	Ca. 95 % *H. trunculus* and *Patella* spp. and *C. glaucum*.
VI Early	KL16/17	69	"Shell midden"	Ca. 65 % *Patella* spp., 30 % *H. trunculus* and a few *C. glaucum*.
VI Early	KL16/17	699	Leveling deposit	Ca. 90 % *S. marginatus* and a few *C. glaucum* and *Patella* spp.
VI Middle	A8	1302	Postholes of a sunken(?) house	Ca. 72 % *S. marginatus*, 12 % *C. glaucum*, 15 % *M. galloprovincialis* and a few *H. trunculus*.
VI Middle	I8	50	Loose soil directly outside of the tower VIh	60 % *H. trunculus*, 35 % *C. glaucum* and a few *M. galloprovincialis*.
VI Late	K4	1040	Deposit immediately outside of Northeast Bastion	Ca. 82 % *H. trunculus*, 13 % *C. glaucum* and a few *O. edulis*, *M. galloprovincialis*, *S. marginatus*.
VI Late	I8	299	Deposit contemporary with and outside of Tower VIh	70 % *H. trunculus*, a few *C. glaucum* and *M. galloprovincialis*.
VI Late	I8	448	Foundation trench of Tower VIh	45 % *H. trunculus*, 45 % *C. glaucum* and a few *M. galloprovincialis* and *S. marginatus*.
VI Late	IK8	752	Topsoil related to Tower VIh	80 % *H. trunculus* and 20 % *C. glaucum*.
VIIa	KL16/17	174	Fill	90 % *H. trunculus* and a few *C. glaucum* and *S. marginatus*.
VIIa	KL16/17	225	Pit	95 % *H. trunculus*, a few *C. glaucum* and *S. marginatus*.
VIIa	KL16/17	236	Pit	Ca. 85 % *H. trunculus*, 10 % *C. glaucum*, a few *O. edulis*, *M. galloprovincialis* and *S. gaederopus*.

Table 9. Some examples of the differences between quantified archaeomalacological results and excavators' accounts of mollusk remains in Troia.

Period	Excavation year	Quadrate	"Behälter"	Context type	Excavation journal	Mollusk content
III	1997	G6	451	Fill	"contains a lot of shells"	1 specimen of *M. galloprovincialis*
IV	1994	D8n	1276	Fill	"consists of burnt bricks and shells"	None
IV	1994	D8n	1752	Fill	"with lime and shells"	None
V	2000	A5/6	485	Fill	"with shells"	1 specimen of *Unio* spp.
V	2001	A5/6	661	Pit	"filled with shells"	None
V	2001	A5/6	670	Pit	"containing shells"	None
V	2001	A7	1496	Fill	"including a lot of charcoal and shells"	None
VI Middle	2001	A5/6	917	Pit	"typical pit fill, loose brown earth with shells"	None
VI Middle	1996	A8	978	Shell layer	"with a lot of shells"	5 specimens of *C. glaucum*

4.3.5. Properties of *C. glaucum* populations in Bronze Age Troia.

4.3.5.1. Morphology of Bronze Age *C. glaucum* populations

A large number (4238) of *C. glaucum* specimens from Troia provided at least one morphometric measurement (Table 30). Morphological attributes show variability between chronologically meaningful units. The standard mean error of the measurements obtained from samples originating from archaeological contexts dating to the Troia III, Troia V and Troia VIIb periods is too large to allow sound interpretations of the morphological properties compared across chronologically different units. They can only be approached with caution. These samples also represent limited excavation areas: the Troia III sample represents a single fill deposit around the megaron wall in Quadrate G6 (Beh. 307) and consists only of nine individuals. The small Troia V sample represents three excavation units: One is a soil deposit from a wall cleaning in Quadrate H6 (Beh. 250) and the other two are found in close proximity underneath the foundations of the Troia VI wall in Quadrate I8. The Troia VIIb sample represents residue from eight different excavation units scattered in Quadrates D9, K4 and KL16/17. The units include pits and post-hole deposits.

The largest specimen of all measured individuals is 45.4 mm in height, while the smallest specimen is 9.9 mm high (Table 30). The valve heights of the bulk of the remains deviate between 21.3 and 30.1 mm (Figure 39). Within the samples that are securely dated to the Early Bronze Age, two are large enough samples to be discriminated: the mean valve height of the *C. glaucum* shells recovered from Troia I deposits is significantly larger (p = <0.0000) than that obtained from the *C. glaucum* shells originating from deposits dating to Troia II (Figure 40a). The mean valve heights of the *C. glaucum* shells recovered from the Early, Middle, and Late Troia VI phases are significantly different from each other (p = <0.0000). While individuals dated to the Early Troia VI have valves smaller than those dated to the subsequent Middle Troia VI and Late Troia VI periods, no other meaningful chronological scheme (e.g., reduction through time or change in accordance with the abundance of *C. glaucum* shells in the assemblages) could be observed between the mean sizes of the samples. Within the Late Bronze Age sample, individuals with the smallest valves are dated to the Troia VIIa period. The mean valve height of the Troia VIIa sample is ca. 25.4 mm. The mean valve height of the Troia VIIb sample is significantly larger than the Troia VIIa sample with a mean valve height of ca. 27.3 mm (Table 29 and Figure 40a).

The difference in valve height between the Early Bronze Age and Late Bronze Age samples is significant (p = <0.0000; Figure 40b). The mean valve height of the Early Bronze Age sample deviates between 20.2 and 28.8 mm, while the mean VH of the Late Bronze Age sample deviates between 22.0 and 30.6 mm (Table 30 and Figure 39). Although increased breadth of gray incremental zones of the Late Bronze Age valves (See Section 4.3.5.2) are partially responsible for the valve height difference between the Early Bronze Age and Late Bronze Age populations, it will be demonstrated in Section 4.3.5.3 that this size difference is actually a co-efficient of a change in the demographic structure of the two populations.

Figure 39. Valve height of *C. glaucum* populations from Troia (Distributions).

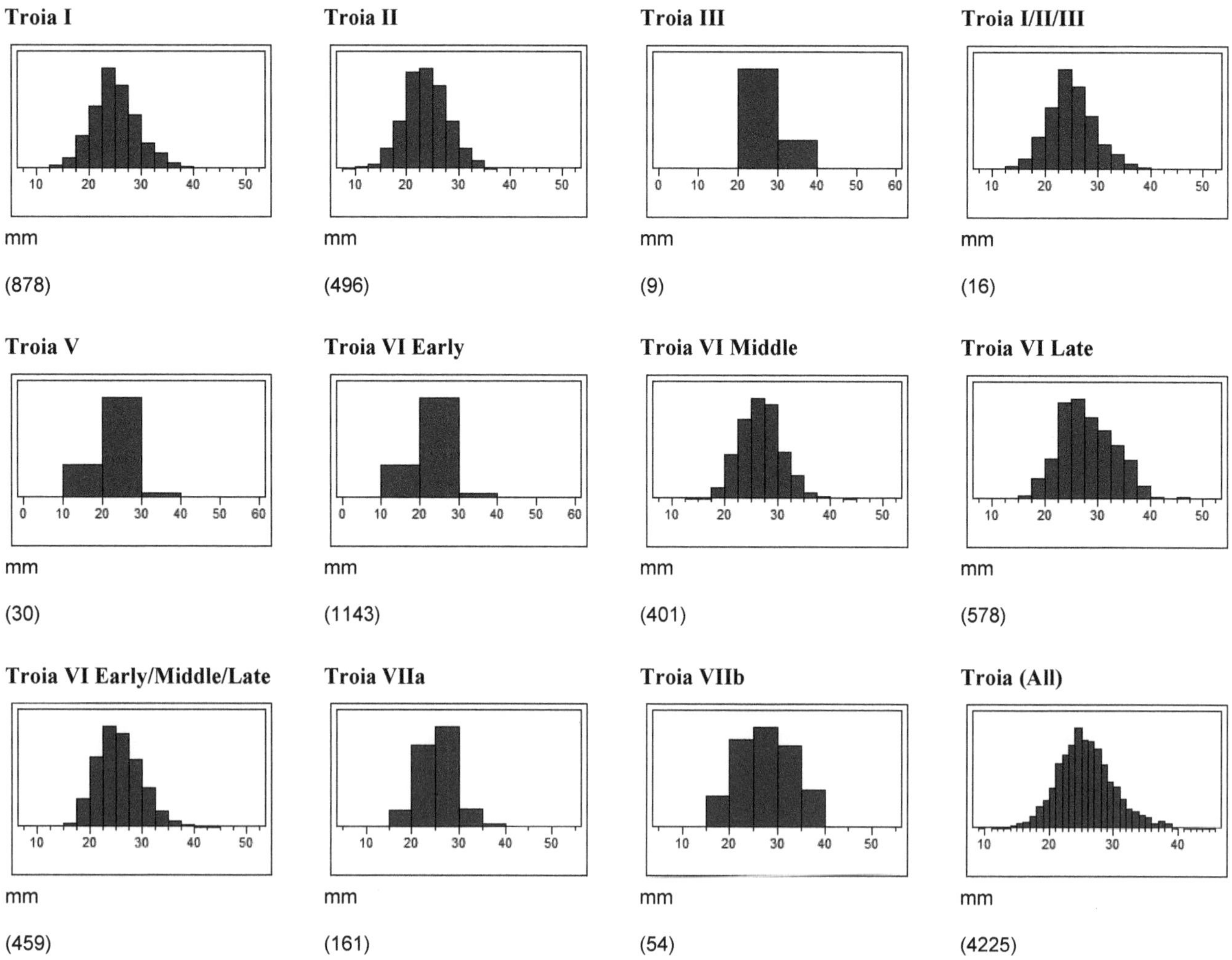

(878) = No. of individuals.

Solidity indices were compiled of measurements from 3524 individuals. The solidity indices of chronologically discrete *C. glaucum* valves vary considerably in equilibrium with the chronological sequence of the periods (e.g., Troia I, II, VI Late et cetera), but they also show variability in positive relationship with the valve heights (Table 30 and Figure 41). The solidity values of Troia I and II are also significantly different from each other (Figure 42a; p = <0.0000). The Early, Middle, and Late Troia VI samples and the Troia VIIa sample are significantly different from each other in terms of solidity (Figure 42a; p = <0.0000), but as in the case of valve height values there are no meaningful chronological patterns of increasing or decreasing solidity through time.

When data from different sub-periods are merged and the Early Bronze Age assemblage is compared with the Late Bronze Age assemblage, the solidity is significantly different (Figure 42b; p = <0.0001): the Late Bronze Age valves are more solid than the Early Bronze Age valves.

Curve indices were produced from measurements taken on 3687 individuals (Table 30 and Figure 43). Indices of curve vary roughly parallel to indices of solidity. The curve of Troia I and II samples is significantly different (Figure 44a; p = 0.0087). The Early, Middle, and Late Troia VI samples are significantly different from each other in terms of shell curve (Figure 44a; p = <0.0000), but as in the case of valve height values there are no meaningful chronological patterns of increasing or decreasing curve through time.

Again, when data from different sub-periods are pooled together into the two larger samples, one dating to the Early Bronze Age and the other dating to the Late Bronze Age, the mean of curve is significantly different (Figure 44b; p = 0.0054). The Late Bronze Age valves are more curved than the Early Bronze Age valves.

Figure 40. Comparison of valve heights in *C. glaucum* populations in Troia according to periods (a) and phases (b) (ANOVA).

a.

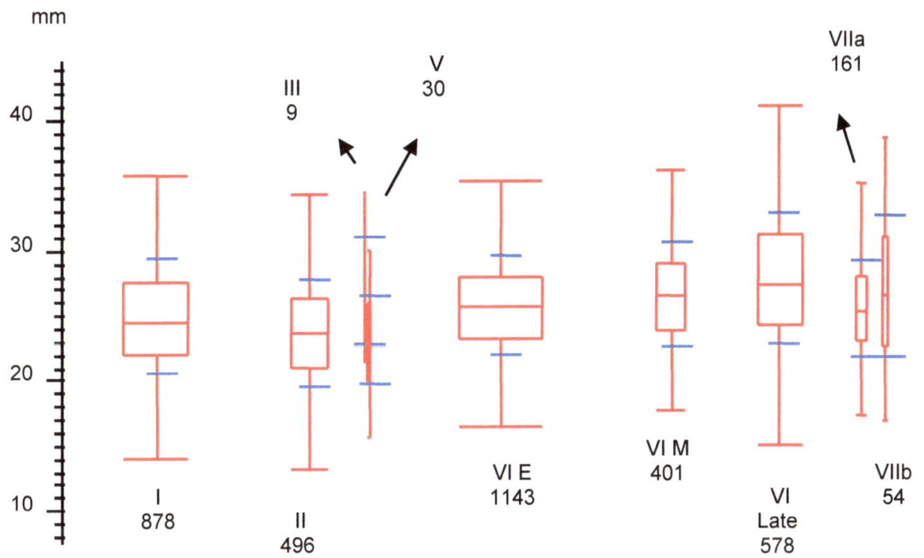

b.

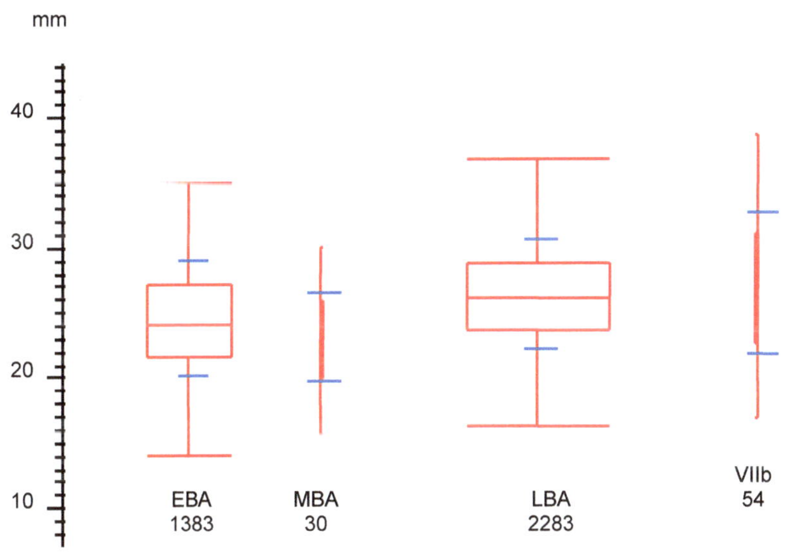

(I) = Chronological period or phase. (878) = No. of individuals.

Figure 41. Solidity of *C. glaucum* populations from Bronze Age Troia (Distributions).

Troia I

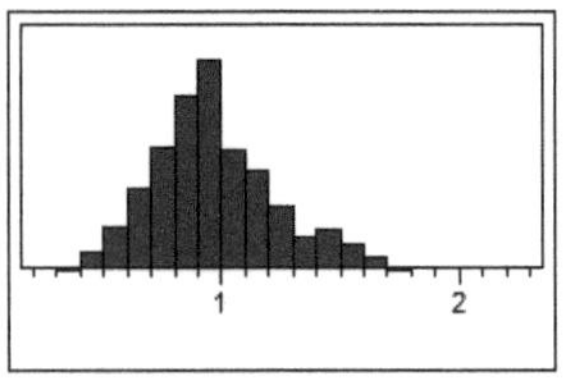

(838)

Troia II

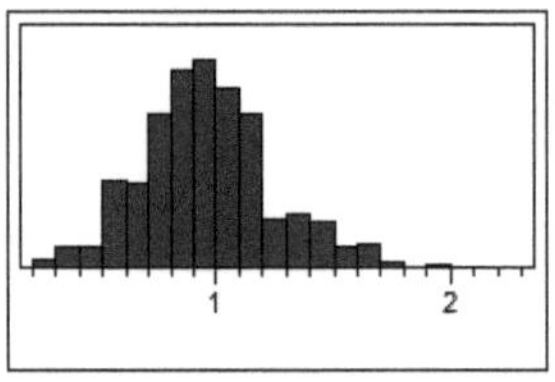

(472)

Troia III

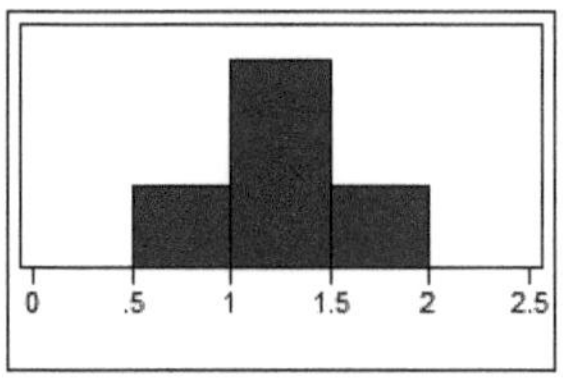

(9)

Troia I/II/III

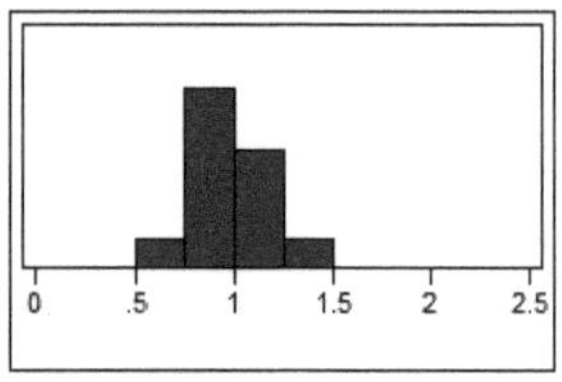

(12)

Troia V

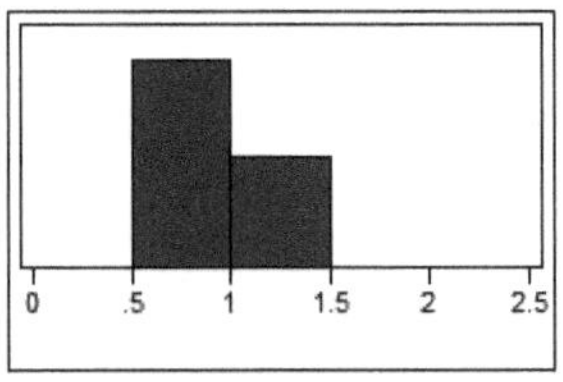

(23)

Troia VI Early

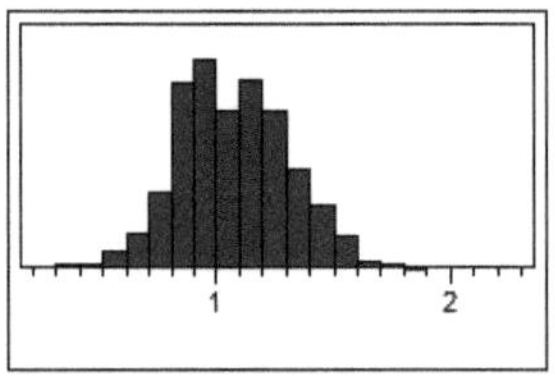

(904)

Troia VI Middle

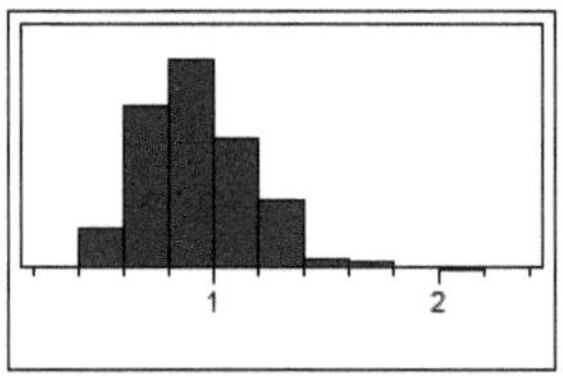

(315)

Troia VI Late

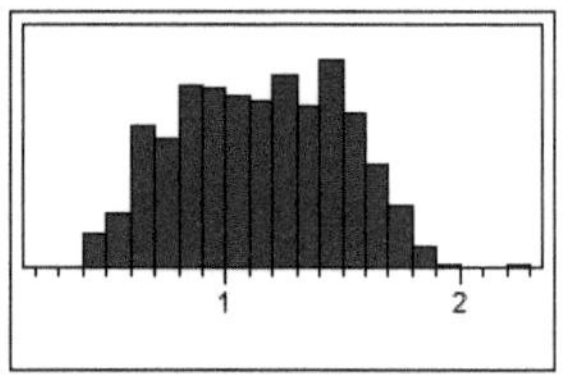

(473)

Troia VI Early/Middle/Late

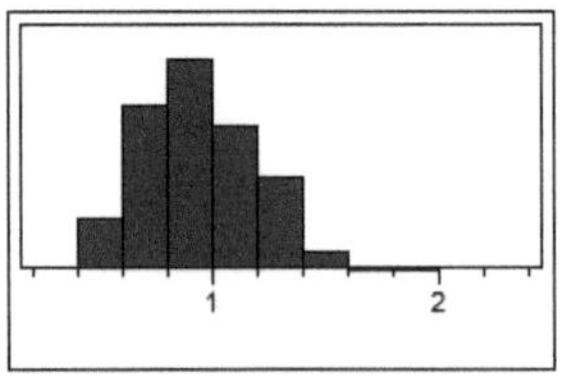

(384)

Troia VIIa

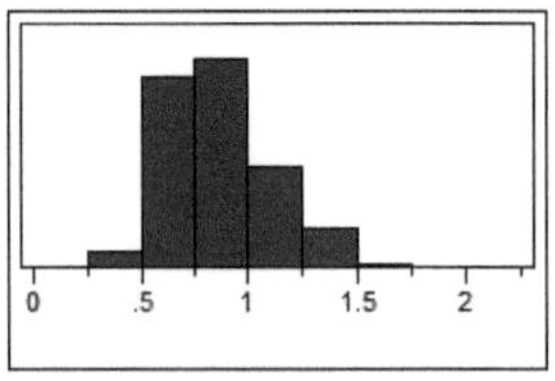

(144)

Troia VIIb

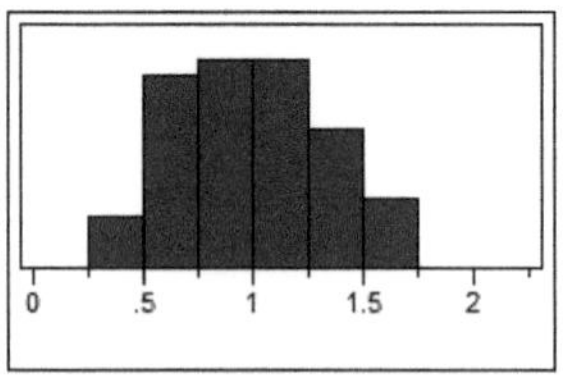

(50)

Troia (All)

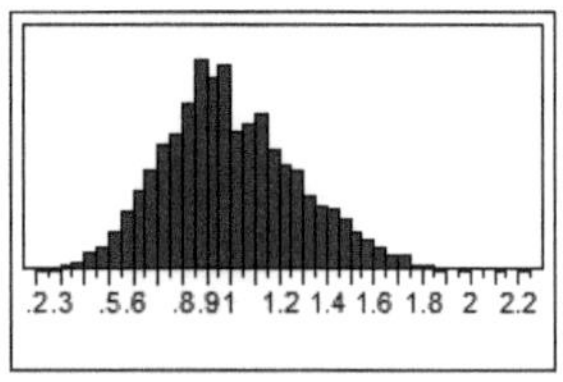

(3624)

(838) = No. of individuals.

Figure 42. Comparison of solidity in *C. glaucum* populations in Troia by period (a) and phase (b) (ANOVA).

a.

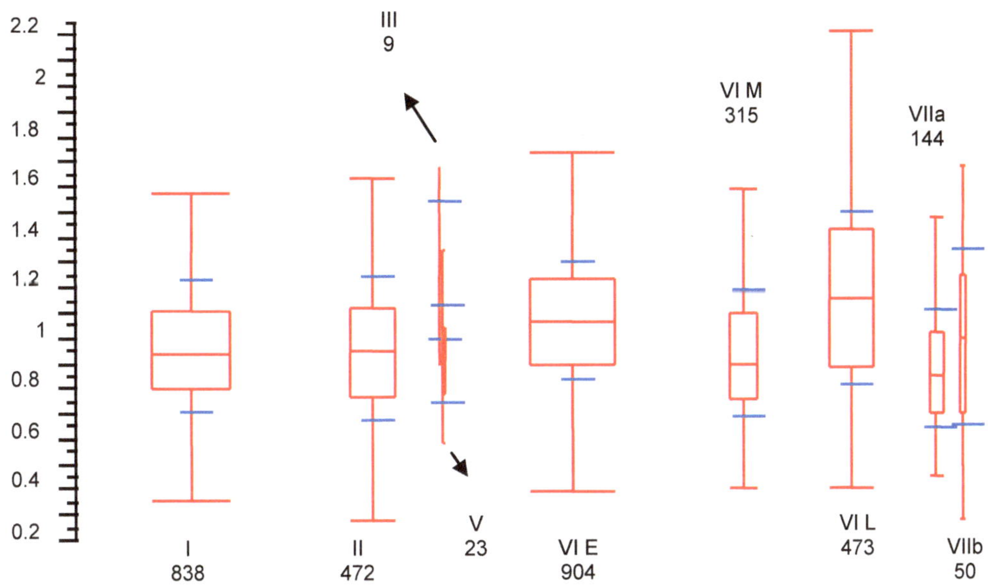

b.

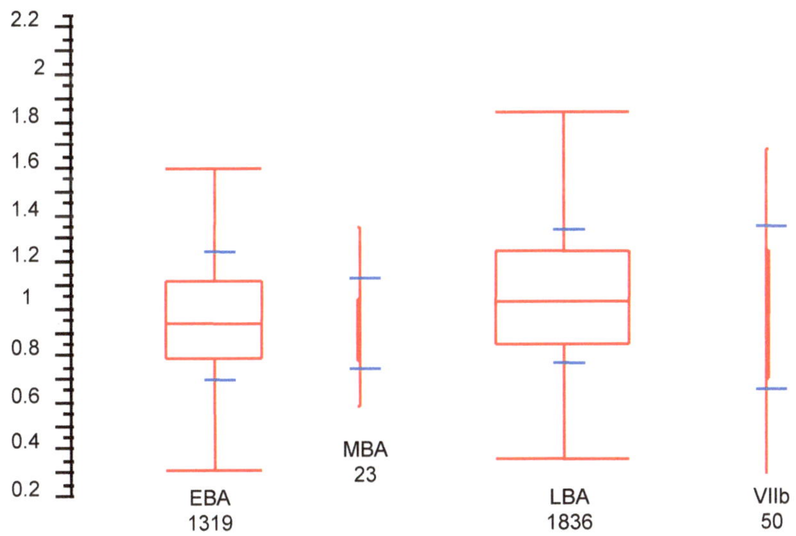

Figure 43. Curve of *C. glaucum* valves from Troia (Distributions).

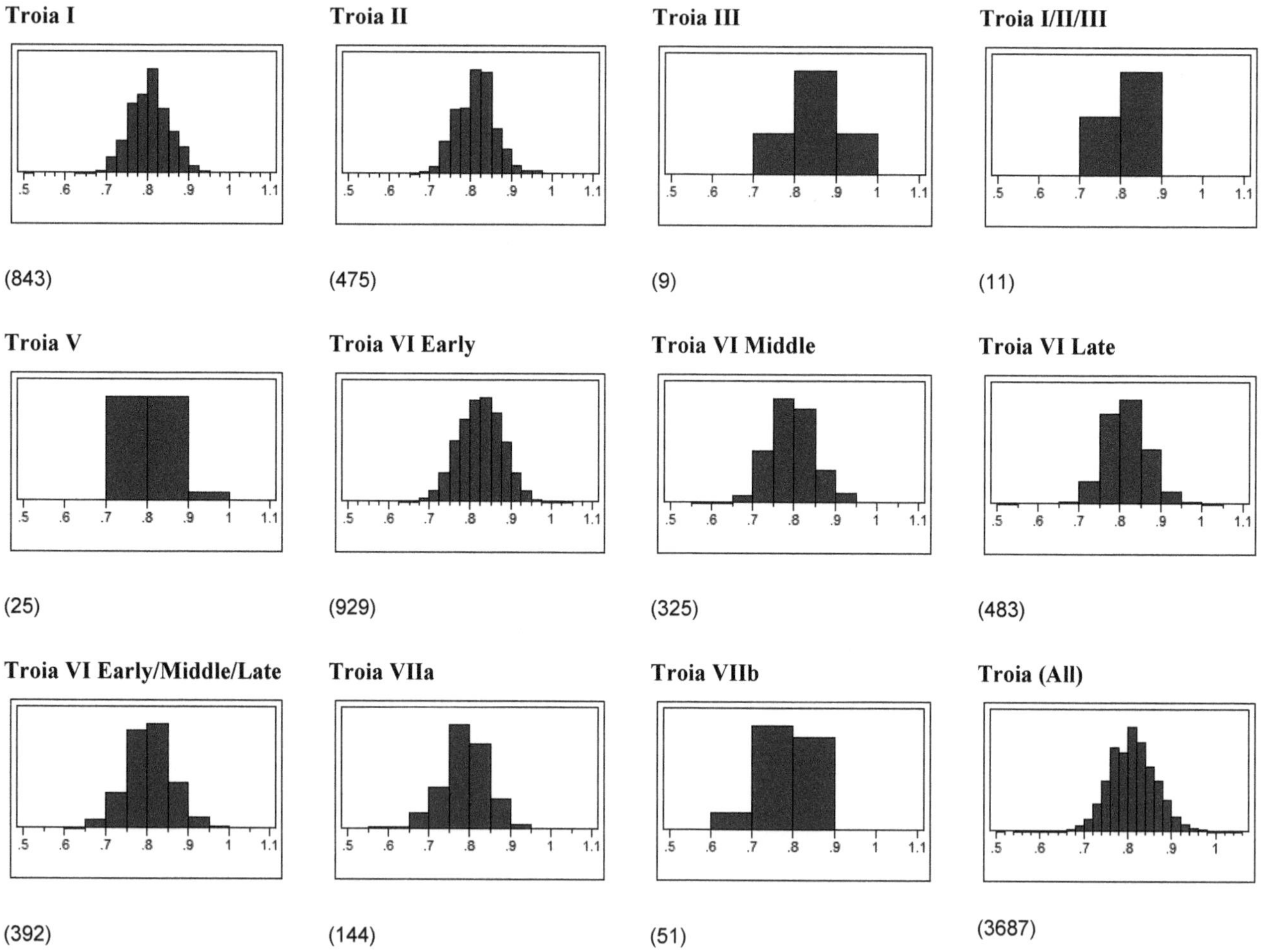

(843) = No. of individuals.

Figure 44. Comparison of curve in *C. glaucum* populations in Troia according to periods (a) and phases (b) (ANOVA).

a.

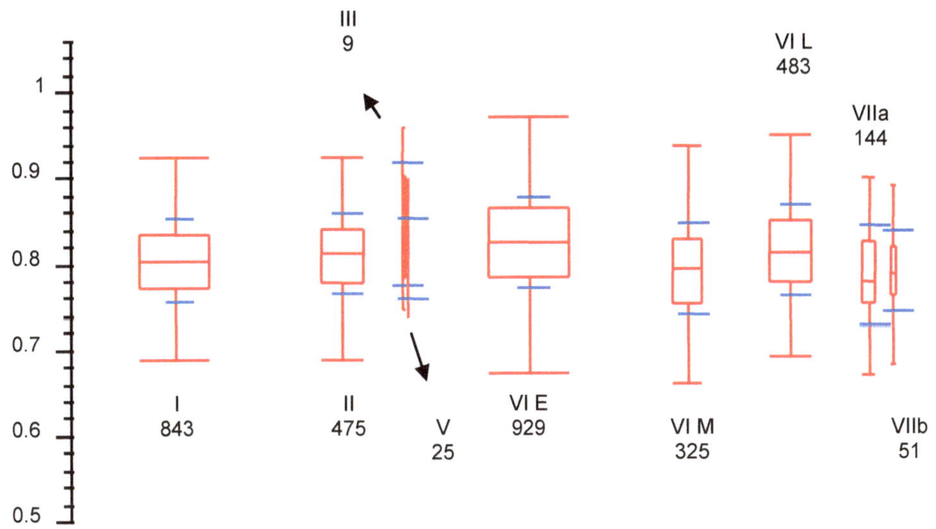

b.

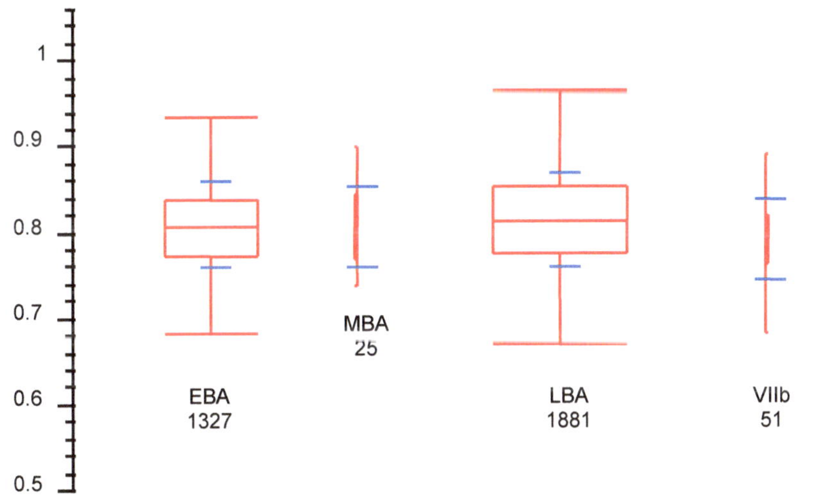

4.3.5.2. Description of incremental shell growth in Bronze Age *C. glaucum* populations

Alterations in the shell growth rates of 2284 individuals from the Bronze Age layers of Troia were observed. Variations in shell secretion rates are visible both on the surface and on the increments of archaeological *C. glaucum* specimens. Similar to modern *C. glaucum* shells, the appearance of external lines are not uniform in appearance. Therefore, they are not considered to be very informative (Figure 35). Alteration between gray and white incremental zones is however clearly visible on the cross-sections of the specimens (e.g., Figures 48 and 49).

Figure 45. Comparison of breadths of white (a.) and gray (b.) incremental zones in archaeological *C. glaucum* (ANOVA).

a.

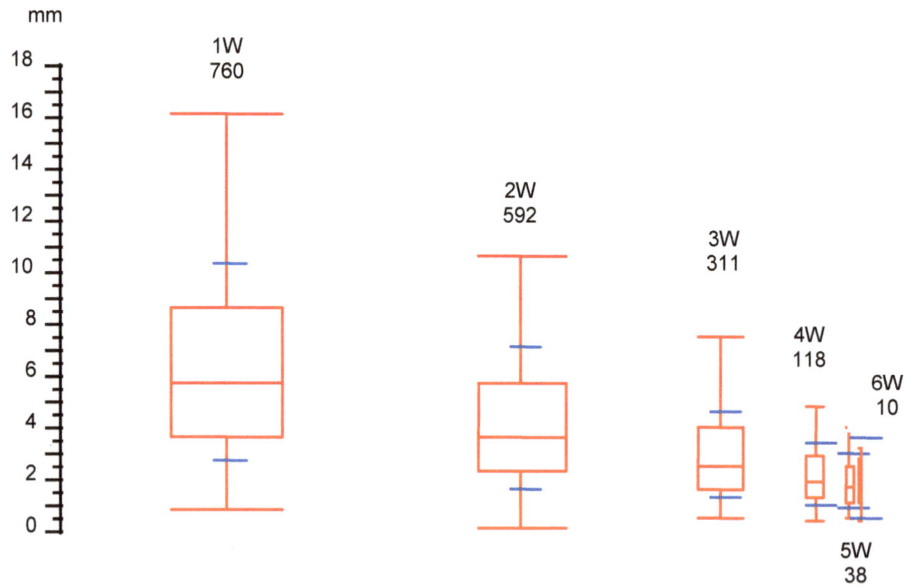

b.

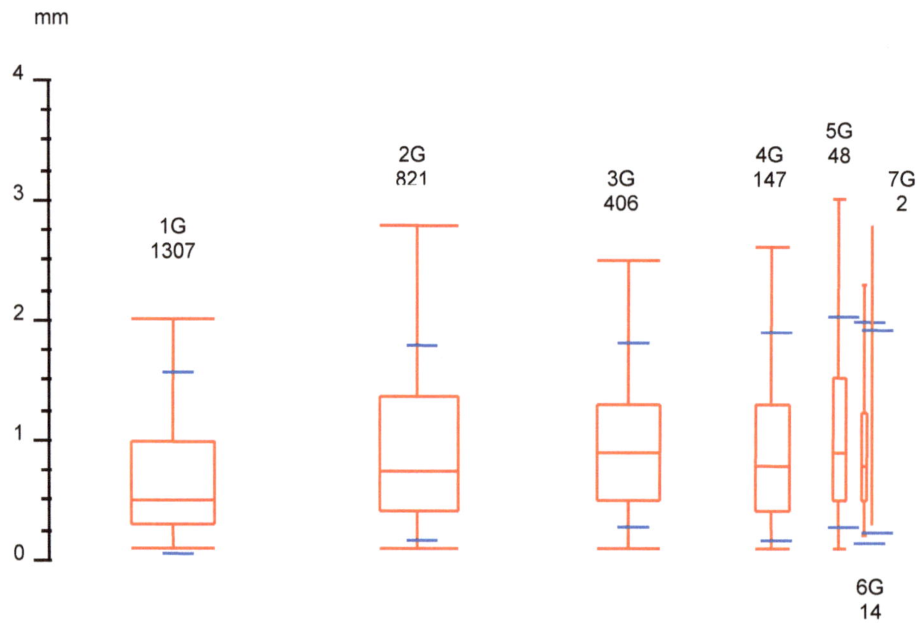

The breadth of white and gray zones vary greatly among individuals, but the mean breadth of successive white incremental zones become narrower as the animal grows (Figure 45). The difference in the mean breadth of incremental zones 5W and 6W is negligible (Figure 45a). Only a few archaeological individuals demonstrate 6W in their cross-sections. The trend of the mean breadth of gray zones is less clear (Figure 45b). Incremental zones 1G, 2G, and 3G are on average broader than the incremental zones 4G, 5G, 6G, and 7G. The narrowest gray increments are only 0.1 mm. The broadest gray incremental zone was a 2G with 8.1 mm, recorded on an individual from Troia VI Early period. This individual is an outlier, which probably survived through unusual environmental or metabolic conditions.

Figure 46. Comparison of completed successive breadths of white incremental zones in *C. glaucum* populations by occupational phase in Troia (ANOVA).

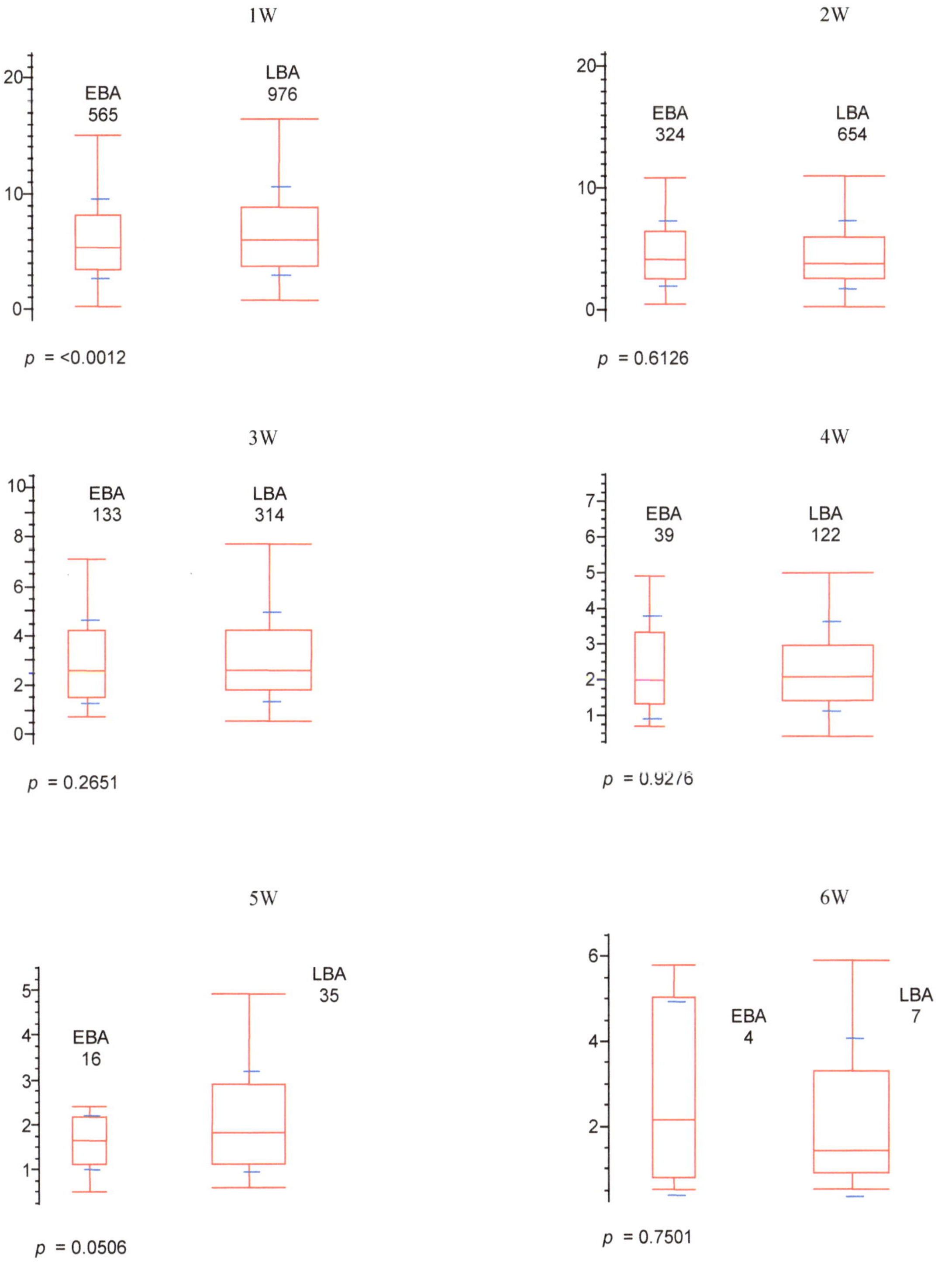

The comparison between the mean breadths of complete successive white incremental zones of individuals from the Early and the Late Bronze Age phases of Troia shows that there is very little difference of significance (Figure 46). While the mean breadth of incremental zone 1W is significantly smaller than the mean breadth of the same incremental zone in Late Bronze Age specimens, the differences between the mean breadths of the white incremental zones that follow 1W are insignificant (Figure 47). The complete successive gray incremental zones of the individuals from the Early Bronze Age are, however, generally narrower than those observed on the Late Bronze Age individuals (Figure 47). The difference between the breadths of 1G is small. The following three gray incremental zones, 2G, 3G, and 4G are broader in the Late Bronze Age samples than they are in the Early Bronze Age samples. The insignificant difference between 5G and 6G is attributable to the sample size and to the ambiguous appearance of these two incremental zones.

Examples of differential growth patterning in archaeological *C. glaucum* are demonstrated in Figures 48 and 49.

Figure 47. Comparison of completed successive breadths of gray incremental zones in *C. glaucum* populations by occupational phase in Troia (ANOVA).

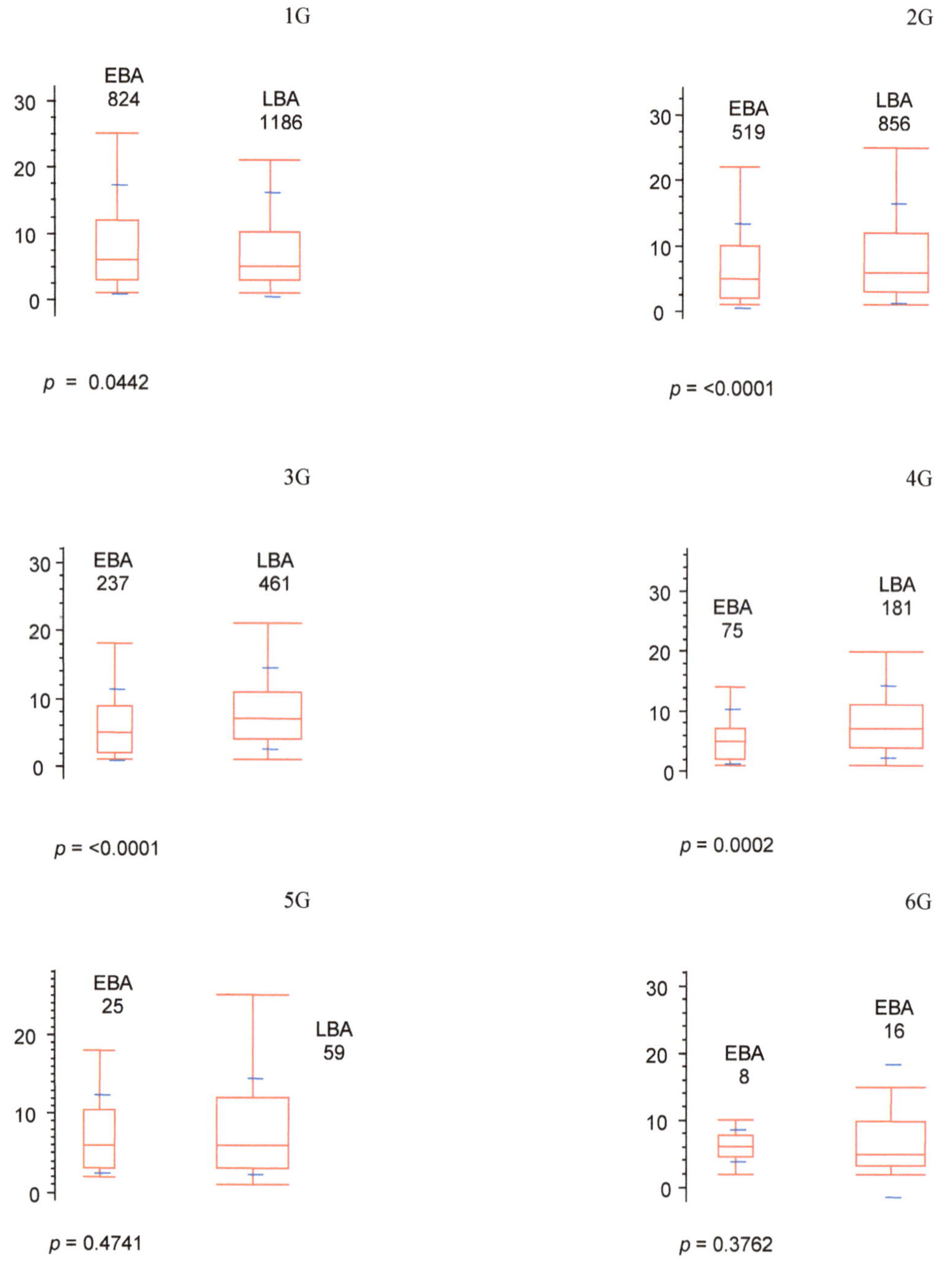

Figure 48. Incremental growth pattern of sample *C. glaucum* individuals from Bronze Age Troia (1).

a. Individual no. 1138-2

b. Individual no. 1254-2

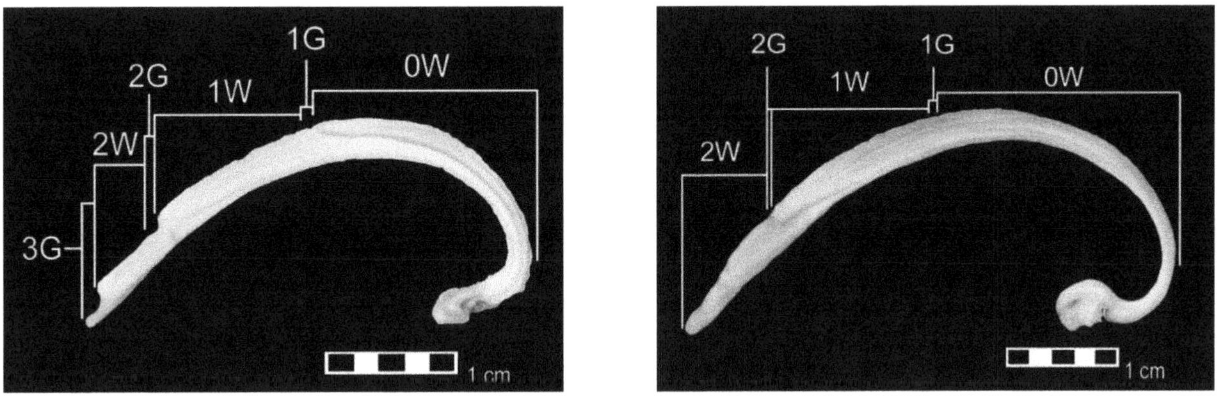

c. Individual no. 2075-4

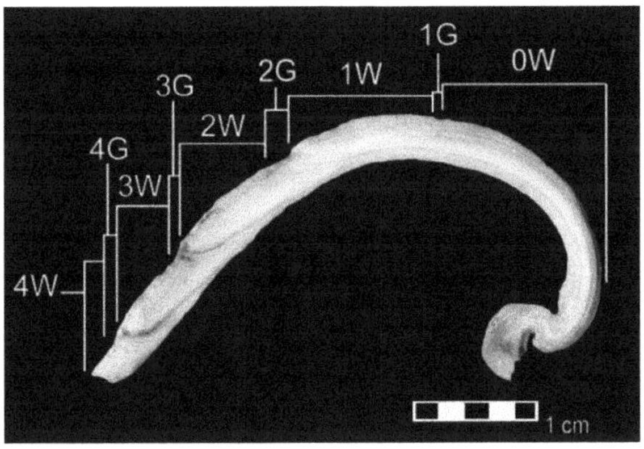

4.3.5.3. Age profiles of Bronze Age *C. glaucum* populations

Using the pattern of shell secretion in the modern *C. glaucum* populations in the Troad as an analogue, an age year was considered to be represented by two successive incremental zones in the archaeological *C. glaucum* populations from Troia. Age profiles of 2284 archaeological individuals from Bronze Age Troia were thus derived from the count of incremental zones they exhibited in their sections. These counts were then translated into age stages as shown in Table 7. In archaeological assemblages too, individuals older than Stage 5 are categorized as senile animals.

The Early Bronze Age demographic profile is represented by 936 individuals. Only very few of these originate from Troia III and mixed I/II/III contexts. The age profiles of these populations are not well distributed (Figure 50). Samples dating to the Troia I period contained very few individuals that were collected during Stage 5 and older. The bulk of the specimens represent individuals between Stages 2 and 4. About 10 % of the individuals were at Stage 1 when they were procured. Although the age profile of Troia II specimens is not very different from that of Troia I specimens, the number of specimens at Stage 1 slightly decreases, as the number of specimens at Stages 2, 3, 4, and 5 slightly increases. The proportion of senile animals remains the same. The pooled age profile of specimens originating from Early Bronze Age deposits indicate that the bulk of the individuals (63 %) were gathered while at Stages 2 and 3. Individuals at Stage 4 are considerably less extant, at about 17 % of all individuals. Slightly more than 10 % of the individuals represent Stage 1. Animals at Stage 5 and older make up only a very small portion of the collected part of the population. Together they represent about 8 % of the assemblage.

The Middle Bronze Age is represented by a limited amount of specimens from various deposits dated to the Troia V period. When this population is compared to the Early Bronze Age population, the number of young individuals collected during Stages 1 and 2 decreases, while the proportion of older individuals increases (Figure 50). The increase in the amount of individuals at Stage 5 and senile individuals may be noteworthy.

Throughout the early, middle, and late phases of the Troia VI period and in Troia VIIa, the age profile of the specimens belonging to the same period is evenly distributed (Figure 50). The majority of the individuals were collected while at Stage 3. The number of individuals at Stage 1 is very small in the Troia VI Early period. In Troia VI Middle, the amount of individuals at Stage 1 decreases, while the number of individuals at Stage 2 increases. Individuals, which originate from mixed deposits roughly dated assigned to the early, middle or late phases of the Troia VI, represent a population younger than the individuals in the other phases of Troia VI. In Troia VIIa, the proportion of young individuals at Stage 1 and the number of adult individuals, which represent Stages 2,3, and 4 increase, while the amount of individuals at Stages 5 and senile animals remain in negligible amounts. The cumulative age profile of all individuals belonging to Troia VI and VIIa periods represents a very small number of individuals collected at Stage 1 (7 %), a very large amount of adult individuals at Stages 2, 3, and 4 (altogether 78 %), and a small amount of individuals at Stage 5 (10 %) and senile individuals (5 %).

Figure 49. Incremental growth pattern of sample *C. glaucum* individuals from Bronze Age Troia (2).

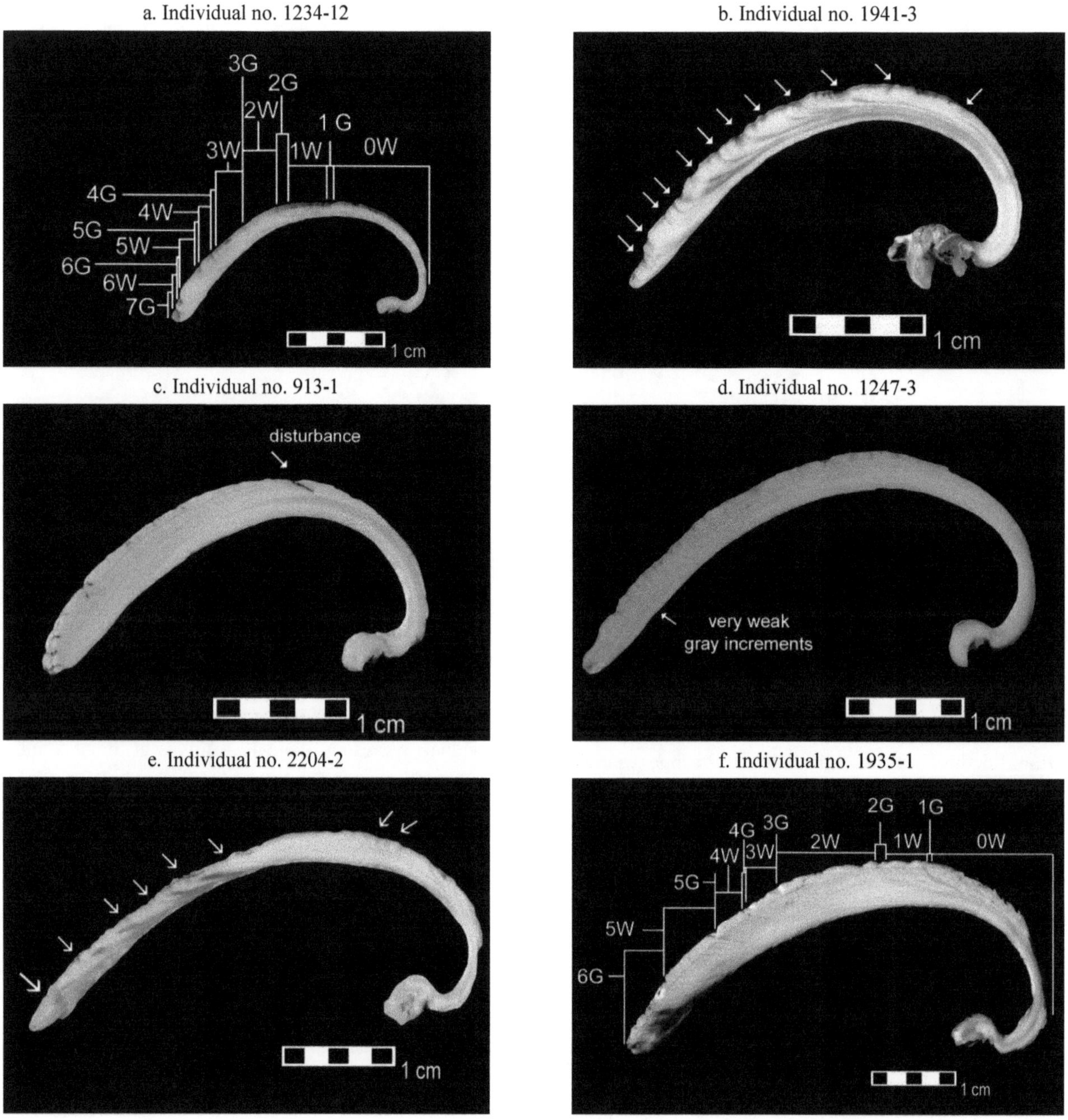

Similar to the age profile derived from individuals from Troia VIIa layers, the majority of the population represented by the shells from Troia VIIb layers consists of individuals at Stages 2 and 3. The number of individuals at Stages 4 and 5, and senile animals are significantly smaller than those represented by the Troia VI.

When the age profiles of the populations from the Early Bronze Age and the Late Bronze Age are observed collectively and compared to each other, it can be seen that the age profile drawn from the Late Bronze Age individuals is more skewed towards older individuals than in the Early Bronze Age (Figure 50). While individuals collected at Stages 1 and 2 are better represented in the Early Bronze Age population, individuals at older stages of their lives (Stages 3, 4, 5, and senile animals) are better represented in the Late Bronze Age population.

In summary, the portion of the *C. glaucum* populations consumed by the Trojans during the Bronze Age consists of individuals belonging to all age stages. The majority of the specimens are individuals collected at Stages 2 and 3, each making up 29 % of the Bronze Age population. Individuals at Stage 4 are slightly less common than those collected at Stages 3 and 4. The number of individuals representing Stages 1 and 5 are small. Senile animals are infrequent.

Figure 50. Age profiles of archaeological *C. glaucum* populations from Troia.

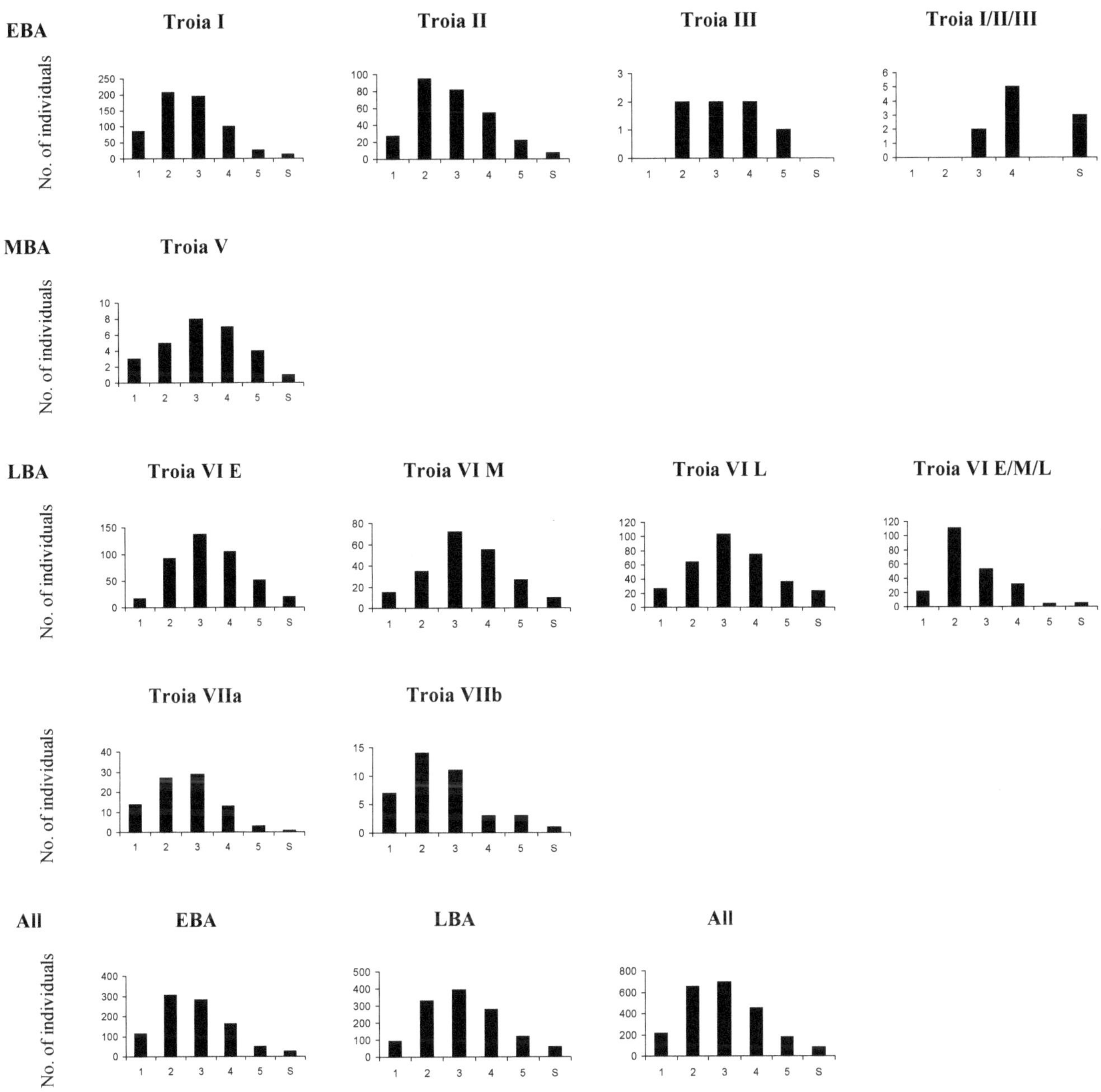

4.3.5.4. Seasonal incremental growth pattern of Bronze Age *C. glaucum* populations

In modern *C. glaucum* populations in the Troad, individuals' shell secretion behavior corresponded to the seasonal changes in their host environment (cf. Section 4.2.2.4). The timing of the change from one type of incremental zone to the other was primarily related to the age of the animal. Based on these observations, in order to delineate the seasonality of *C. glaucum* gathering behavior in the Bronze Age Troia, individuals originating from different periods and different contexts of the site have been classified into age groups (Figure 51). The amount of individuals demonstrating gray and white incremental zones at their margins are represented within their age group.

The individuals collected while secreting white incremental zones and the individuals collected while secreting gray incremental zones are both represented by the studied assemblage from Bronze Age Troia (Figure 51). Individuals with gray incremental zones at their margins have generally lower representation than the individuals secreting white incremental zones at the time of death. The relative abundance of these individuals changes, however, between the samples from different chronological periods.

The proportion of the white incremental zones to the gray incremental zones recorded at the margins of the individuals from Troia I is very similar to the proportion of the same elements for the individuals from Troia II. Very few individuals at Stage 1 demonstrate gray incremental zones at their margins. The quantity of the individuals with gray incremental zones increases with age groups but remains below 40 % at all age stages, with the exception of the senile age group. Although the distribution of the age groups are not normal for Troia III and Troia I/II/III samples due to limited sample size, it should be noted that the individuals at Stage 3 during these periods had been all collected as they were secreting white incremental zones (Figure 51).

Figure 51. Marginal incremental zones of archaeological *C. glaucum* valves from different occupational phases in Troia.

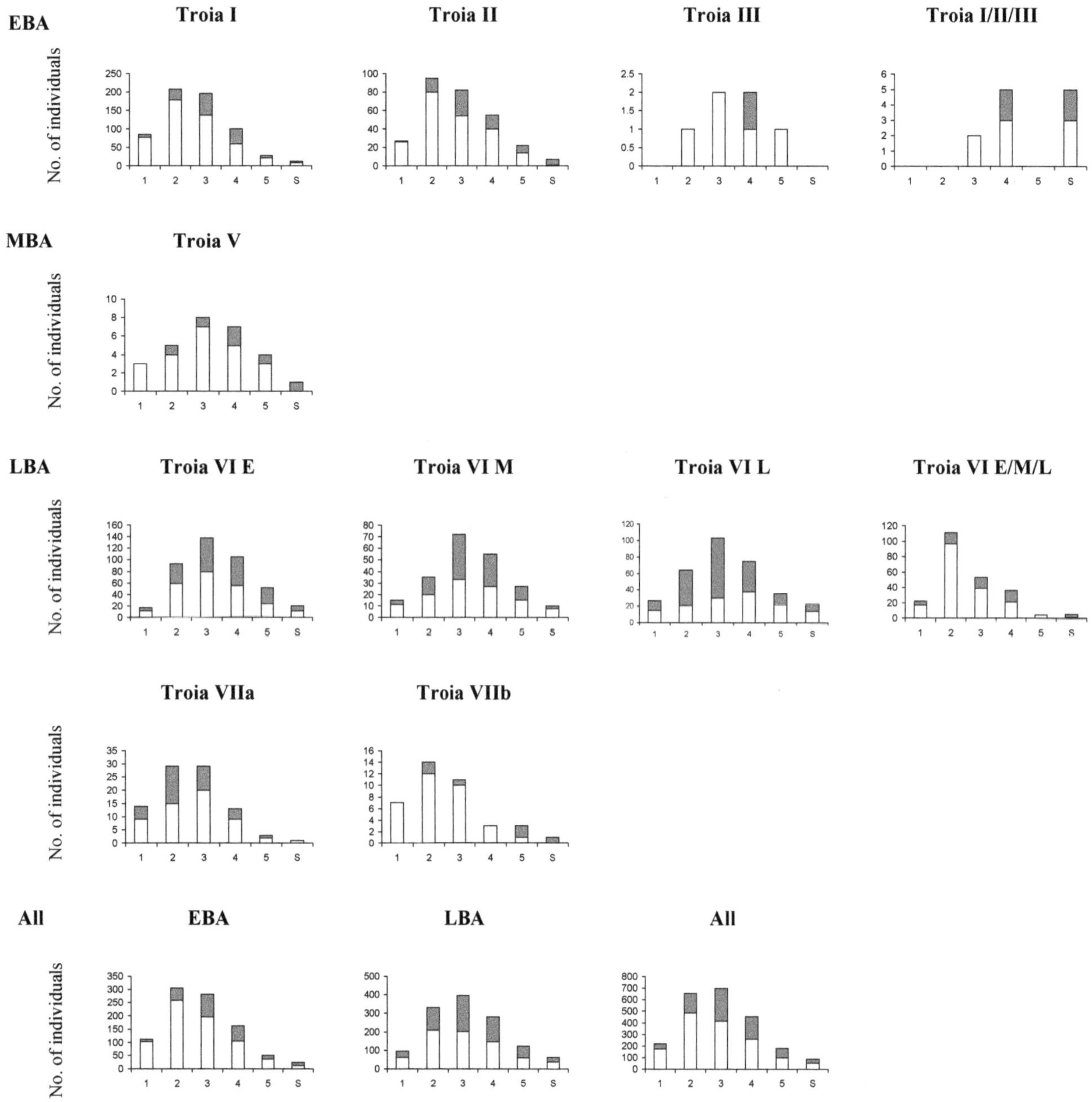

Likewise, the Middle Bronze Age is also represented by a limited amount of individuals. In this sample, the proportion of individuals showing gray incremental zones at their margins is smaller compared to the Early Bronze Age assemblages (Figure 51).

The proportion of individuals with gray incremental zones at their margins is much larger in the Late Bronze Age assemblages than it is in the assemblages of Early and Middle Bronze Ages (Figure 51). The percentage of young and adult animals that were secreting gray increments at the time of collection increases up to 27-77 % per age group. In

Figure 52. Marginal incremental zones of archaeological *C. glaucum* individuals from different excavation units in Troia.

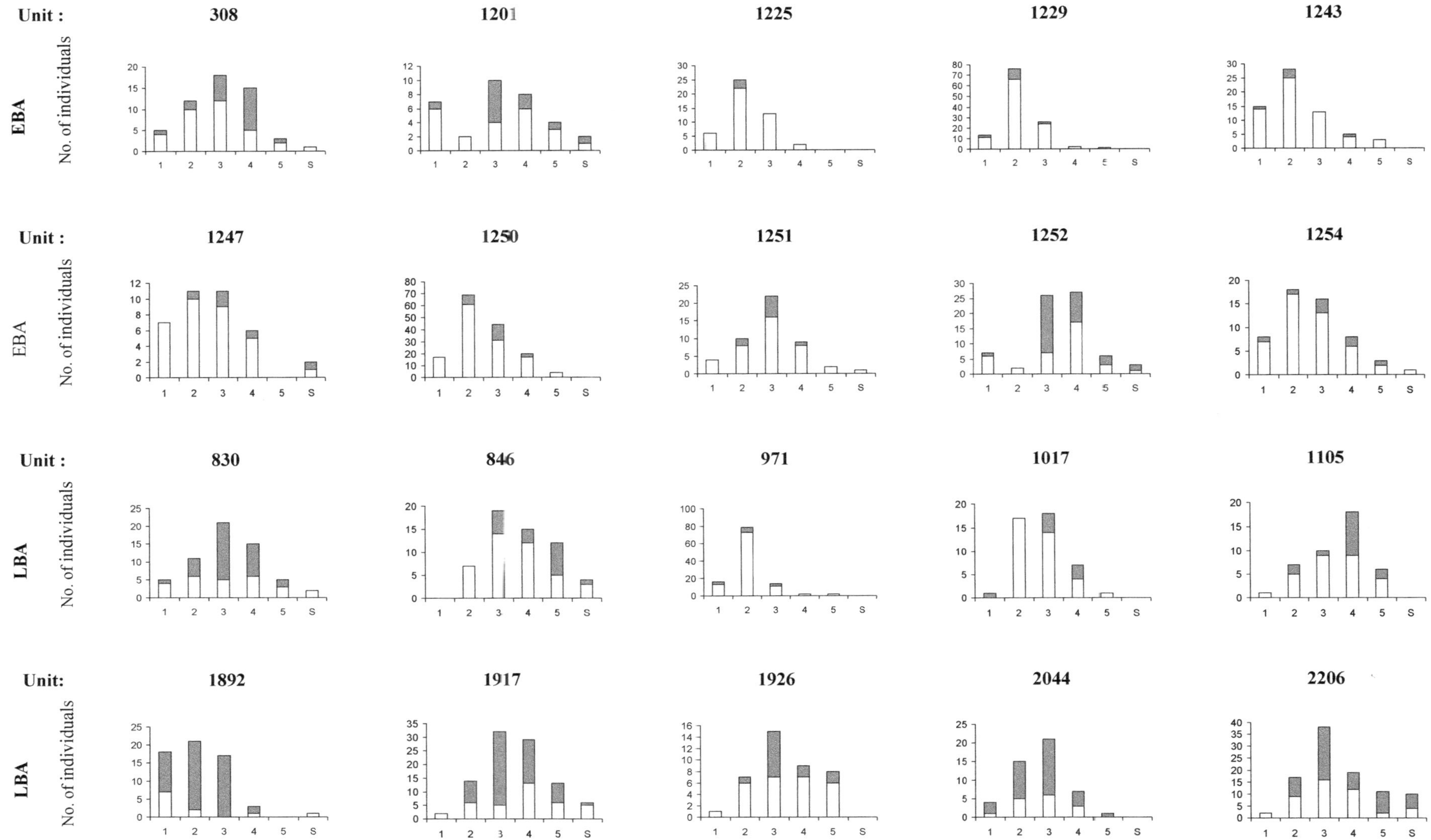

the assemblages roughly dated to any one of the occupational layers of Troia VI, the picture is completely different than that represented by more exactly dated assemblages. Here, the percentage of individuals with gray incremental zones is relatively lower for all age groups except senile animals. Individuals with gray zones at their margin are slightly less present in the sample representing the Troia VIIa period, but their proportion is more in line with the trends in Troia VI assemblages than the assemblages from Early Bronze Age Troia.

The assemblage of *C. glaucum* shells from Troia VIIb deposits mainly represents individuals collected during the fast-growth period (Figure 51). Small numbers of individuals at Stages 2 and 3 were collected as they were growing slowly. Furthermore, almost all older individuals (Stage 5 and older) were collected when they were secreting gray incremental zones.

In order to detect variations in the character of incremental zone secretion at the time of collection within chronologically grouped *C. glaucum* assemblages, samples containing more than 30 individuals from different archaeological units were observed for their growth-frequency profiles (Figure 52). None of the samples containing more than 30 individuals come from archaeological contexts that may represent discrete single-event domestic activities, such as pit, hearth or bin deposits. All of the units represent fill deposits accumulated during the disuse periods of architectural features or layers. All 10 of the Early Bronze Age units containing more 30 complete valves of *C. glaucum* originate from Quadrate D5 (9 units) and a related fill deposit excavated in Quadrate DE4. Together they make up a large shell layer clearly visible in the section of the excavation areas. Five of the units representing Late Bronze Age assemblages originate from Quadrates I8 and K8 immediately outside of the Troia VI citadel and are dated to various occupational periods of Troia VI. The other five units were recovered from KL16/17 and they, too, represent fill deposits dating to Troia VI Early till Late periods. It can be seen in the graphic (Figure 52) that while most of the units (Units 1225, 1229, 1243, 1247, 1251, and 1254) consist mainly of individuals collected during fast-growth (81-93 %), other units (Units 308, 1201, and 1254) contain relatively more individuals that were collected during slow-growth phase (33-49 %). Likewise, among the units representing the Late Bronze Age, the growth-sequence profiles of different excavation units demonstrate that eight of the illustrated units (Units 830, 846, 1105, 1892, 1917, 1926, 2044, and 2206) consist of 28-82 % of individuals secreting gray incremental zones at the time of collection. Meanwhile only in two of the units observed from Troia VI levels (Units 971 and 1017) contain only 11-18 % slow-growing individuals.

When the assemblages discussed above are merged thereby creating two larger assemblages –one dating to the Early Bronze Age and the other dating to the Late Bronze Age– it becomes clear that the Late Bronze Age assemblages contain more individuals collected while secreting gray incremental zones than the Early Bronze Age assemblages do (Figure 52). Such is the situation for all age groups from Stage 1 to Stage 5. Conversely, in the Late Bronze Age assemblage, slightly less senile animals demonstrate gray incremental zones at their margins, in comparison with the senile animals which demonstrate gray incremental zones at their margins during the Early Bronze Age. This situation, however, may be a statistical effect caused by the differential quantity of the senile animals in the two assemblages: 25 senile animals were recorded in the Early Bronze Age assemblage as opposed to 61 in the Late Bronze Age assemblage.

4.3.5.5. Isotopic compositions of Bronze Age *C. glaucum* specimens

Two valves from the same archaeological unit excavated in the lower town (Quadrate KL16/17, Beh. 561, Individuals 2206-2 and 2206-6) were selected for sampling of the *C. glaucum* carbonate (Figure 53a and 53b). They are archaeologically dated to the last occupational phase of the Late Bronze Age (Troia VI Late, ca. 1350-1300 BC). Their valve heights measure 32.8 mm (2206-2) and 32.0 mm (2206-6). They are good representatives of the solid and curved *C. glaucum* shells typical of the Bronze Age. Specimen 2206-2 demonstrates incremental zones 0W-3G and specimen 2206-6 demonstrates incremental zones 0W-2W. Exact locations of the carbonate samples taken along the growth axis are indicated in Figure 53. In 2206-2, carbonate samples represent incremental zones 1G, 1W (2 samples), 2G, and 2W (Table 10). In 2206-6, carbonate samples represent incremental zones 0W, 1G, 1W, and 2G.

The $\delta^{18}O$ values of all samples are more negative than the standard (PDB) sample. The $\delta^{18}O$ values of the samples range between -2.07 and -0.49 ‰ PDB (Table 10). Although these $\delta^{18}O$ values appear to be close to the range of oceanic water, meteoric water in low altitudes close to river mouths also display similar values (Hoefs 2004: 61-62). Only two out of nine $\delta^{13}C$ values are positive relative to the PDB standard. $\delta^{13}C$ values range between -2.21 and 0.16 ‰ PDB (Table 12). The range of these values assuredly shows that no contamination with the archaeological soil took place during the sampling of the shell carbonate (cf. Göbel 2005 and Section 2.4.1). The mean $\delta^{13}C$ isotopic content of the archaeological *C. glaucum* shells are also within the range of the published $\delta^{13}C$ values from *C. glaucum* specimens taken from geomorphological cores in the flood plain of the Karamenderes and the Dümrek (Rapp et al. 1982: 188). Rapp et al.'s (1982) $\delta^{13}C$ values from two samples were determined on whole shells during ^{14}C analyses; they are -1.55 ‰ and -0.7 ‰ PDB respectively. Although Rapp et al. (1982: 188) interpreted the range of these two $\delta^{13}C$ values as indicative of marine environments, this is a simplistic approach to the interpretation of carbon isotopic content of mollusk shells for two reasons; one involving the difficulties in explaining the fractionation amount between the DIC in the host water and the $CaCO_3$ formation of the mollusk shell, and the other related to the overlapping range of the $\delta^{13}C$ compositions of the DIC in ocean and meteoric water reservoirs (Hoefs 2004: 48). While the mean $\delta^{13}C$ compositions around -1 ‰ PDB are close to the range of δ

^{13}C marine carbonates, deducing that such values represent marine environments are complicated by the fact that the δ ^{13}C composition of fresh water DIC can be as high as +1 ‰ when the meteoric water originates from carbonate rich areas, because carbonates are isotopically heavier than organic matter (Hoefs 2004: 48, Schöne et al. 2006). Moreover, although the δ ^{13}C composition of mollusk $CaCO_3$ should be lighter than that of the DIC source, the equation of the physiological fractionation process is unknown (Kennedy et al. 2001). Therefore, neither the oxygen nor the carbon isotopic contents of the samples analyzed for the present study are helpful in designating their growth environment as belonging to the marine, brackish, or freshwater realms with a high level of certainty.

Table 10. Oxygen and carbon isotopic composition of carbonate samples from two Bronze Age *C. glaucum* specimens.

Individual no.	**2206-2**		**2206-6**	
Growth stage	δ ^{13}C ‰ PDB	δ ^{18}O ‰ PDB	δ ^{13}C ‰ PDB	δ ^{18}O ‰ PDB
2W	-0.72	-1.58		
2G	-2.13	-1.83	-0.94	-1.39
1W	-0.25	-1.11	0.16	-0.49
1W	0.32	-0.90		
1G	-2.21	-2.06	-1.72	-2.07
0W			-0.70	-1.79

Figure 53. Incremental growth pattern of the *C. glaucum* specimens sampled for isotopic analysis.

a. Individual no. 2206-2

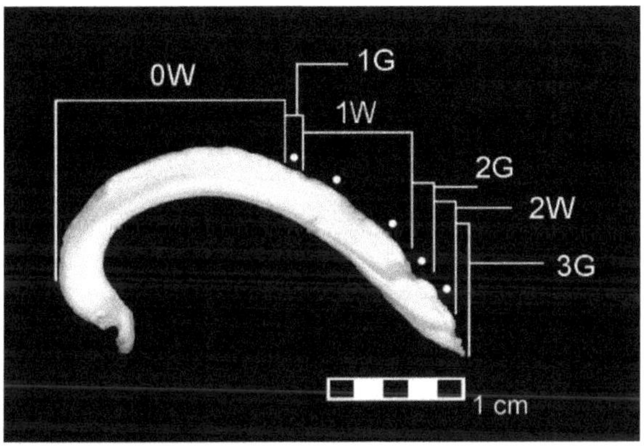

b. Individual no. 2206-6

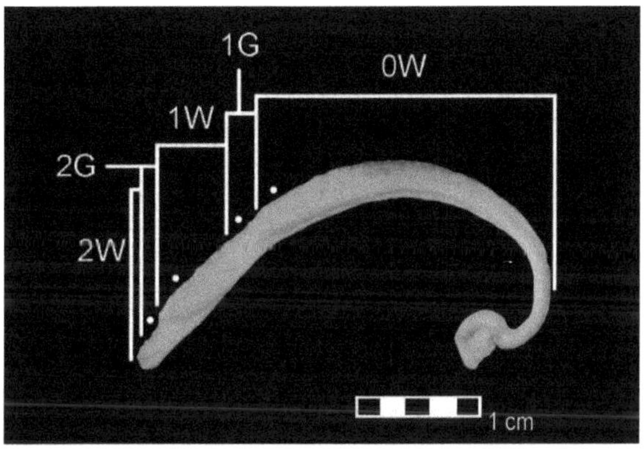

White points indicate the location of carbonate samples.

There are three apparent trends in the ranges of isotopic compositions: firstly, there is a significant co-variation between the δ ^{18}O and δ ^{13}C values. When δ ^{18}O record of a sample is depleted in heavier isotopes, the δ ^{13}C record of the same sample is also depleted, and vice versa (Figure 54a). Such a positive relationship between the δ ^{18}O and δ ^{13}C compositions of carbonate samples from *C. glaucum* shell has been recorded elsewhere in the Aegean on an archaeological *C. glaucum* (cf. Section 4.6.2 below; Shackleton 1970: 944, Figure 2)[29]. Secondly, the isotopic variations correlate well with the optical variations of the incremental zones. The isotopic compositions of the carbonate samples representing white incremental zones are enriched in heavier oxygen and carbon isotopes, in relation to the isotopic compositions of the adjacent gray incremental zones (Figures 54a and 54b). For example, the δ ^{18}O and δ ^{13}C compositions of 2G of 2206-6 are lighter than those of the carbonate sample representing 2W and the δ ^{18}O and δ ^{13}C compositions of the 1G carbonate sample of the same specimen are lighter than those of the adjacent 0W. Finally, despite the fact that the samples are time-averaged, the amplitude of the isotopic compositions between different incremental zones ($\Delta_{\text{White carbonate (2206-2) - Gray carbonate (2206-2)}}$ and $\Delta_{\text{White carbonate (2206-6) - Gray carbonate (2206-6)}}$) are large enough to represent seasonal changes in the isotopic content of the host water.

These trends are explicable in a variety of ways:

If salinity variations and evaporation constitute important factors influencing the host water of these two individuals, then the higher δ ^{18}O values might be indicating higher salinities during dry seasons, hence placing the formation of white incremental zones to seasons with decreased freshwater influx and increased evaporation, roughly between June and November –based on the seasonality of present time pluvial regimes. Evaporation is the chemical

[29] Shackleton's (1970) account is the only publication where both the carbon and oxygen compositions along the growth axis of a *C. glaucum* specimen from the Aegean (modern or archaeological) have been demonstrated. This study does not indicate the locations of the carbonate samples with respect to the character of the incremental zones.

process that causes this isotopic fractionation, through which the lighter oxygen isotopes in the seawater evaporate to create meteoric water in the atmosphere, leaving heavier oxygen isotopes in the seawater (Hoefs 2004: 116-117). But the δ ^{18}O composition of the host water of these two specimens may also have been determined by temperature effects. In that case, more negative δ ^{18}O values should signify high SST values occurring during warm seasons, placing the formation of gray increments within summer.

Figure 54. Oxygen and carbon isotopic composition of incremental zones from two *C. glaucum* specimens from Troia.

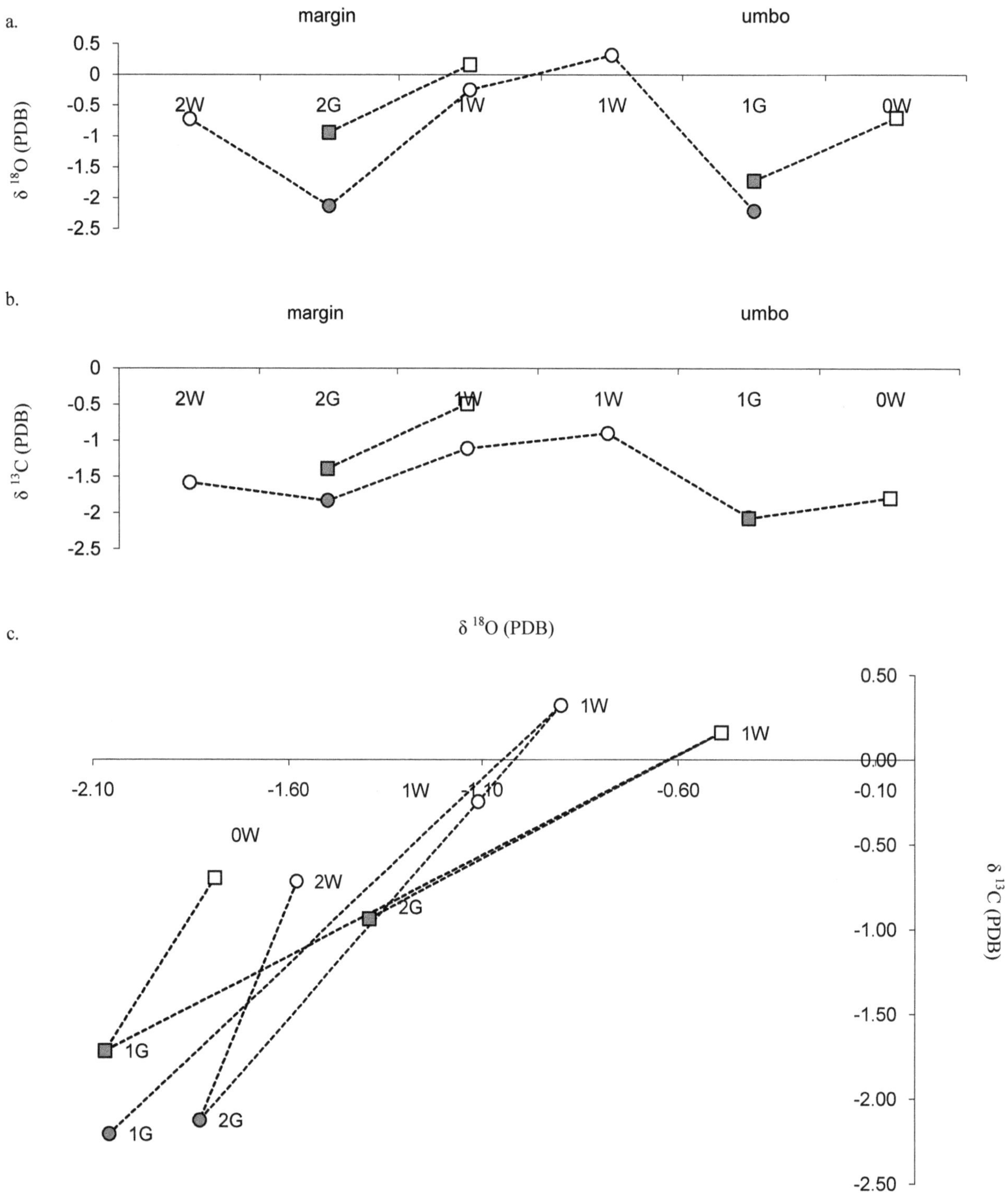

(○) = 2206-2 (□) = 2206-6.

The correlation of the positive relationships between the $\delta^{18}O$ and $\delta^{13}C$ compositions of these two specimens implies that metabolic effects had little if any influence on the fractionation process of the $\delta^{13}C$ content of the *C. glaucum* shell. The same trend observed on Shackleton's (1970) specimen might be confirming this suggestion. If this suggestion is true, then it should be considered that the mechanism that leads to the $\delta^{13}C$ fractionation in the *C. glaucum* shell is directly related to the $\delta^{13}C$ of the DIC in the ambient water. What kind of natural phenomena cause the coeval depletion of $\delta^{18}O$ of water and $\delta^{13}C$ of DIC in the water?

There are two possibilities: the $\delta^{13}C$ composition of productive waters is generally more positive than waters with low productivity (e.g., surface water versus deep water). In the northern Aegean, it was shown that the level of primary production is higher between April and September than it is between October and March, i.e., during the winter (Field et al. 1998). This should mean that the $\delta^{13}C$ of the surface DIC influencing the formation of the *C. glaucum* shell in the Troad is more positive during the summer months. But warm waters contain less CO_2 because of decreased solubility with temperature. Since CO_2 is an important constituent of oceanic DIC, increasing SST may cause more negative $\delta^{13}C$ compositions of DIC. Thus, even if it is assumed that the ambient water of the specimens was dominated by marine waters, it is difficult to predict the amount of DIC and its $\delta^{13}C$ composition.

Upwelling events in the Dardanelles further complicate attempts to predict the seasonal variation of $\delta^{13}C$ of DIC in the water of the Trojan Bay. Although upwelling events generally bring isotopically light DIC to the surface (Richardson et al. 2001), because of the lack of nutrients in the Dardanelles below ca. 25 m (Jönsson 2002), it is difficult to estimate the effect of upwelled DIC on the $\delta^{13}C$ composition of the surface waters in the Dardanelles.

The influence of freshwater DIC could also be significant on the shell formation of these *C. glaucum* specimens. It was mentioned in Section 3.7 that while a number of factors affect the isotopic composition of freshwater DIC, the weathering of carbonates in carbonate rich areas significantly enrich the $\delta^{13}C$ of riverine DIC (Schöne et al. 2006). The immediate geological surrounding of Troia can best be described by limestone and sandstone ridges whose compositions are dominated by calcium carbonate (Kayan et al. 2003). If the weathering of these carbonates significantly affected the freshwater inflowing into the Trojan Bay in the past, the positive $\delta^{13}C$ of the freshwater DIC, contrary to the common opinion that the mixing of freshwater to estuaries depletes the $\delta^{13}C$ of estuarine water mass (e.g., Geist et al. 2005), may have enriched the estuarine $\delta^{13}C$ on a seasonal basis. In the Troad, such a process would have produced positive $\delta^{13}C$ values in the *C. glaucum* carbonate precipitated during winter months, when the freshwater coming from the Karamenderes and Dümrek Rivers is more significant. But there are as mentioned before, other factors affecting the freshwater DIC.

In summary, the fact that hinders the interpretation of $\delta^{13}C$ compositions of the samples taken from archaeological *C. glaucum* carbonate is that data on the seasonal variation of the $\delta^{13}C$ composition of both marine and meteoric waters in the Troad are not known. Additionally, there is the possibility that the main cause of the positive correlation between $\delta^{13}C$ and $\delta^{18}O$ values in the *C. glaucum* shell might be metabolic. It is evident from the correlation between the growth pattern and $\delta^{18}O$ isotopic composition of the mollusk shell that natural phenomena producing variations in the $\delta^{18}O$ composition of the *C. glaucum* shell also changes the metabolism of the organism. This change in the organism's metabolism may very well be causing the preferential uptake of ^{12}C isotopes and hence the depletion of $\delta^{13}C$ values in the shell carbonate. Similar trends of $\delta^{13}C$ values in the carbonates of marine mollusks have been observed in other studies (e.g., Kennedy et al. 2001).

The only result of the isotopic analyses which is in good accord with the seasonal nature of incremental zone formation observed on the modern *C. glaucum* populations in the Troad is the well-fitted correlation between the changes in the character of the incremental zones and coeval changes in $\delta^{18}O$ and $\delta^{13}C$ compositions. Although the amplitude of these changes is large enough to signify seasonality, it is not possible to relate the formation of different incremental zones to certain seasonal events concretely, based solely on their isotopic content. However, further analyses of both archaeological and modern *C. glaucum* carbonate sampled at shorter intervals along the growth axis of the valves have strong potential to elucidate all the points that are raised in the discussion above.

4.3.6. Summary

A large amount of faunal remains (probably the largest archaeozoological assemblage from the Aegean) was studied from the New Excavations in Troia. The faunal samples are unevenly distributed among the stratigraphical layers of the site. Mollusk remains are most common in the Early Bronze Age deposits, particularly in Troia II. They are unusually rare among the Troia III layers. Although their proportion slightly picks up in Troia IV and V, their contribution to the faunal assemblages almost returns to the level it was at the beginning of the occupation only during Early Troia VI. Mollusks constitute about one fourth of the faunal assemblages dating to the Late Bronze Age. The overall ratio of the mollusk remains in the faunal assemblages of Post-Bronze Age Troia is lower than it is in the Bronze Age assemblages. In general, there is a proportional decrease in the amount of mollusk remains throughout the occupational history of the site.

Patella spp., *H. trunculus*, *M. galloprovincialis*, *O. edulis*, *S. marginatus*, *T. decussatus*, and *C. glaucum* are the most prominent mollusk taxa represented in Troia. *C. glaucum* is the prevailing species throughout the Bronze Age, except in Troia V. Along with *C. glaucum*, *M. galloprovincialis*, *O. edulis*, and *T. decussatus* are significant features of the Troia I assemblages. Species other than *C. glaucum* lose their importance in the Troia II layers. In Troia V, *S. marginatus* remains constitute the leading element in the NIS and WIS figures. Gastropods, such as *Patella* spp. and Monodontinae, and particularly *H. trunculus,* make their first significant appearances in the Troia VI Early period. *C. glaucum* is still the most abundant species, making about three fourth of the archaeomalacological assemblage dating to the Late Bronze Age. While the species composition of Troia VIIb does not contrast with the preceding Troia VI and VIIa assemblages in general, it is remarkable that the proportion of *H. trunculus* remains is significantly smaller during this period. In the Post-Bronze Age, *C. glaucum* is gradually replaced by *O. edulis.* Pectenidae also becomes an important component of the mollusk assemblages during the Post-Bronze Age. *Unio* spp. appears very infrequently in all assemblages.

The taphonomic condition of the mollusk remains can be best described as species-specific. The majority of *H. trunculus* were found in crushed form. *C. glaucum* shells are fairly intact; few are burnt and few others are fragmented. *M. galloprovincialis* and *O. edulis* shells are exfoliated. The majority of the rarely encountered species appear to be artificially modified or they were collected as empty shells.

The studied mollusk remains originate from diverse archaeological deposits across the site. Most of them belong to fill deposits, some of which are only arbitrarily defined. In general, the mollusk taxa are homogeneously distributed among the deposits. Exceptions include the *M. galloprovincialis*-rich "shell layers" in Troia I deposits of Quadrate D5, the large quantities of *H. trunculus* remains recovered in Quadrate IK8, and the *H. trunculus* and *Patella* spp. accumulations found in Quadrate KL16/17 in the lower Late Bronze Age town.

The valve heights of the bulk of the Bronze Age *C. glaucum* populations deviate between 21.3 and 30.1 mm. The mean size of *C. glaucum* valves changes between periods. Most notably, the Late Bronze Age valves are significantly larger than the Early Bronze Age valves. Age profiles of chronologically grouped assemblages show parallelism with observations on the size profiles. The proportion of older individuals is greater in the Late Bronze Age assemblage than it is in the Early Bronze Age assemblage. The Late Bronze Age valves are also more curved and solid than the Early Bronze Age valves. Correlation between smaller successive chronological units and morphological patterns is, however, less clear.

Similar to modern specimens, the external lines of the *C. glaucum* valves are not uniform in appearance. The breadths of the incremental zones demonstrate high variability. The complete successive gray increments of the individuals coming from the Late Bronze Age deposits are generally broader than their Early Bronze Age counterparts. Meaningful differences have not been observed between the successive white incremental zones of valves coming from different phases of the Bronze Age.

Incremental growth analyses accounting the differences in the demographic structure of the two populations show that the proportion of the individuals gathered during fast growth is larger in the Early Bronze Age assemblage than it is in the Late Bronze Age assemblage. Isotopic analyses of archaeological *C. glaucum* carbonate demonstrate that the differences between the $\delta^{18}O$ and $\delta^{13}C$ content of visually differentiated incremental zones are of seasonal amplitude, thus confirming the observations pertaining to the incremental growth pattern of extant *C. glaucum* populations. Yet, the results of the isotopic analyses do not point to the season of incremental zone formation (i.e., as summer or winter) unambiguously. The results obtained from the incremental analyses of archaeological *C. glaucum* are thus evaluated in light of observations made on extant populations.

4.4. Assemblages of Yenibademli

The studied faunal samples from Yenibademli are divided into two stratigraphic groups: one including the faunal remains yielded from excavation units belonging to the earlier occupational layers (Yenibademli III, IV, V, VI and VII); and, the second consisting of remains derived from later occupational layers (Yenibademli I and II). In this study, the former is tentatively called "Early Yenibademli", and the latter is called "Late Yenibademli." Both groups together are dated to between ca. 2900 and 2600 BC (Hüryılmaz 2001).

4.4.1. Proportion of mollusk remains in faunal assemblages

The number of identified faunal specimens from Yenibademli is 14728 (Table 31). 14036 of these specimens originate from 212 stratified excavation units. The specimens' total weight amounts to 118257.2 g (Table 31). Other than mollusks, the faunal assemblage includes domesticated mammals, wild mammals, wild birds, reptiles, amphibians, and fish. Ca. 14.1 % (NIS)/ 6.7 % (WIS) of the vertebrate remains could not be assigned to any faunal taxa (Table 31).

Figure 55. The proportion of mollusk remains in the faunal assemblages of Yenibademli.

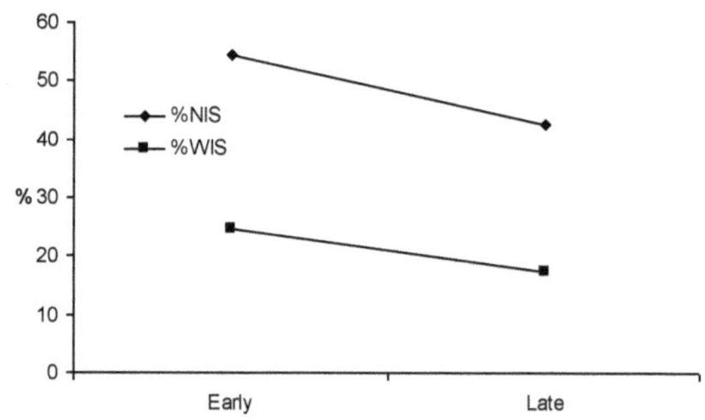

Mollusks remains make up 39.7 % of the NIS and 18.6 % of the WIS of the entire faunal assemblage (Table 31). The ratio of mollusk remains is higher in the faunal samples retrieved from the Early Yenibademli deposits with 54 % in NIS and 24.6 % in WIS, whereas samples from younger phases include 42.6 % in NIS and 17.3 % in WIS of mollusk remains (Table 31 and Figure 55). The share of mollusk remains in the faunal assemblage from Late Yenibademli is replaced by the increasing share of domestic mammals. A larger amount of faunal samples from the different layers of the site need to be analyzed to test whether this change in the proportion of mollusk remains is archaeologically significant.

4.4.2. Archaeomalacological assemblages

At least 12 gastropod and 12 bivalve species are represented in Yenibademli (Table 32). All species are marine organisms. The freshwater clam, *Unio* spp. is absent in the studied assemblages.

Figure 56. Changes in the relative abundance of mollusk species represented in Yenibademli (NIS).

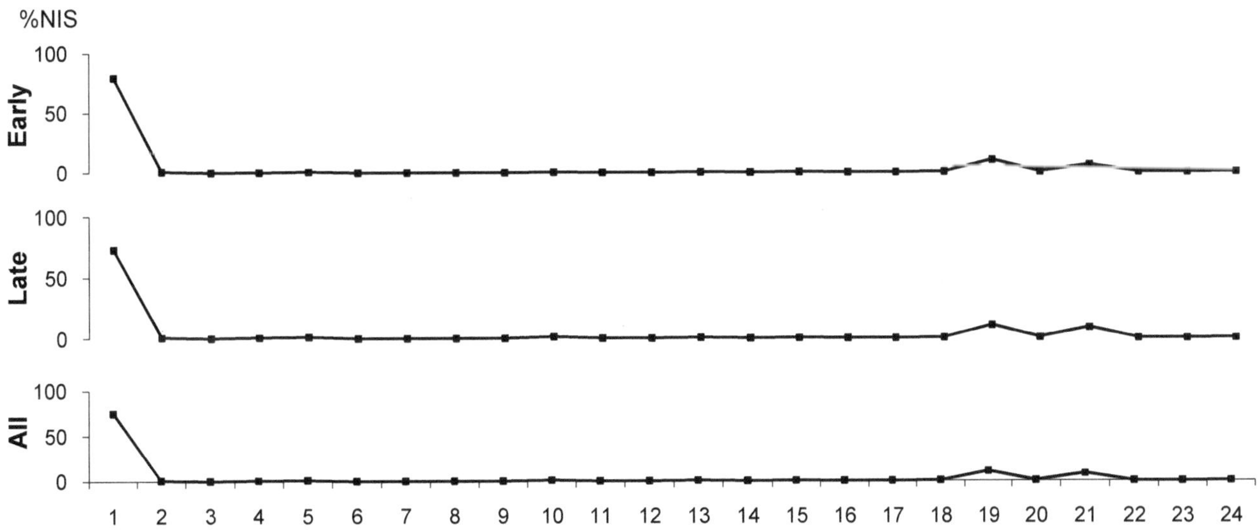

1. *Patella caerulea*
2. *Patella rustica*
3. *Haliotis* spp.
4. Monodontinae
5. *Cerithium* spp.
6. Cypraeidae
7. *Phalium granulatum*
8. *Charonia tritonis*
9. *Bolinus brandaris*
10. *Hexaplex trunculus*
11. *Fasciolaria lignaria*
12. *Conus mediterraneus*
13. *Arca noae*
14. Glycymerididae
15. *Mytilus galloprovincialis*
16. *Pinna* spp.
17. Pectenidae
18. *Spondylus gaederopus*
19. *Ostrea edulis*
20. *Acanthocardium tuberculatum*
21. *Cerastoderma glaucum*
22. *Solen marginatus*
23. *Donax trunculus*
24. *Tapes decussatus*

The most frequent species in Yenibademli is *P. caerulea*. The remains of this species make ca. 75.3 % of the NIS and 43.2 % of the WIS of the mollusk assemblage (Table 32, Figure 56). Limpets are represented with one more species, *P. rustica*, whose remains are much fewer than those of *P. caerulea*, making up less than 1 % of the NIS and about 0.3 % of the WIS with 45 specimens in total. *O. edulis* is the second most frequently represented species after *P. caerulea*. *O. edulis* remains contribute 621 specimens to the assemblage, comprising 10.6 % of the NIS and 34.6 % of the WIS of the total assemblage (Table 32, Figure 56). The only other species whose remains have any significant contribution to the mollusk assemblage at the site is *C. glaucum*. *C. glaucum* remains constitute 8 % of the NIS and 9.9 % of the WIS. Remains of Monodontinae, *Cerithium* spp., *H. trunculus*, *A. noae*, *S. gaederopus*, and *A. tuberculatum* are fairly common (i.e., more than 20 specimens) (Table 32, Figure 56).

Only minor differences exist between the mollusk assemblages of Early and Late Yenibademli. The proportion of *P. caerulea* decreases slightly in the later phase, whereas the proportions of *C. glaucum* and *H. trunculus* vaguely increase (Table 32, Figure 56). Species, such as *Haliotis* spp., *F. lignaria, C. mediterraneus*, *Pinna* spp., Pectenidae, which are represented with single specimens in the assemblage of the later phases are absent from the spectrum of the earlier phases. This is probably only due to the smaller size of the mollusk assemblage coming from the earlier phases. At the time of writing, the archaeologists working on the stratigraphy of the site were not able to assign the single specimen of *Charonia tritonis* to any of the six phases of occupation.

4.4.3. Taphonomic description of mollusk specimens

Because most forms of post-depositional fragmentation are species-specific, the taphonomic condition of the mollusk assemblage from Yenibademli does not differ largely from that of Troia (cf. Section 4.2.4). The similarity of the taphonomy of *P. caerulea* specimens in Yenibademli with those of Troia nicely illustrates this point (Figure 57, cf.

Figure 34a). Below, an account of those conditions that differ from Troia will be given.

One of the *C. mediterraneus* specimens is almost complete, but appears to have been abraded by wave action (Figure 57b). It also bears a hole on the apex. This hole was probably made by a predatory gastropod and subsequently smoothened by wave action. The shell was already empty when it was brought to the settlement.

One *L. lurida* (57c) and one *E. spurca* specimen (Figure 57d) were found almost intact. Both specimens bear holes near the apex. The similarity between the location and appearance of these holes suggests that they were made artificially.

At least two of the *A. noae* specimens have been abraded and encrusted by marine organisms.

Figure 57. Typical taphonomy of various mollusk specimens from Yenibademli.

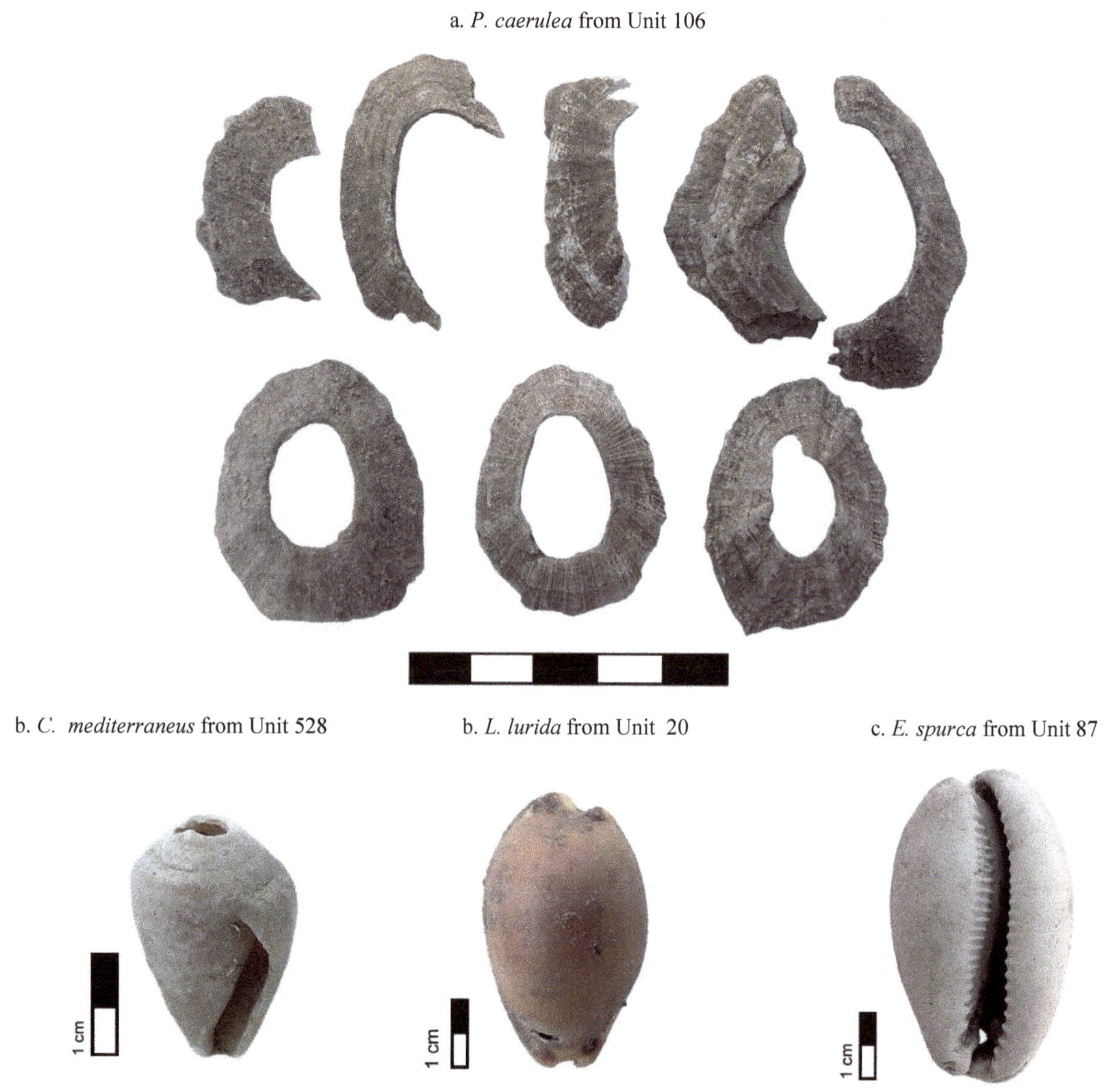

Two specimens of Glycymerididae shells with slightly smoothened margins have been abraded by natural agents (Figure 58a). One Glycymerididae specimen consists solely of an abraded ventral margin (Figure 58b). Another one has lost its color completely and bears a hole near the umbo (Figure 58c).

Seven specimens of *S. gaederopous* have corroded shells. In some cases the shell surface is pitted, in others the shell was smoothed completely by wave and sand action (Figure 59a). These examples were collected dead near or on the coast.

Some *O. edulis* specimens are found still cemented to the stones they dwelled upon (Figure 59b). One *O. edulis* specimen appears to have been collected when already dead, as encrustation on the inner shell surface indicates.

Three specimens of *A. tuberculatum* bear holes on their umbones. At least two of these appear to have been artificially perforated.

Specimens of *C. glaucum* are largely intact, a few with both valves still attached together. A peculiar characteristic of some of the *C. glaucum* valves of Yenibademli is the abraded shell surfaces on the posterior margins and umbones (Figure 59c). This form of abrasion can occur on empty shells that are exposed to post-mortem abrasion, though it can also be observed on bivalves that live in coarse-sandy substrates (per. obv., Figure 60). The smooth margins and umbones may also be of human production, pertaining to burnishing or polishing on a softer material like clay or leather. Altogether, 46 specimens were found to be abraded in this manner. Two specimens bear irregular smooth-edged holes on their umbones. One example bears a regular and sharp-edged hole on the same part of the valve (Figure 59d). This hole was probably made by scrubbing the umbo against a hard surface.

Figure 58. Glycymerididae specimens from Yenibademli.

a. Outer (left) and inner (right) surface of a Glycymerididae valve from Unit 494.

b. Water-worn Glycymerididae valve from Unit 003

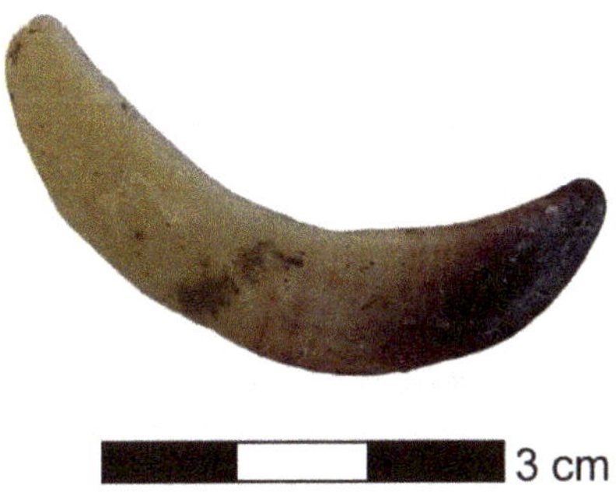

c. Water-worn Glycymerididae valve from Unit 536.

4.4.4. Archaeological context of mollusk remains

The studied sample of the faunal assemblage from Yenibademli represents a randomly chosen collection of excavation units. The excavated remains in Yenibademli mainly consist of fairly uniform rectangular buildings of domestic character (Hüryılmaz 2001). The studied sample of faunal remains was yielded from fill deposits inside and outside of these buildings in Quadrates H7, G8, H8, E9, H9, E10, and H10. Two units (Units 137 and 138) were extracted from inside a hearth in Quadrate G8. Eight others (Units 77, 108, 111, 513, 515, 552, 571, and 572) come from the vicinity of outdoor hearths. All faunal units contain mollusk remains, with the exception of four units which include very few faunal specimens. There are no notable differences among individual faunal units that belong to separate archaeological contexts. The faunal composition of each unit is fairly uniform in terms of species spectra and proportions.

Figure 59. Taphonomy of various mollusk specimens from Yenibademli.

a. A *S. gaederopous* valve from Unit 554

b. *O. edulis* specimens found attached on rock in Unit 526

c. *C. glaucum* valves with abraded surfaces from Unit 106

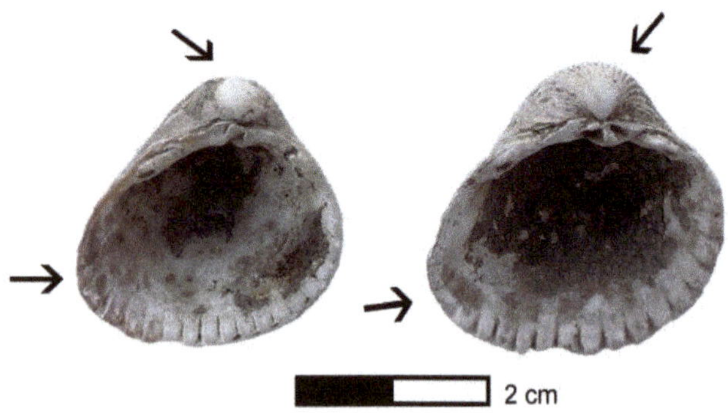

d. Artificially modified *C. glaucum* from Unit 530

Figure 60. Live collected (Beşik Bay, 2004) *A. tuberculatum* specimen with abraded umbo.

4.4.5. Morphology of *P. caerulea* populations

The length, height, and width of *P. caerulea* specimens from the early and late phases of Yenibademli were compared. Specimens coming from earlier layers of the site are significantly longer (p = 0.0001), higher (p = 0.0005), and wider (p = 0.0001) than specimens from later layers of the site (Table 33, Figures 61 and 62). The height/length and length/width relationships of specimens from earlier and later layers were also compared. There are no significant differences for either case (p = 0.2623 for height/length, p = 0.1876 for length/width) (Figure 61). The mean shell length, width, and height of the archaeological *P. caerulea* populations from Yenibademli are similar to the corresponding morphometric attributes shown for modern populations along the coast of the Saros Bay (Öztürk and Egemen 1999: 517; Table 1, cf. Table 4 this study).

Figure 61. Morphology of the *P. caerulea* populations in Yenibademli (Distributions).

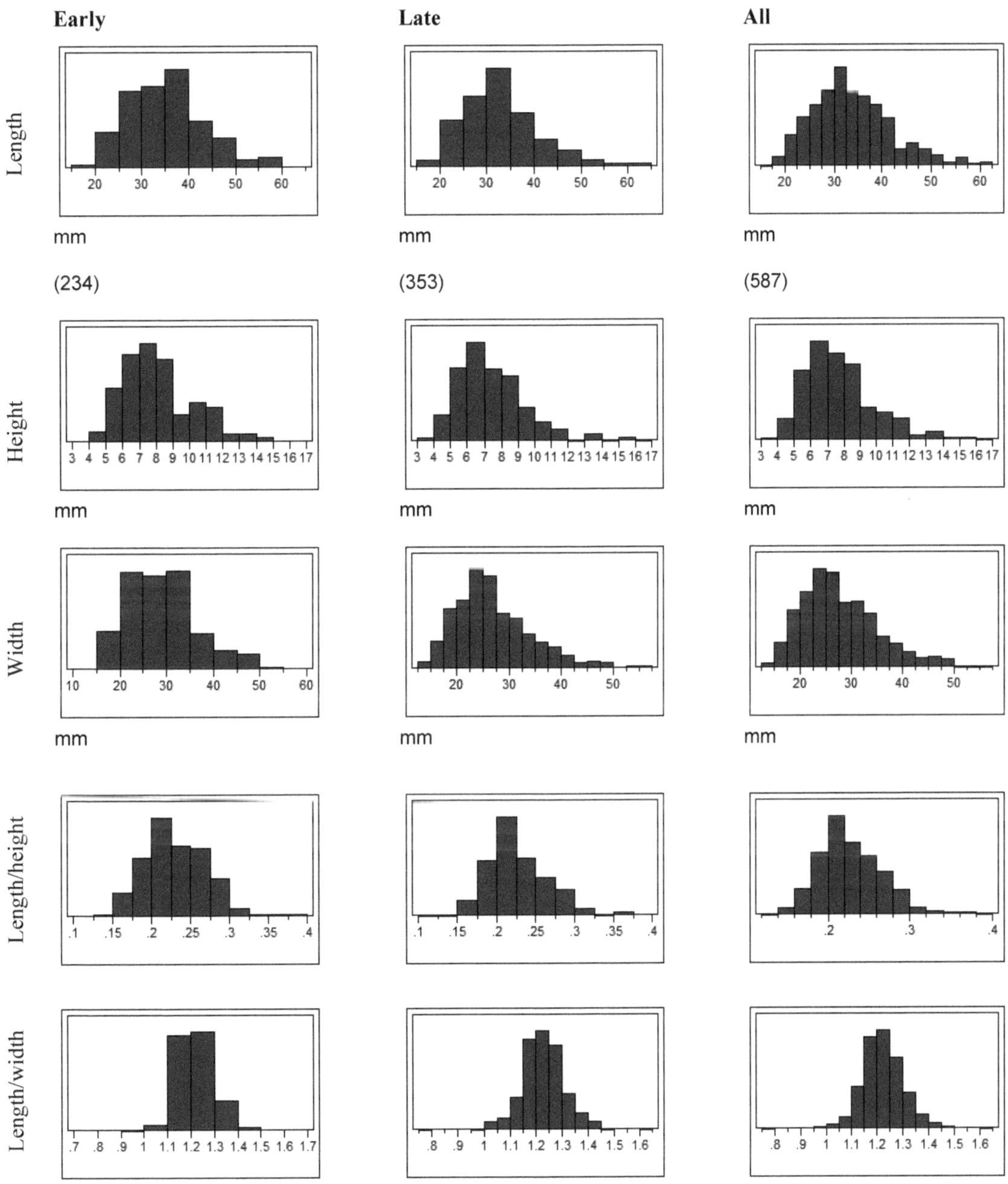

(234) = No. of individuals.

4.4.6. Morphology of *C. glaucum* populations

The valve heights of the *C. glaucum* specimens from Yenibademli range between 15.7 mm and 51.2 mm, with the bulk of the remains clustering between 23.8 mm and 32.4 mm (Table 34, Figure 63). The mean valve height of *C. glaucum* specimens does not differ between the earlier and later layers of the site ($p = 0.6191$) (Figure 64). The solidity and curve of the specimens are also not significantly different between the earlier and later layers ($p = 0.2135$ for solidity and $p = 0.4821$ for curve) (Figure 64).

4.4.7. Summary

Stratified excavation units of Yenibademli are tentatively divided into two chronological groups, as early and late phases. Mollusk remains make up 39.7 % of the NIS and 18.6 % of the WIS of the entire faunal assemblages. The contribution of the mollusk remains to the faunal assemblages is slightly larger in Early Yenibademli than it is in the later phases. The most abundant mollusk species are *P. caerulea*, *O. edulis*, and *C. glaucum*, in order of frequency. *P. caerulea* become slightly less abundant in the assemblages from Late Yenibademli, whereas the proportion of *C. glaucum* and *H. trunculus* remains increase.

Figure 62. Morphology of the *P. caerulea* populations in Yenibademli by occupational phase (ANOVA).

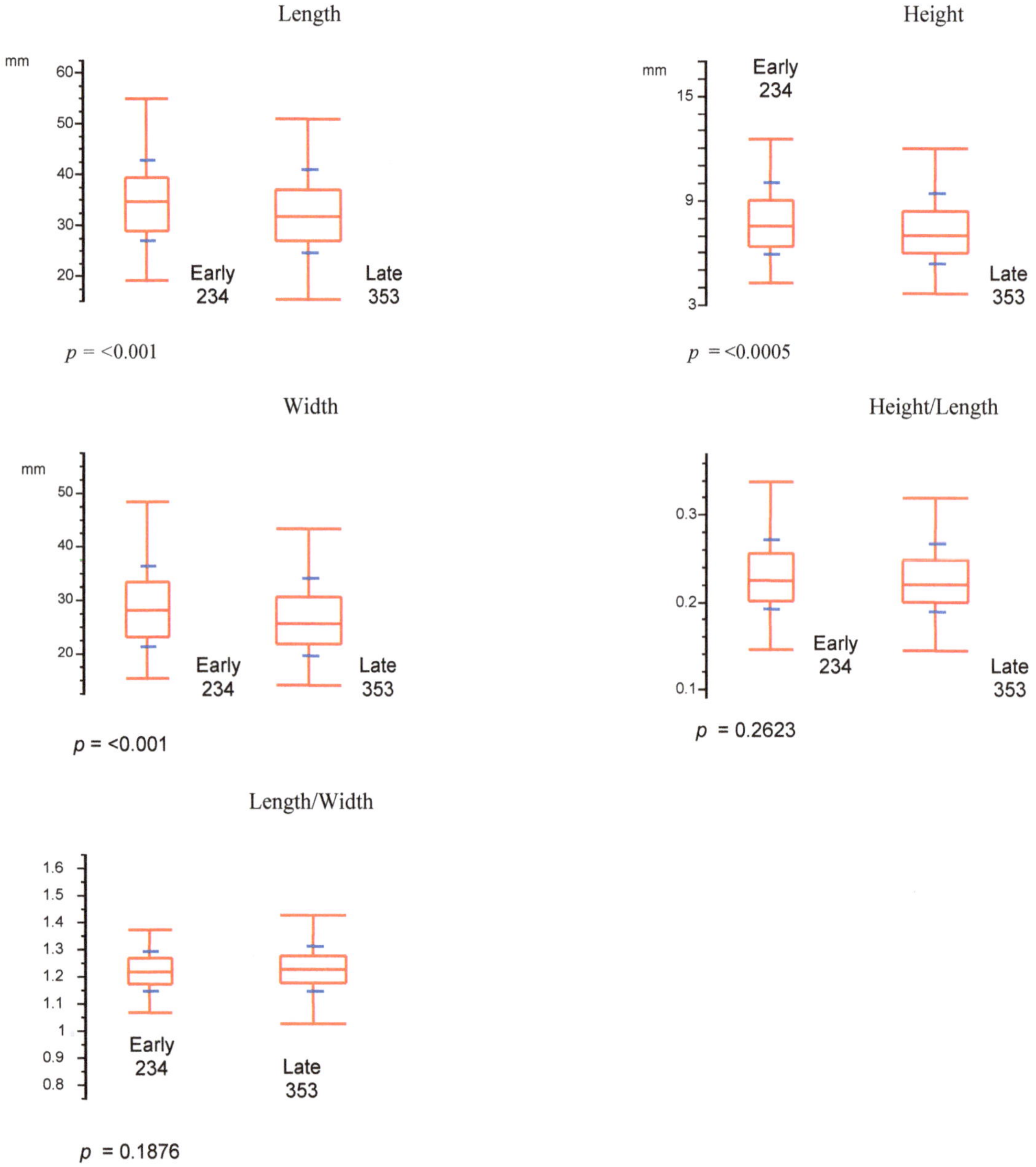

Figure 63. Morphology of the *C. glaucum* populations in Yenibademli (Distributions).

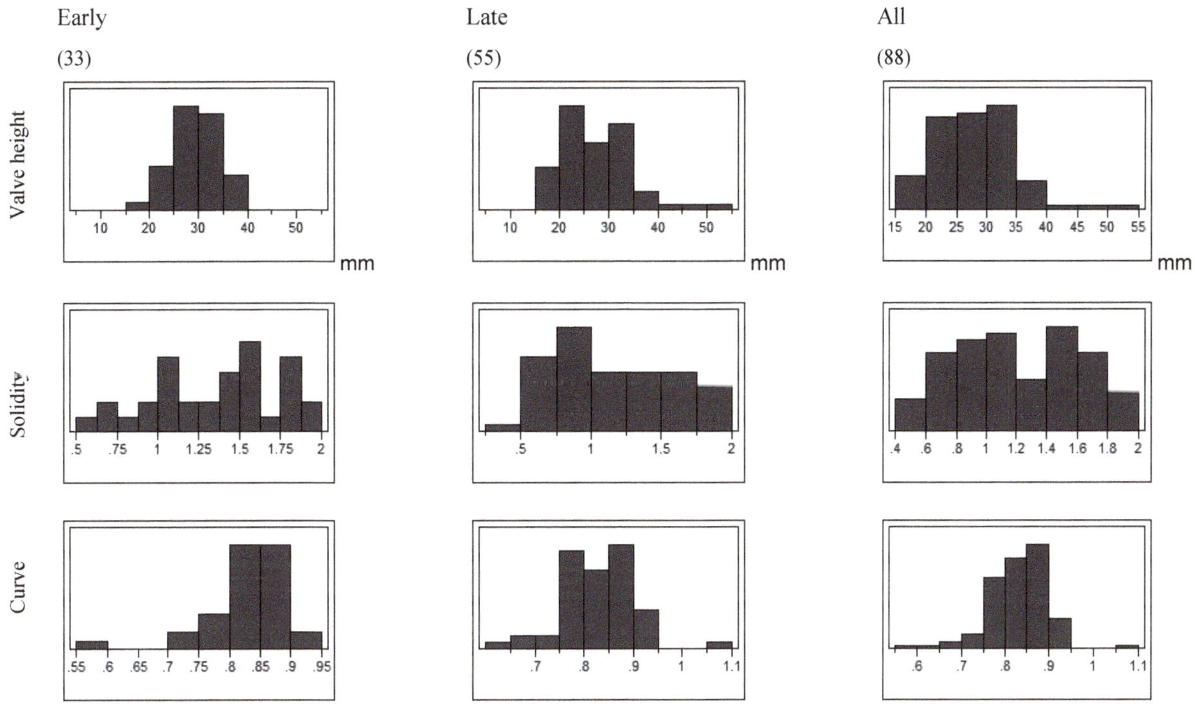

Figure 64. Morphology of the *C. glaucum* populations in Yenibademli by chronological period (ANOVA).

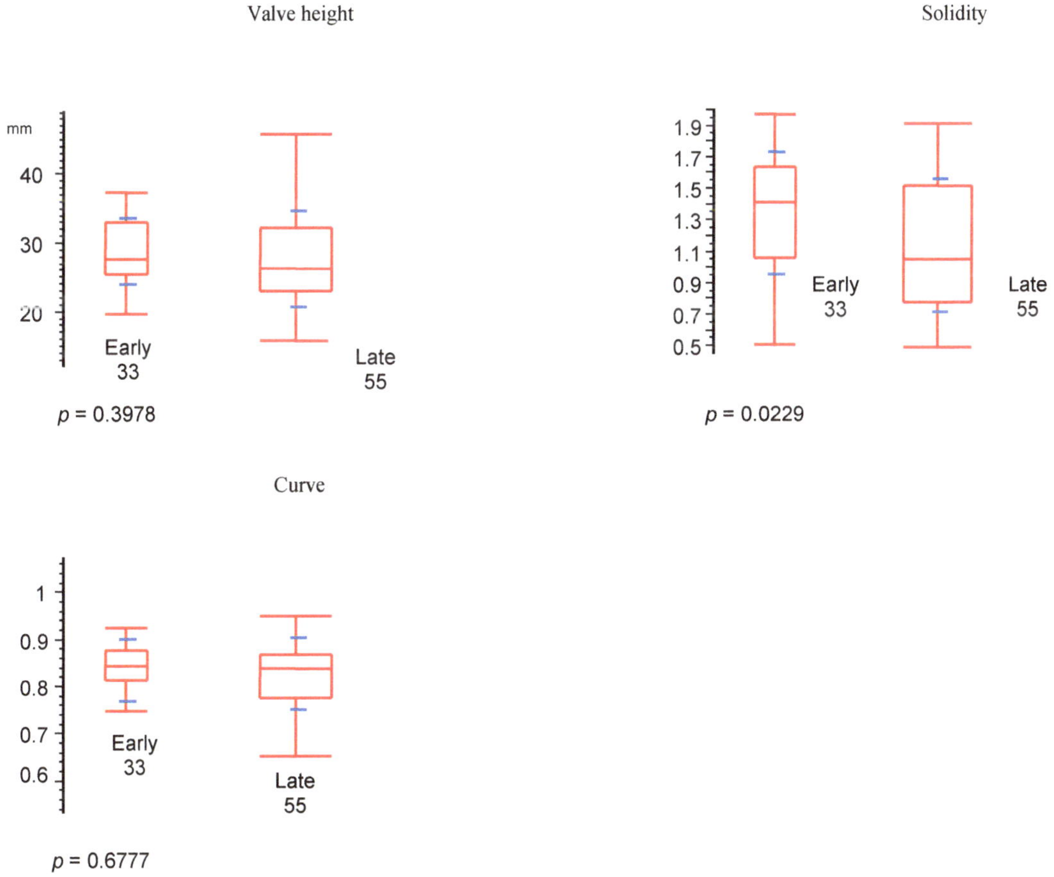

The taphonomy of *P. caerulea* and *O. edulis* shells is very similar to that observed for the specimens of these species in Troia. Species which are represented only by a small amount of specimens appear to be collected as empty shells. Some of them were subsequently modified. Abraded posterior margins and umbones of 46 specimens of *C. glaucum* is a peculiar feature of this find group.

Almost all excavation units contain mollusk remains. There are no notable differences between the mollusk samples from different archaeological deposits.

While *P. caerulea* specimens originating from Early Yenibademli are significantly longer, higher, and wider than the specimens from the later phase of occupation, no significant differences are observed between the height/length and length/width relationships of the specimens coming from different phases. The morphological properties of the archaeological *P. caerulea* from Yenibademli demonstrate similarities with the extant populations in the Saros Bay. The valve height of the bulk of the *C. glaucum* population from Yenibademli measures between 23.8 and 32.4 mm. There are no significant differences between the mean valve heights of the *C. glaucum* populations belonging to different layers of the settlement.

4.5. Assemblages of Ulucak

Archaeozoological analyses of faunal remains dating from the Chalcolithic Age and the Early Bronze Age, excavated between 1995 and 2003, have been completed. Although all mollusk remains retrieved from this excavation from its beginning until 2003 have been studied, only those remains from the Early Bronze Age are discussed here in detail. A short account of the mollusk assemblage of the Chalcolithic levels is also provided, in an attempt to underline the characteristics of the mollusk assemblage from the Early Bronze Age layer of the site by contrasting them with the archaeomalacological patterns of the period prior to it.

4.5.1. Proportion of mollusk remains in faunal assemblages

Altogether 5507 faunal specimens, weighing 48667.3 g, have been studied (Table 35). They originate from 49 different excavation units. Twenty-one units date from the Chalcolithic Age, including 3183 specimens weighing 26,193 g. Twenty-eight units belonging to the Early Bronze Age levels of the site produced 2324 specimens weighing 22,474.7 g. The assemblage includes domestic and wild mammals, wild birds, reptiles, amphibians, and fish. The preservation state of the remains allowed 71.4 % of the NIS to be identified to the family, genus or species level. Mollusk remains amount to 13.2 % of the NIS and 3.4 % of the WIS of the entire faunal assemblage (Table 35).

There is a significant difference in the amount and proportion of the mollusk remains retrieved from the Chalcolithic and the Early Bronze Age layers of the site. Mollusks are represented by 48 specimens in the Chalcolithic layers, making up only 1.5 % in NIS and 0.5 % in WIS of this older assemblage from Ulucak. Early Bronze Age levels, on the other hand, yielded 681 mollusk specimens comprising 29.3 % in NIS and 6.8 % in WIS of the faunal assemblage dating to this period (Table 35, Figure 65).

Figure 65. Proportion of the mollusk remains in the faunal assemblages of Ulucak.

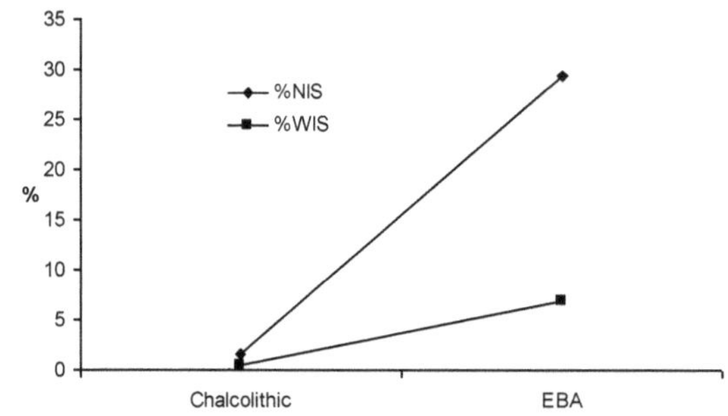

4.5.2. Archaeomalacological assemblages

The aquatic mollusk spectrum of Ulucak is represented by at least 11 different taxa (Table 36). Eight of these are present in the assemblage dating to the Chalcolithic Age. *C. glaucum* is the most commonly represented species with 37 specimens making up 77.1 % of the mollusk assemblage in NIS. Other than the two specimens of Monodontinae and four specimens of *A. noae*, all taxa are represented by single specimens. *B. brandaris*, *C. rustica*, and *Unio* spp. are present only in the Chalcolithic assemblage (Table 36, Figure 66).

C. glaucum dominates the Early Bronze Age assemblage as well. With 668 specimens weighing about 1416.1 g, *C. glaucum* makes up 98.1 % of the NIS and 92.9 % of the WIS (Table 36, Figure 66). *A. noae* is represented with six specimens and *O. edulis* is represented with two specimens, while all other taxa are represented extremely rarely, each with single specimens. *Cerithium* spp., *C. tritonis* and *S. marginatus* appear only in the Early Bronze Age assemblage.

There are no important differences in the proportions of the represented taxa between the Chalcolithic and Early Bronze Age mollusk assemblages: *C. glaucum* predominates in the assemblages of both periods and *A. noae* is the only other species worth of mentioning in terms of abundance (Table 36).

4.5.3. Taphonomic description of mollusk specimens

The single specimen of *C. tritonis* is represented by an aperture lip. *B. brandaris* is represented solely with a body whorl. One of the *C. rustica* specimens is holed on the apex, probably by natural agents, while the other is a complete example (Figure 67a). Single intact valves make up 75 % of the *A. noae* remains. *Pinna* spp. is represented by a fragment of the valve body. One *O. edulis* specimen is a complete shell with two valves found together, albeit with splintered off shell margins. *C. glaucum* remains are very well preserved, occasionally with specimens represented by both valves still clinging to each other at the umbo (Figure 67b). Only 43 specimens consist of less than ca. 30 % of the valve.

Figure 66. Changes in the relative abundance of mollusk species represented in Ulucak (NIS).

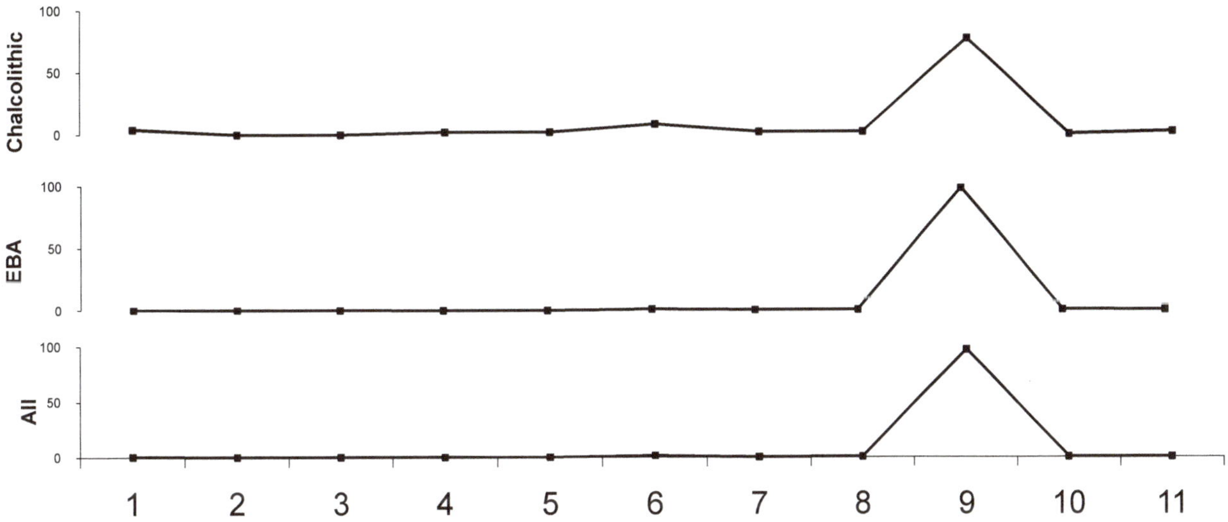

1.	Monodontinae	5.	*Columbella rustica*	9.	*Cerastoderma glaucum*
2.	*Cerithium* spp.	6.	*Arca noae*	10.	*Solen marginatus*
3.	*Charonia tritonis*	7.	*Pinna* spp.	11.	*Unio* spp.
4.	*Bolinus brandaris*	8.	*Ostrea edulis*		

Figure 67. A *C. rustica* and a *C. glaucum* specimen from Ulucak.

a. *C. rustica* (Context 530)

b. *C. glaucum* (Unit 523)

4.5.4. Archaeological context of mollusk remains

All faunal units that were excavated between 1995 and 2003 and securely attributed to either the Chalcolithic or the Early Bronze Age contexts have been analyzed.

Mollusk remains were found mainly as scattered finds in the archaeological contexts of the Chalcolithic layers. Most of the *C. glaucum* shells dating to the Early Bronze Age were recovered from five contexts (CSP, CUK, COV, CTG, and CUV) in different parts of Quadrate O13. Context CSP consisted almost entirely of *C. glaucum* specimens, altogether 582 in NIS (Figure 68).

Figure 68. Context CSP in Ulucak (Courtesy of the Ulucak Höyük Excavations).

4.5.5. Morphology of the *C. glaucum* assemblages

The valve heights of *C. glaucum* specimens from the Chalcolithic and the Early Bronze Age deposits at Ulucak range between 15.4 mm and 36.4 mm (Table 37, Figure 69). The bulk of the individuals have valve heights between 19.3 mm and 23.8 mm (Figure 69). The amount of Chalcolithic *C. glaucum* specimens (only 7) is not sufficient to draw a normal valve height distribution. The valve height distribution of the Early Bronze Age specimens is skewed towards smaller specimens (Figure 69). The valve height index of *C. glaucum* from the Chalcolithic layers of the site is not significantly different from those of specimens coming from the Early Bronze Age layers (Figure 70; $p = 0.543$). Neither the solidity nor the curve differ between the two periods ($p = 0.0112$ for solidity and $p = 0.3821$ for curve; Table 37, Figure 70). This situation results from the imbalance between the sizes of these chronologically defined samples.

4.5.6. Summary

The faunal assemblage of the Chalcolithic and Early Bronze Age Ulucak represents the smallest primary assemblage presented in this study. The quantity of the assemblage reflects the limited amount of archaeological deposits recovered from these two chronological periods. Mollusk remains constitute a very small portion of the faunal remains yielded from the Chalcolithic deposits, whereas the mollusk remains found in the Early Bronze Age layers make up more than one fourth (in NIS) of the faunal remains dating to this period. *C. glaucum* is the most abundant species in both periods. Bulk of the Early Bronze Age *C. glaucum* remains comes from a single excavation unit. Other mollusk remains were found scattered. Most of them are represented by single specimens. *C. glaucum* shells are well-preserved; intact specimens with both valves were occasionally found. The valve heights of the *C. glaucum* shells cluster between 19.3 mm and 23.8 mm. The distribution of the valve height values demonstrates the predominant presence of smaller individuals among the *C. glaucum* assemblage.

4.6. Archaeomalacological data from other sites in the Aegean

Published archaeomalacological data from sites located in the Aegean basin shows that mollusk remains constitute a prominent feature of the archaeological deposits, indicating the recurrent exploitation of mollusk populations across the basin. While there is great disparity in the quality and quantity of published information on the mollusk remains retrieved from the sites in the Aegean, the available quantified data is commonly in the form of number of identified mollusk specimens (NIS) counts, allowing for deductions of the most abundant mollusk species recorded from the sites. Archaeomalacological accounts rarely include information on the weight of identified specimens (WIS) or taphonomic descriptions. For some archaeological sites, the NIS of mollusks is reported along with the detailed accounts of other faunal remains. Although some of these are very useful in assaying the proportion of mollusk remains within the total faunal assemblages from these sites, such quantitative data is difficult to interpret. Differential excavation, identification, and registration techniques, as well as taphonomic conditions are all effective in shaping the published quantitative data. These effects are largely unpredictable.

Figure 69. Morphology of the *C. glaucum* specimens in Ulucak by chronological period (Distributions).

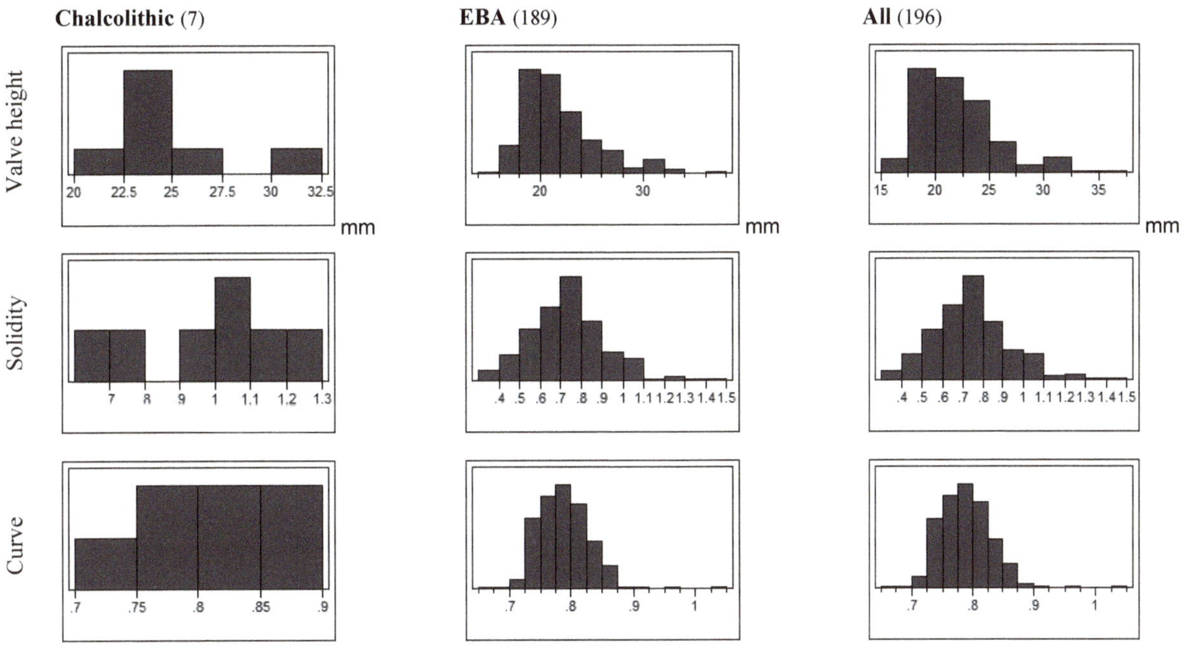

Figure 70. Comparison of the morphology of the *C. glaucum* specimens in Ulucak by chronological period (ANOVA).

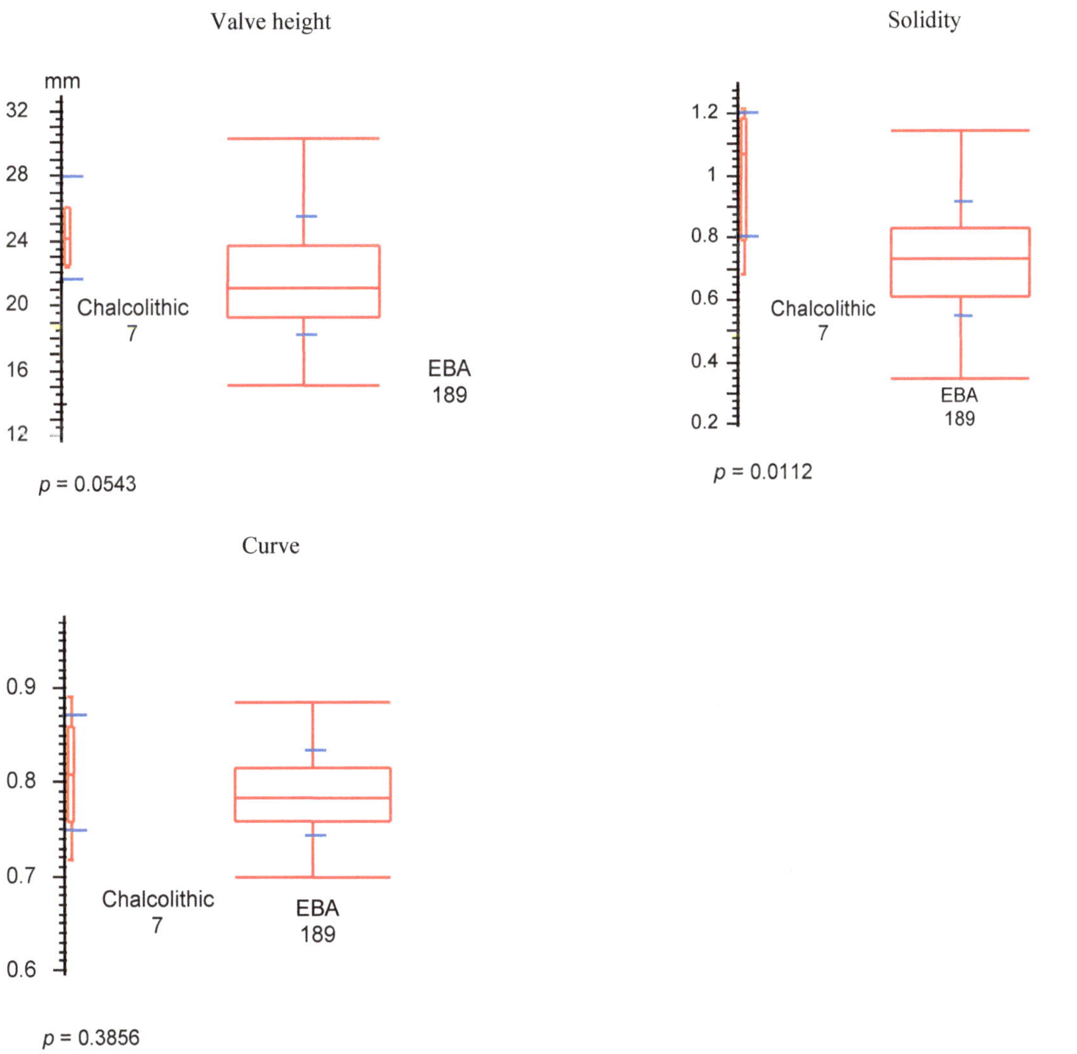

4.6.1. Species spectra

Table 11 summarizes the most frequently represented mollusk species from 28 archaeological sites in the Aegean Basin. Conventional and absolute dates of the mollusk assemblages and generalized geomorphological background to the sites are also provided. Sites in this table include those for which information on the relative abundance of mollusk species is available. Sites, for which only a list of mollusk species has been presented, such as Thermi (Lamb 1936: 216-217), have been excluded from this review list.

The species spectra of the mollusk assemblages reported from the archaeological sites in the Aegean consist of species which are already familiar from the primary sites discussed in the present study. The most common species represent eleven different genera, all of which are represented in the archaeomalacological record of Troia and Yenibademli (See Sections 4.3.2 and 4.4.2). The relative abundances of species differ between the sites.

One of the earliest mollusk assemblages of the Aegean prehistory comes from the 9400 year old deposits of the Franchthi Cave in Greece. This assemblage is dominated by *Patella* spp. and *Monodonta* spp. remains (Shackleton 1988a, 1988b). As the eustatic sea-level rises during the Late Glacial aftermath and seawater inundates the inner reaches of the Franchthi embayment (Shackleton 1988a: 37-41, Figures 9-12), *C. neritea* and *C. vulgatum* start dominating the mollusk assemblages (ca. 8500 BP in the Mesolithic and Neolithic periods). Another shell-bearing Mesolithic site of the Aegean is the Cyclope Cave of the Youra Island (Sampson 1998). Youra Island was connected to the mainland at the time of Mesolithic occupation. The Cyclope Cave is located on a rocky coast and the most common mollusks found in the deposits of the cave belong to *Patella* spp. and *Monodonta* spp.

The Neolithic and Chalcolithic mollusk exploitation in the Aegean is represented by numerous sites, including Hocaçeşme (located on the Turkish side of the Meriç River), Makriyalos, Nea Nikomedia, and Magula Pevkakia in the North Aegean, the Neolithic layers of Lerna in the Argolid, Saliagos in the Cyclades, and Kumtepe, Sivritepe, and Alacalıgöl in the Troad (Andreasen 1998, Boessneck 1986, Coy 1977, Falkner 1975, Gejvall 1969, Higgs and Clench 1962 and per. com. M. Uerpmann). Among these sites, Saliagos and Kephala represent settlements located close to rocky shores. The mollusk assemblages of these two sites are dominated by supralittoral-mesolittoral rocky-shore gastropods like *Patella* spp. Sites that were located near lagoons or estuaries, such as Hocaçeşme, Lerna, M. Pevkakia, Makriyalos, and Nea Nikomedia contain abundant *C. glaucum* remains. At others, such as Kumtepe and Sivritepe –sites that are on the shore of small bays but not nearby substantial freshwater sources– *O. edulis* and *M. galloprovincialis* are the prevailing species (Boessneck 1986, Gejvall 1938, and per. com. M. Uerpmann). A noteworthy feature of Sivritepe is the proportion of mollusk remains in the faunal assemblage of the site, which consists of 5366 mollusk specimens and 167 animal bones (Boessneck 1986). Similarly, abundant mollusk remains were the most conspicuous feature of Alacalıgöl, another pre-Bronze Age site (5th millennium BC) (Gabriel et al. 2004). Similar to the archaeomalacological species spectra of Kumtepe (A and B) and Sivritepe, the mollusk remains at this site consist mainly of *O. edulis* specimens (per. obv.).

In the Early Bronze Age, remains of *C. glaucum* become a staple feature of the archaeological deposits of the Aegean. It is the most abundant mollusk species in settlements located near small bays, estuaries or lagoons, such as Beşik-Yassıtepe, Eleusis, Franchthi, Kumtepe, Lerna, and Limantepe (Cosmopoulos et al. 2003, Gejvall 1969, Reese 2006, Shackleton 1988, Sorrentino 1997, von den Driesch 1999, and per. obv. at Kumtepe). At those sites, which are located on promontories overlooking a rocky shore and a lagoon or a small bay, such as Yassıtepe and Poliochni (on Lemnos), *Patella* spp. constitutes a large portion of the archaeomalacological assemblages (Sorrentino 1997, von den Driesch 1999). Additionally, *A. noae, O. edulis, S. gaederopous, T. decussatus* make their appearances in relatively larger amounts during the Early Bronze Age. Another observation that needs to be noted is the relatively more common presence of *A. noae* and *S. gaederopous* remains in the archaeological deposits of the southern Aegean (Cosmopoulos et al. 2003, Gejvall 1969).

The ubiquitous presence of *C. glaucum* remains at sites near lagoons and estuaries is a recurrent phenomenon. Also in the Middle Bronze Age, *C. Glaucum* remains are the most common elements of the mollusk remains in Ayios Mamas, Lerna, Magoula Pavlina, and Panaztepe. *A. naoae* and *S. gaederopous* continue to be common species in the mollusk assemblages of southerly sites such as Lerna and Eleusis (Cosmopoulous et al. 2003, Gejvall 1969). A perplexing feature of the Lerna assemblage is the complete absence of *O. edulis* remains, despite an outstanding abundance of *S. gaederopous* shells (Gejvall 1969: 7, Table 4). While the archaeological remains of *O. edulis* and *S. gaederopous* shells may appear undistinguishable to the untrained examiner, the otherwise consistent and precise archaeomalacological species list provided by Gejvall (1969) does not imply a possible identification failure. In Palaikastro, the most abundant species are *Patella* spp. (Reese 1987a). This settlement is located in close proximity to a rocky promontory.

Kastanas represents the only archaeological site in the Aegean where freshwater clams *Unio* spp., constitute a very large portion of the archaeomalacological remains. Kastanas is located on the Axios River, which slowly silted the coastal embayment where it met the Aegean Sea during the course of Middle and Late Holocene. In the Middle Bronze Age, *Unio* spp. is the most abundant mollusks in the mollusk assemblage of Kastanas (Becker 1986: 232-234, Abbildung 78).

The Middle Bronze Age marks the first appearance of crushed *H. trunculus* remains in the archaeomalacological assemblages of the Aegean. They are found in large amounts in Palaikastro, at the northeastern coast of Crete (Reese

1987a). They are also present in crushed form in the Middle Bronze Age assemblages of Eleusis near Athens, and Ayios Mamas in Thessaly (Becker 1996, 2001, Cosmopoulous et al. 2003). The small quantity of crushed *H. trunculus* remains in Ayios Mamas led Becker (2001) to interpret these finds as remnants of small-scale domestic purple dye production. Their relative abundance is much more significant in Eleusis (Cosmopoulos et al. 2003).

The mollusk assemblages of Eleusis and Lerna in the southern Aegean, Kastanas in the northern Aegean, and Skala Sotiros on the Thassos Island represent Late Bronze Age mollusk consumption. The proportion of *H. trunculus* remains in Eleusis and Lerna increases during this period (Cosmopoulos et al. 2003, Gejvall 1969). In Kastanas, *C. glaucum* becomes the most common species in the deposits dated to the time period between 1600 and 1400 BC (Becker 1986: 233, Abbildung 78). The *Unio* spp. becomes once again predominant in the archaeomalacological assemblages of Kastanas during the subsequent transitional period into the Iron Age (1300-1000 BC).

A few Post-Bronze Age assemblages are also included in the presentation of mollusk data from other sites in the Aegean. Not much change seems to occur in the mollusk assemblages. Similar to the situation in the Early Bronze Age, *C. glaucum, Patella* spp., and *O. edulis* are the most common species in the Archaic-Hellenistic deposits of Beşik-Yassıtepe. In Eleusis, the mollusk remains of the Classical and Roman periods are in insignificant amounts, but crushed *H. trunculus* shells and *A. Noae* shells are still present. In Kalabaktepe near Miletos, the Archaic deposits contain large amounts of *C. glaucum* shells and some *O. edulis* remains (Peters 1992). The mollusk procurement ground of Kalabaktepe is the nearby estuary of the Menderes River. In New Halos, located only ca. 1 km from the Middle Bronze Age site of Magoula Pavlina, the most common species present in the mollusk assemblages are the same as in the Middle Bronze Age site of M. Pavlina. These are namely *C. glaucum, O. edulis, and S. gaederopous.*

At least two prehistoric inland sites –Pradeisos, located on the River Nestor near Thessalonica in mainland Greece; and, Baklatepe in Anatolia– yielded large amounts of *C. glaucum* remains (Reese 1987c, Reese 2006). Baklatepe is today located ca. 33 km up the Büyük Menderes river basin.

4.6.2. Morphometric, incremental, and isotopic data

Morphometric and isotopic data on the archaeomalacological assemblages of the Aegean is scarce. Data on the incremental growth of any archaeological mollusk population is not in the literature.

One of the most descriptive accounts of archaeological mollusk populations in the Aegean has been provided for Kastanas (Becker 1986: 230-236). Becker measured the valve heights and valve lengths of *Unio* spp. and *C. glaucum* specimens from Kastanas layers 1-14a (1300-200 BC). The definitions of these measurements have not been illustrated or otherwise provided in this publication. Becker recognized two forms of *C. glaucum* shell: curved ("geschnäbelten Schalen" or "Normalform") and long ("lange Form") (Becker 1986: 231). The morphometric measurements taken on both the long and normal ("Normalform") shells of *C. glaucum* were illustrated (Becker 1986: Abbildung 79b). The valve heights of the "long" shells range from ca. 29 mm to 36 mm. The same measurement on the "normal" shells ranges from ca. 26 mm to 30 mm. The valve lengths of the "long" shells vary between ca. 35 and 42 mm, whereas the valve lengths of the "normal" shells vary between ca. 29 and 33 mm. The mean difference between the valve heights and valve lengths is larger in the "long" *C. glaucum* shells. Both the valve heights and lengths of the "long" and "normal" forms of *C. glaucum* shells decrease with time (Becker 1986: 233-234, Abbildung 79b). The reduction in the shell size of *C. glaucum* in the Iron Age coincides with the increase in the contribution of *C. glaucum* shells to the mollusk assemblage of the site (over 90 %, Becker 1986: 234). Becker interprets this size depletion as an indicator of either environmental deterioration or a change in the mode of shellfish procurement (Becker 1986:234). Size decline was also observed in the valve heights and lengths of the *Unio* spp. specimens recovered throughout the stratigraphical layers of the site (Becker 1986: 236, Abbildung 79a).

Another morphometric account of archaeological *C. glaucum* shells in an Aegean site comes from the Late Neolithic (5400-4500 BC) layers of Makriyalos in Greece. The Neolithic architecture at the site is characterized by round pit dwellings (Andreasen 1998: 8). The presence of a nearby lagoon at the time of occupation at Makriyalos has been detected by geomorphological studies (Andreasen 1998: 6). Andreasen compared the valve heights of *C. glaucum* shells from two different phases of the site: Phase I, earliest Late Neolithic; and, Phase II, latest Late Neolithic, probably contemporary with "classical" Dimini. The mean height of *C. glaucum* specimens significantly decreases from 22.1 mm in Phase I to 21.3 mm in Phase II (Andreasen 1998: 28, 46, Table 5 and 49, Table 6, $p = <0.0001$). Andreasen attempted to correlate his observations on shell size with changes in the age structure of the Makriyalos population and the seasonality of *C. glaucum* foraging. After examining 85 acetate replicas of sectioned *C. glaucum* valves (1998: 20), Andreasen concluded that the sampled group demonstrated no clear pattern of incremental shell growth (1998: 27).

Prummel (2005) recently published the results of comparative morphometric analyses of mollusk shells from the Middle Bronze Age site at Magoula Pavlina, the Hellenistic town of New Halos and the Hellenistic dwellings at the southeast gate of the same site; all located ca. 2.5 km from each other in Thessaly (Prummel 2005). At all three locations the most abundant species is *C. glaucum*, followed by *S. gaederopous* and *O. edulis.* Although a summarized description of height, length, and width measurements of various gastropods and bivalves are provided in the publication (Prummel 2005: 113-114, Table 3), only the *C. glaucum* assemblage consists of an adequate number of specimens to speak of an archaeological population. Thus only the morphometric analysis of this group will be discussed here. The mean valve height and length of the *C. glaucum* population from the Middle Bronze Age settlement are significantly larger than those from the archaeological deposits recovered at the Hellenistic Southeast Gate Dwellings of New Halos. The size of the *C. glaucum* shells yielded from the Hellenistic town of New Halos is even smaller than those found at the Southeast Gate Dwellings (Prummel 2005: 118, Figure 8, $p = <0.0001$). The valve heights of the Middle Bronze Age specimens range from 17.2 mm to 45.5 mm with a mean of 31.6 mm, whereas the valve heights of the specimens from the Hellenistic town range between 14.6 mm and 45.5 mm, with a mean of 24.9 mm. The mean valve height value obtained from the Southeast Gate Dwellings of New Halos is 27.8 mm (Prummel 2005: 114, Table 3). Prummel (2005: 118-119) relates this analytical result to the optimal conditions that existed in the nearby lagoon during the Middle Bronze Age, which "possibly resulted in a rapid growth of the species." Additionally, she notes that the Middle Bronze Age specimens had "thicker and more stable valves than those found in the two Hellenistic sites" and suggests that the former lagoonal conditions must have been responsible for such shell morphology (Prummel 2005: 119).

A morphometric analysis of a sizable archaeological *Patella* spp. population is provided for the Late Neolithic site of Saliagos in the Cyclades. The specimens' specific identification has not been determined (Shackleton 1968: 129). Shackleton (1968) compared the lengths of *Patella* spp. shells from a pit assemblage and from a "shell concentration" in another area of the excavations. The bulk of the *Patella* spp. specimens from the pit measured more than 25 mm in length, whereas more than half of the specimens measured from the shell concentration were smaller than 25 mm in length (Shackleton 1968a: 129-130, Figure 28). Shackleton (1968a: 129) interpreted the small-sized *Patella* spp. of the shell-concentration representative of "a meal collected from a rather unfavorable spot or perhaps a period of scarcity."

Additionally, Boessneck (1986) published some measurements he took on the archaeomalacological material from Sivritepe. Yet, since no morphometric indices and only a few measurements have been provided, their use is limited. Stable isotopic analysis on the mollusk remains from the Aegean sites has been conducted as a means of determining past coastal environmental parameters and seasonality of mollusk procurement; as well as for resolving the direction of prehistoric long distance exchange routes and the origin of shell artifacts found in inland sites (Deith and Shackleton 1988, Shackleton 1970, Shackleton and Renfrew 1970). Here, the focus will be on the results of stable isotopic analyses of *C. glaucum* remains from Nea Nikomedia and Franchthi, which were conducted for purposes of palaeoenvironmental and palaeoeconomic reconstruction.

In Nea Nikomedia, isotopic analyses determined the $\delta\ ^{18}O$ and $\delta\ ^{13}C$ compositions of carbonate samples taken from the margins of thirteen *C. glaucum* specimens (Shackleton 1970). The isotopic compositions ranged from -1 to -5 ‰ for $\delta\ ^{18}O$ and from -3 to -7 ‰ for $\delta\ ^{13}C$. These values suggest the presence of an estuarine environment near Nea Nikomedia (Shackleton 1970). Additionally, the $\delta\ ^{18}O$ and $\delta\ ^{13}C$ compositions of carbonate samples taken along the growth axis of a single *C. glaucum* specimen has been illustrated (Shackleton 1970: 944, Figure 2). The values have been interpreted in terms of water salinity, with lighter isotopic compositions suggesting lower salinity, presumably related to winter floods; and higher isotopic compositions suggesting higher salinity, presumably related to reduced freshwater inflow during summer (Shackleton 1970).

In 1988, a series of $\delta\ ^{18}O$ profiles of archaeological *C. glaucum* specimens from Franchthi and of a beach-collected modern specimen were published (Deith and Shackleton 1988). The $\delta\ ^{18}O$ values of the modern specimen varied between 1.5 ‰ and -1 ‰. The $\delta\ ^{18}O$ values of the archaeological specimens were generally more positive, ranging between 2 ‰ and -1.5 ‰, consistent with the profile of a habitat with no marked differences in salinity (Deith and Shackleton 1988: Appendix F). The authors interpreted the more positive $\delta\ ^{18}O$ values as indicative of cold temperatures and the more negative ones as indicative of warm temperatures, arguing that in estuarine environments of the Aegean, the effect of low water temperatures would lead to enriched oxygen isotopic values in the winter (Deith and Shackleton 1988: 151). Based on the stable isotopic profiles of Franchthi mollusk specimens (including *Monodonta* and *Cerithium* specimens), it was suggested that the seasonal pattern of shellfishing differed in the Neolithic, with increased activity during spring and summer, in relation to the older periods of the site (Deith and Shackleton 1988, also see Deith 1988a).

4.6.3. Summary

According to quantified data from 28 sites across the Aegean basin, aquatic mollusk remains are ubiquitous in the archaeological deposits of the region. Yet, most abundant mollusk remains represent no more than eleven taxa. The species identification of most abundant mollusk remains at any site is related to the geomorphological properties of the coastline, i.e., habitats available, near the site in question. Chronological patterns do not seem to have played a significant role in the formation of archaeomalacological assemblages. The appearance of crushed *H. trunculus* remains in the Middle Bronze Age levels of various sites constitutes an exception.

Data pertaining to the morphometric, incremental, and stable isotopic properties of archaeomalacological mollusks is very scarce. Such analytical data is provided for Kastanas, Makriyalos (unpublished), Magoula Pavlina, New Halos, Saliagos, Sivritepe, Nea Nikomedia, and Franchthi. The datasets from these sites, with two exceptions, namely Makriyalos and Kastanas, involve small numbers of mollusk specimens. Most analytical techniques, such as morphometric and isotopic analyses, have been employed mainly on the *C. glaucum* remains. There are many shortcomings in the employment, presentation, and the interpretation of these results. For example, although Becker (1986) and Prummel (2005) observe the morphological differences (i.e., differences in solidity and curve) between the archaeological and modern populations of *C. glaucum,* they do not pursue an analytical method to demonstrate their empirical statements objectively. The absence of raw data in the published accounts hinders the reproducibility of the results. Despite these hindrances, when appropriate, the data summarized above (Sections 4.6.1 and 4.6.2) is integrated into the discussions presented in the following sections of this study.

4.7. Comparative account of analytical results

4.7.1. Proportion of mollusk remains in faunal assemblages

Among the primary faunal assemblages that are dealt with in this study, only the faunal assemblages of the Troia I settlement and Yenibademli can be compared adequately in terms of the contribution of mollusk remains to the total faunal assemblage. The locations of the settlements in relation to the coast line, the preservation condition of the faunal assemblages, excavation techniques that were used during their retrieval and the methods that were used during their analysis are similar. Despite these similarities one must approach the contributions of the mollusk remains to the total of the faunal assemblage with caution, since differences between the archaeology of the two sites are also present, such as the species identity of the most abundant species and the differential characters of the settlement structures. Also present are chronological problems such as the large and unresolved overlap between the absolute dates of Troia I and Troia II settlements and questions that might be posed about the contemporaneity of Yenibademli with Troia I.

The mollusk remains make up 37.9 % in NIS and 24.7 % in WIS of the faunal assemblage originating from Troia I deposits in Troia (cf. Section 4.3.1). In Yenibademli, the percentages are very similar to Troia I in Troia. The mollusk remains constitute 39.7 % in NIS and 18.6 % in WIS of the total of the faunal remains (cf. Section 4.4.1). The larger difference between the percentages obtained from WIS calculations is related to the differential species-specific fragmentation levels and shell weight per individual. In average, *Patella* spp. individuals have shells lighter than those of *C. glaucum.* They are also more fragile, and thus more vulnerable to post-depositional fragmentation. The result of this situation is relatively larger differences between specimen counts and weights.

4.7.2. Comparison of archaeomalacological assemblages

The species spectra represented by the mollusk assemblages in Troia, Yenibademli, and Ulucak are very similar. All the species identified in Troia are represented in the assemblage from Yenibademli (Tables 24, 32, and 36, cf. Sections 4.3.2, 4.4.2 and 4.5.2). Altogether, they are also not very different from the species spectra reported from coastal settlements of the Bronze Age Aegean. The relatively long species list produced from the mollusk assemblage of Yenibademli is a consequence of identification methods. The short species list produced for Ulucak should be related to the distance this settlement lies from the coast and the human behavior they represent, but also to the quantity of the archaeomalacological assemblage studied from this site (cf. Section 6.1 for a detailed discussion).

The relative abundance of the species represented by the mollusk assemblages differ between sites and individual chronological periods. This is due to a variety of palaeoenvironmental and palaeoeconomic factors, which will be discussed in Section 6.1.

4.7.3. Comparative morphology of archaeological and modern *C. glaucum* populations

The morphometric properties of the archaeological and modern *C. glaucum* populations examined for this study have been described in Sections 4.3.5.1, 4.4.6, and 4.5.5. The available morphometric data from other prehistoric sites in the Aegean has been reviewed in Section 4.6.2. It is now appropriate to return to these results with a comparative approach.

The mean valve heights of archaeological and modern *C. glaucum* populations are significantly different (p = <0.0001, Figure 71). The mean valve height of the shells of *C. glaucum* specimens collected in 2005 and 2006 in the delta area of the Karamenderes is 33.3 mm. This value is much larger than those of the archaeological *C. glaucum* shells from Troia, Yenibademli, and Ulucak (Tables 17, 30, 34, 37). The *C. glaucum* population from Yenibademli constitutes the *largest C. glaucum* populations among the archaeological populations with a mean valve height of 28.0 mm (Table 34). The Bronze Age (all phases and periods) populations from Troia have a mean valve height of 25.7 mm and they are significantly smaller than Yenibademli populations, but larger than the mean valve height of the population examined from Ulucak, which has a mean valve height of 21.9 mm (Table 37).

When the comparison of the archaeological assemblages is confined to roughly contemporary assemblages dating to the Early Bronze Age, the mean valve heights of the populations from Troia, Yenibademli, and Ulucak are again significantly different (p = <0.0001, Figure 71). The mean valve height of the Yenibademli population is, at 28.0 mm, larger than the mean valve heights of the Early Bronze Age populations in Troia, at 24.5 mm, and Ulucak, at 21.8 mm.

The mean valve heights of the archaeological *C. glaucum* assemblages examined in the present study can be compared to the mean valve heights provided in other studies. The size of the two chronologically defined *C. glaucum* populations from Makriyalos are –with 22.1 mm for the older phase and 21.3 for the younger phase– are similar to the mean valve height of the Chalcolithic and Early Bronze Age assemblages of Ulucak. The mean valve heights of Hellenistic populations from New Halos –24.9 mm and 27.8 mm– are within the range of the mean valve heights of the populations from Troia and Yenibademli. The mean valve height of the Middle Bronze Age population from Magoula Pavlina is similar to the Early Bronze Age population of Yenibademli.

The solidity indices of all the *C. glaucum* populations examined for this study are significantly different from each other (p = <0.0001, Figure 72a). The *C. glaucum* shells in Yenibademli represent the most solid population, with a mean solidity of 1.2. The Bronze Age populations of Troia are significantly more solid, with a mean of 1.0, than the modern populations of the Troad, whose solidity values have a mean of 0.82. The *C. glaucum* populations of Ulucak have the least solid valves with a mean solidity of 0.74.

Likewise, the solidity indices of the Early Bronze Age populations represented by the *C. glaucum* assemblages in Troia, Yenibademli, and Ulucak are different from each other (p = <0.0001, Figure 72b). The Early Bronze Age population of Yenibademli is more solid (1.2) than the Early Bronze Age population of Troia (0.96). The Early Bronze Age *C. glaucum* populations from Ulucak represent the least solid of all Early Bronze Age populations, with a mean solidity of 0.73.

Figure 71. Comparison of the valve heights of *C. glaucum* from Troia, Yenibademli, Ulucak, and the modern populations from the Troad (ANOVA).

a. All individuals

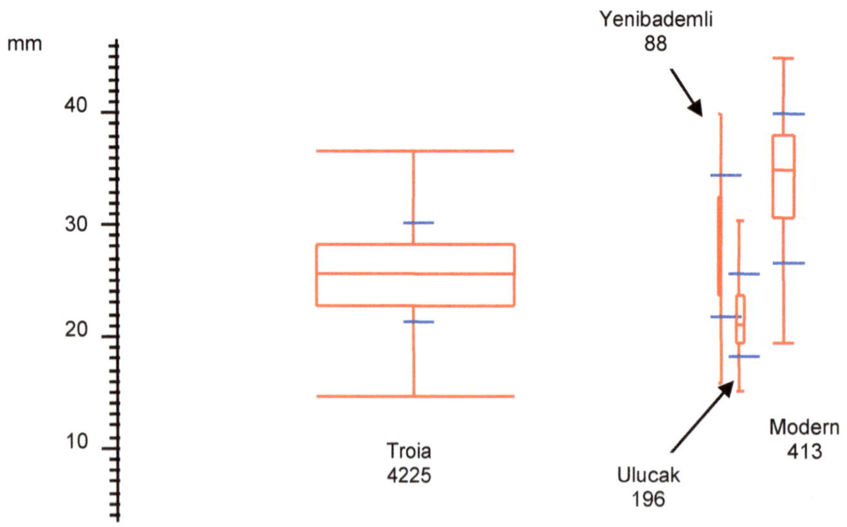

b. Individuals dating to the EBA

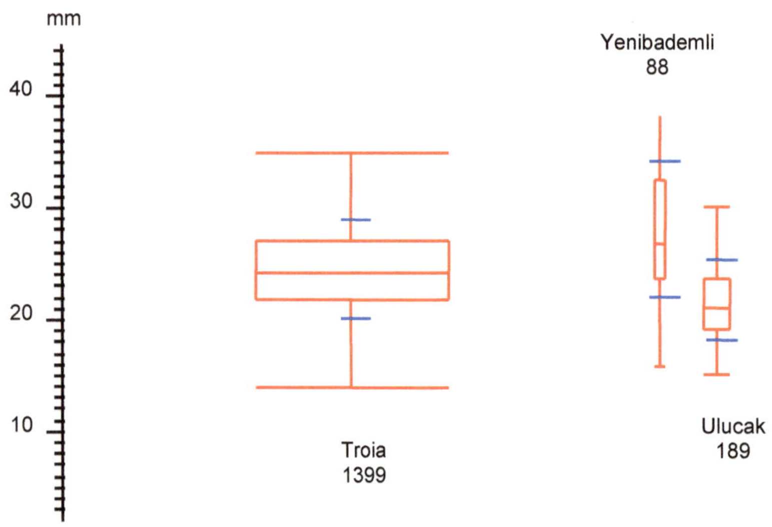

The curve indices –closely related to the solidity indices– are also different for all archaeological and modern populations (p = <0.0001, except for the difference between the mean curves of Yenibademli and Troia populations where the difference is slightly smaller at p = 0.0004. Figure 73a). The valves from Yenibademli represent the most curved population, with a mean curve of 0.83. The populations of Troia are less curved than this population, with a mean curve of 0.81. The *C. glaucum* population represented by the remains in Ulucak, with a mean curve of 0.79, is less curved than the archaeological populations but more curved than the modern populations, with a mean of 0.70.

Likewise, the mean curves of populations limited to the Early Bronze Age are significantly different (p = <0.0001, Figure 73b). The mean curve of the Early Bronze Age population in Yenibademli is larger, at 0.83, than the mean curve of *C. glaucum* valves from the Early Bronze Age deposits of Troia, which have a mean of 0.80. The mean curve of the Ulucak population is the smallest among the Early Bronze Age populations with 0.79.

Figure 72. Comparison of the solidity of *C. glaucum* from Troia, Yenibademli, Ulucak, and the modern populations from the Troad (ANOVA).

a. All individuals

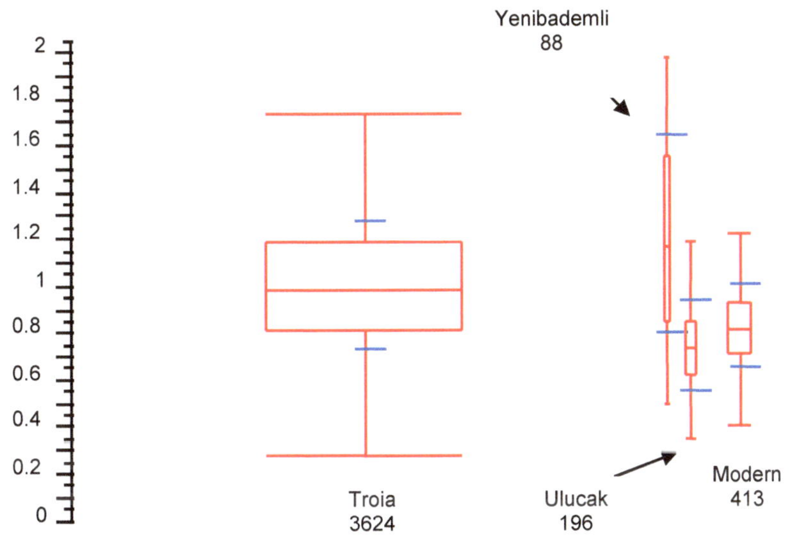

b. Individuals dating to the EBA

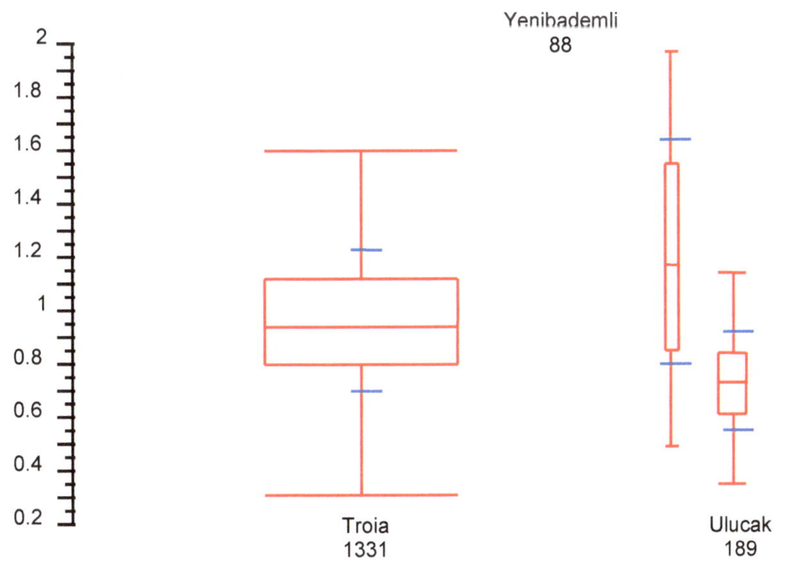

Figure 73. Comparison of the curve of *C. glaucum* from Troia, Yenibademli, Ulucak, and the modern populations in the Troad (ANOVA).

a. All individuals

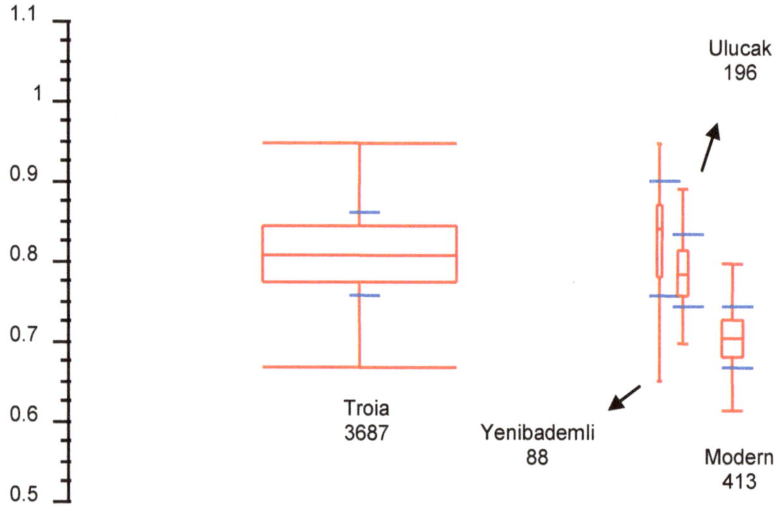

b. Individuals dating to the EBA

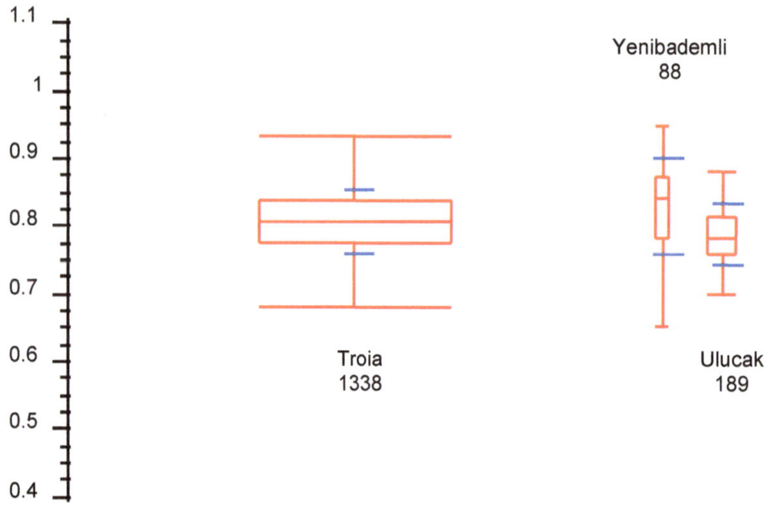

The relationship between the mean solidity and curve indices of modern and archaeological populations plotted in Figure 74. For this analysis, the solidity and curve indices of the populations living in different areas of the Karamenderes Delta Area have been pooled. The curve and solidity indices of these separate communities do not act co-dependently or in correlation with the differences in average SST and salinity values recorded at the stations from which they were obtained. This situation is not in agreement with Mariani et al.'s (2002) observations on modern *C. glaucum* populations, where mean solidity and curve of *C. glaucum* shells alter together and in positive relationship with recorded SST and salinity values. Because this situation is caused by the imbalance between age and size profiles of the sampled portion of the populations, it is considered best to treat these separate communities as a single population representing the modern conditions in the Karamenderes Delta. It is evident in this plot that even though the solidity and curve ranges of archaeological populations differ from one another, the relationship between their solidity indices and their curve indices show similarity. Contrarily, the modern population from the Troad appears to have assumed morphological features dissimilar to the archaeological populations. This situation shows similarities with other archaeomalacological descriptions which draw attention to the thicker and oblique (or curved) shells of the archaeological *C. glaucum* populations from Kastanas and New Halos in relationship to modern populations (cf. Section 4.6.2; Becker 1986: 231,

Prummel 2005: 119). It should be added that although von Martens, who was the first to examine mollusk remains from Schliemann's excavations in Troia, did not compare his findings about the morphology of archaeological *C. glaucum* at this site with modern specimens; he did note their oblique form and even provided measurements (1879).

Figure 74. The relationship between curve (y) and solidity (x) indices for *C. glaucum* populations from Troia, Yenibademli, and Ulucak, and modern *C. glaucum* populations in the Troad.

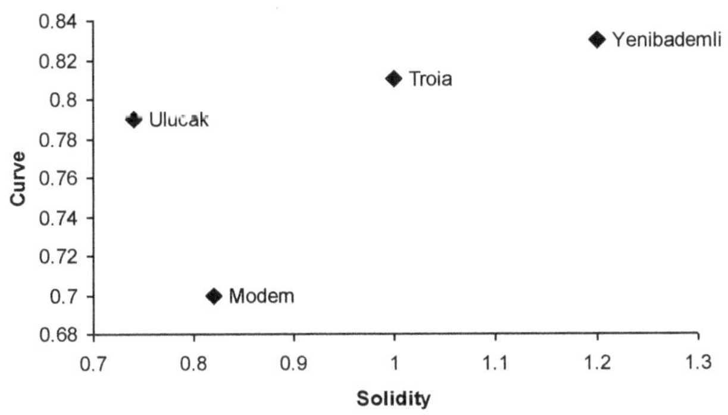

4.7.4. Comparative incremental growth pattern of archaeological and modern *C. glaucum* populations

Different types of data were compiled from the examination of incremental growth patterns of archaeological and modern *C. glaucum* populations: (1) type of incremental zone being secreted at the time of death (gray or white); (2) frequency of the alteration between gray and white incremental zones; (3) demographic structures of populations; and, (4) mean amount of shell growth per age category and added growth rates.

According to periodic observations on the incremental growth pattern of the *C. glaucum* populations living in the delta area of the Karamenderes, a majority of the individuals in the extant populations secrete gray incremental zones during warm months (cf. Section 4.2.2.4). Annual growth consists of one pair of white and gray incremental zones (cf. Section 4.2.2.3 and 4.2.2.4).

The age profiles of the Bronze Age populations from Troia and of the extant populations show the following differences (Figure 75): animals collected at Stage 1 are proportionally less abundant in the archaeological assemblages than they are in the modern collections, whereas animals collected at Stage 2 are much better represented in the archaeological assemblages than in the modern collections. The presence of individuals at Stages 3 and 4 is similar in proportion. Individuals at Stage 5 and senile individuals are relatively more numerous in the modern collections. In summary, the age profile of archaeological populations is skewed towards younger individuals (Stages 1-3), whereas the age profile of modern collections is biased towards older individuals (Stages 4-senile).

The difference in the added growth per age year between the archaeological and modern populations is illustrated in Figure 76, where the mean valve heights of the individuals at the same age stage are compared. The mean valve heights of individuals at Stage 1 of modern and archaeological populations are similar, indicating that the frequency of spawning periods of the Bronze Age *C. glaucum* populations in the Troad was similar to the present. The mean valve heights of the modern individuals at subsequent age stages are greater than the mean valve heights of the individuals at the corresponding age stages represented by the archaeological *C. glaucum* assemblages in Troia. Extant populations grew faster –i.e., added more to the valve height– than the archaeological populations. In order to understand the biometrical reason behind this difference, the breadths of incremental zones observed on the two populations were compared. A meaningful difference between the breadths of corresponding incremental zones of archaeological and modern populations was not observed. While some incremental zones are significantly broader in the archaeological populations, others are broader in the modern samples. For this reason, it is not possible to pinpoint which aspect or aspects of growth are directly related to the fast growth of modern populations. Differential shell morphology (i.e., curve and solidity) are also influential in these results.

Analogy between the seasonal timing of changes in the shell secretion rates of modern populations and the changes in the shell secretion rates of archaeological populations is considered meaningful. Although the present aquatic environment in the Troad is considerably different from the aquatic environment that prevailed during the 3rd and 2nd millennia BC (cf. Chapter 2), the most important factor determining the annual patterns is the seasonality of climate –the effect of which is similar in both environments. While anthropogenic regulation of the flood waters of the Karamenderes, fluctuations in the global and regional climate, and the greater variety of past habitats should be considered as important factors affecting the timing of the shift from slow growth to fast growth and the duration of growth episodes, the essential seasonal alternation between warm and more saline conditions and cold and less saline conditions must have dominated the shell growth pattern of the *C. glaucum* in the Troad in the past as it does today. Therefore, it can be accepted that the formation of the gray incremental zones in the shells of the Bronze Age populations represent slow growth during the warmer periods (from June to November or shorter, depending on the age of the individual) of the year, whereas the white incremental zones correspond to the colder periods of the year. When the results of the observations on the marginal incremental zones of modern and archaeological populations are compared, the collection time of the archaeological populations appear to be skewed towards warmer months of the year (Figure 77). If *C. glaucum* gathering in Bronze Age Troia took place primarily during the colder months, then the proportion of the archaeological individuals demonstrating gray marginal incremental zones would have been much smaller. On the other hand, the proportion of the individuals with gray marginal incremental zones in the Late Bronze Age assemblage is probably inflated by two factors: (1) the increase in the amount of older individuals (cf. Section 4.3.5.2); and, (2) the increase in the breadths of gray incremental zones observed in the Late Bronze Age assemblages (cf. Section 4.3.5.1).

Figure 75. Comparison of the demographic structures of modern and archaeological *C. glaucum* populations.

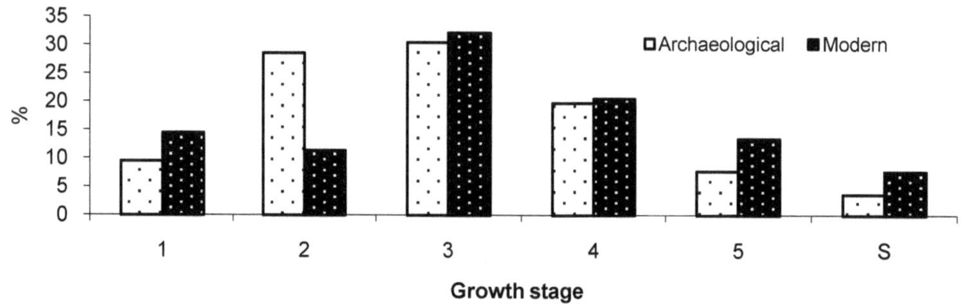

Figure 76. Comparison of shell growth achieved at the completion of each incremental zone in modern and Bronze Age *C. glaucum* populations in the Troad.

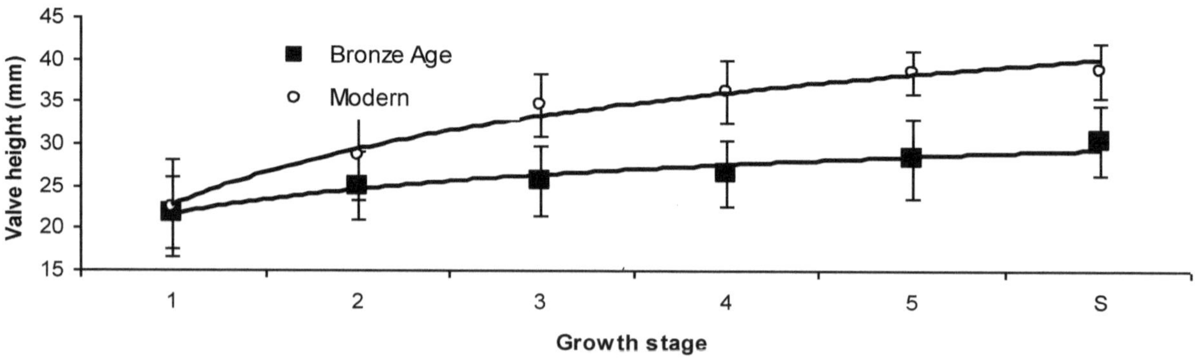

Figure 77. Comparison of the character of the marginal incremental growth zone in modern and archaeological *C. glaucum* populations.

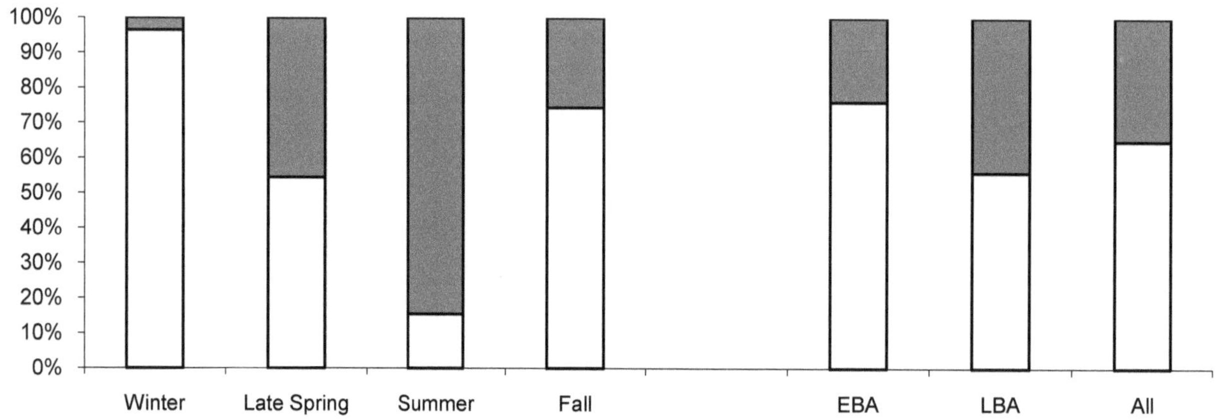

The following chapter, Chapter 5, will discuss mollusk procurement and consumption practices in Troia, Yenibademli, and Ulucak.

Table 11. Overview of archaeomalacological assemblages from other sites in the Aegean.

Site	Date/Period[30]	Nature of coastline	Most common species	Source
Pre-Bronze Age				
Franchthi	Ca. 9400 BP	Closed bay	*Patella* spp., *Monodonta* spp.	Shackleton 1988a, 1988b
Franchthi	Ca. 8500 BP	Closed bay	*C. neritea*, *Cerithium vulgatum*	Shackleton 1988a, 1988b
Franchthi	Ca. 6900 BP	Closed bay	*Cerithium vulgatum*	Shackleton 1988a, 1988b
Hocaçeşme	6500-5700 BC	Deltaic lagoon	*C. glaucum*, *O. edulis*	Buitenhuis 1995
Kephala[31]	5000-4500 BC	Rocky shore	*Patella* spp.	Coy 1977
Kumtepe	5th millennium BC	Estuary and small marine embayment	*O. edulis*, *M. galloprovincialis*, *C. glaucum*	Per. com. M. Uerpmann
Lerna	(Lerna I) Neolithic	Lagoon and estuary	*C. glaucum*	Gejvall 1969
Magula Pevkakia	"Rachman Period" and Early Bronze Age, 3000-2000 BC	Low promontory at a bay	*C. glaucum*	Falkner 1975
Makriyalos	Late Neolithic	Lagoon	*C. glaucum*	Andreasen 1998
Nea Nikomedia	Early Neolithic	Lagoon	*C. glaucum*	Higgs and Clench 1962
Saliagos	5300-4500 BC	Rocky inlet	*Monodonta* spp., *Patella* spp., *C. Glaucum*	Shackleton 1968°
Sivritepe	5th millennium	Bay and lagoon	*O. edulis*, *C. glaucum*, *M. galloprovincialis*, Muricidae	Boessneck 1986
Youra	Neolithic/Mesolithic	Rocky shore	*Patella* spp., *Monodonta* spp.	Sampson 1998
Early Bronze Age				
Beşik-Yassıtepe	2900-2400 BC	Rocky inlet by a lagoon	*C. glaucum*, *Patella* spp., *O. edulis*	von den Driesch 1999
Eleusis	Early Helladic II	Closed bay	*C. glaucum*, *A. noae*	Cosmopoulos et al. 2003
Franchthi	5000 BP	Closed bay	*C. glaucum*, *T. decussatus*	Shackleton 1988a, 1988b
Kumtepe	C/Troia I	Lagoon and estuary	*C. glaucum*	Per. obv.
Lerna	Lerna II/III	Lagoon and estuary	*C. glaucum*, *H. trunculus*, *S. gaederopous*	Gejvall 1969
Limantepe	Early Bronze Age	Small bay	*C. glaucum*	Reese 2006
Poliochni	2900-2400 BC (Troia I-II)	Rocky inlet by an estuary	*Patella* spp., *C. glaucum*, *O. edulis*	Sorrentino 1997
Middle Bronze Age				
Ayios Mamas	2000-1700 BC	Estuary	*C. glaucum*, *A. noae*, *H.trunculus*, *P. caerulea*	Becker 1996, 2001
Eleusis	Middle Helladic	Closed bay	*A. noae*, *S. gaederopus*,*H. trunculus*	Cosmopoulos et al. 2003
Kastanas	2150-1600 BC	Estuary	Unionidae, *C. glaucum*	Becker 1986
Lerna	Lerna IV	Lagoon and estuary	*C. glaucum*, *H. trunculus*, *S. gaederopous*, *A. noae*	Gejvall 1969
Magoula Pavlina	2200-1700 BC	Closed lagoon and estuary	*C. glaucum*, *O. edulis*, *S. gaederopous*	Prummel 2005
Magoula Pevkakia	2000-1550 BC	Low promontory at a bay	*C. glaucum*	Falkner 1975
Panaztepe	Middle Bronze Age-Late Bronze Age	Estuary	*C. glaucum*	Reese 2006
Palaikastro	Middle Minoan	Embayment near a rocky promontory	*Patella spp.*, *H. trunculus*	Reese 1987a

[30] The chronology summarized in this column has been adopted from the publications that are cited for each site.

[31] This assemblage includes finds from the settlement's graveyard.

Table 11. Overview of archaeomalacological from other sites in the Aegean (Continued).

Site	Date/Period[32]	Nature of coastline	Most common species	Source
Late Bronze Age				
Eleusis	Late Helladic	Closed bay	*H. trunculus, A. noae, C. glaucum*	Cosmopoulos et al. 2003
Kastanas	1600-1400 BC	Estuary	*C. glaucum,* Unionidae	Becker 1986
Kastanas	1300-1000 BC	Estuary	Unionidae, *C. glaucum*	Becker 1986
Lerna	Lerna V-VI-VII	Lagoon and estuary	*S. gaederopous, H. trunculus, C. glaucum*	Gejvall 1969
Skala Sotiros	1400-1300 BC	Rocky shore	*Patella* spp.	Karali 1999
Post-Bronze Age				
Beşik-Yassıtepe	Archaic-Hellenistic	Rocky inlet by a lagoon	*C. glaucum, Patella* spp., *O. edulis*	von den Driesch 1999
Eleusis	Classical and Roman periods	Closed bay	*H. trunculus, A. noae*	Cosmopoulos et al. 2003
Kalabaktepe	7th-5th c. BC	Estuary	*C. glaucum, O. edulis*	Peters 1992
New Halos	300-220 BC	Closed lagoon and estuary	*C. glaucum, O. edulis, S. gaederopous*	Prummel 2005
Inland sites				
Baklatepe	Early Bronze Age I	33 km inland on the Büyük Menderes	*C. glaucum*	Reese 2006
Paradeisos	L. Neolithic	On the Nestor	*C. glaucum, Glycymeris* spp.	Reese 1987c

[32] The chronology summarized in this column has been adopted from the publications that are cited for each site.

Chapter 5. Shellfish procurement and consumption

In Chapter 4, primary data produced through the application of a number of analytical methods both on the archaeomalacological material stemming from human behavior and the modern mollusk populations were described. Additionally, a review account of the published archaeomalacological data from other sites in the Aegean was presented. The rest of this study is dedicated to explaining the behavioral (cultural and economic) and environmental implications of this data. Because the foremost aim of this study is to evaluate the role of shellfish as a dietary resource and shellfishing as an economic activity, Chapter 5 will address a number of topics: first, the food species within the archaeomalacological assemblages will be designated; next, the environments exploited and techniques used for gathering, as well as possible methods of shellfish processing, will be described; then, the nutritional values of important mollusk species represented in the assemblages will be discussed; further, factors that may have influenced decisions concerning harvest time in Troia will be assayed; and finally, an attempt to assess the nature of purple dye production in Troia will be made.

5.1. Food species

Mollusk remains can be found in archaeological sites for a variety of reasons. While terrestrial and small pond gastropods can represent naturally induced intrusions in an archaeological deposit, it is almost always certain that aquatic mollusk remains are remnants of human activity. Exceptions can be the shells of certain species carried by hermit crabs to archaeological (mainly shell middens) sites that lay directly on the shore (per. com. K. Szabo). Although a large majority –if not all– of aquatic mollusks are edible, humans are not always interested in their dietary or other economic value, even when the mollusk populations are within easy reach. Ethnological and archaeological evidence demonstrates that mollusk-related human activity can represent activities other than those of a dietary or otherwise economic nature. Therefore, before analyzing the economic role and nature of shellfishing in societies represented by the archaeological sites that are dealt with here, it is important to determine the reasons for the occurrence of each mollusk taxon represented in the archaeological assemblages. Distinguishing the species that were used principally as human food is of primary concern.

There are few circumstances under which humans avoid shellfish as food:

Toxicity is the only known natural factor that negatively affects the edibility of a mollusk species. Outbreaks of toxic algae, commonly known as “red tides” or “toxic algal blooms,” occur throughout the coastal areas of the world and have also been recorded in the Aegean (Buchanan 1988: 83-85, Claassen 1998: 33-36, per. com. P. Leontarakis). Red tides are harmful to mollusks and cause poisoning and even death in humans who consume them (Buchanan 1988: 84). Today when their occurrence is detected by local coastal authorities in Greece, mollusk exploitation is prohibited temporarily (per. com. P. Leontarakis). There is no archaeological or ethnographic evidence (e.g., records of mass mortalities) suggesting that ancient shellfish gatherers of the world were not aware of this danger (Buchanan 1988: 83-85). Red tides occur irregularly (Buchanan 1988: 83-85, Moss 1993, per. com. P. Leontarakis) and can only influence the edibility of shellfish temporarily, unless catastrophic events stimulate taboos dictating shellfish avoidance.

Although it is difficult to trace the reasons behind them, taboos prohibiting shellfish consumption altogether or prohibitions on certain species are common cultural phenomena. Modern examples manifest how such food taboos are powerful determinants of edible food species in many cultures. For example, Judaism, some schools of Islam, and some disciplines of Christianity do not permit shellfish consumption, while others, such as Orthodox Christianity, encourage its consumption during certain periods of the religious calendar. Such religious beliefs have strong authority on common food practices: for example, while mollusk species other than *M. galloprovincialis* are abhorred by the Muslim population in Turkey, all available shellfish species are exploited as human food in Morocco; another predominantly Muslim country. This is simply due to the fact that the present population in coastal Turkey is dominated by the followers of the Hanefi School, which prohibits shellfish consumption, whereas the most commonly practiced school of Islam in Morocco, the Maliki School has not issued such a ban. Although shellfish consumption is not regulated in the taboo level in Greece, a predominantly Orthodox Christian country; Greeks are generally reluctant against consuming shellfish other than *M. galloprovincialis* (per. com. P. Leontarakis). In this country, shellfish becomes an important part of the daily menu during the Easter fast. Other cultural and demographic factors affecting shellfish gathering patterns on the Turkish coast of the Aegean have been discussed in Section 3.8.

Conversely, reluctance against consuming certain shellfish can transform into prejudice against peoples who consume those species, in the process of the formation of group identities: Sir Walter Scott writes in 1814 in his diary that the people from the Orkney Islands used to shun the people of Swona for eating limpets (1972, quoted in Waselkov 1988).

While cultural and natural edibility criteria are not readily visible in the list of species represented by the archaeomalacological record of the prehistoric peoples, it is possible to distinguish important food species from non-food species based on their remains with a high level of certainty. The quantitative and taphonomic variability of the archaeomalacological assemblages imply that a variety of behaviors other than food consumption are represented by the remains of aquatic mollusks. Economically important species in Troia, Yenibademli, and Ulucak have been identified by evaluating the relative and absolute abundances of mollusk species, the taphonomy of the specimens, and the analogies made from ethnographic and archaeological record. Only those mollusk taxa that are represented by large quantities of mollusk remains are considered to have been valued significantly as human food. This consideration is based on the fact that the majority of the mollusk taxa represented in the archaeological assemblages is small in size and the individual mollusks have limited nutritional value (See Section 5.5 below). For shellfish to contribute to human nutrition at a significant level, it was necessary to collect considerable quantities of individual specimens. Mollusk remains bearing traces of wave or sand abrasion are not considered to have been part of the menu.

Altogether, at least 13 gastropod and 13 bivalve species are represented in Troia, Yenibademli, and Ulucak.

Among the gastropods, only *Patella* spp. and *H. trunculus* are represented in sufficient amounts to suggest that they were frequently harvested. *Patella* spp. is the most abundant mollusk species in Yenibademli.

Patella spp., or limpets, though not an important commercial product in the Aegean today, is still exploited as occasional "snacks" in Greece (Davidson 1972: 190, per. com. P. Leontarakis). Shackleton (1968a: 129) also notes that fishers on Antiparos consumed limpets as food in the 20th century. Their remains are among the most common gastropod remains in the archaeological sites in the Aegean (cf. Section 4.6.1; Becker 1996, Boessneck 1986, Coy 1977, Karali 1999, Reese 1987a, Shackleton 1968, 1969, Sorrentino 1997, von den Driesch 1999), in the eastern Mediterranean (Ridout-Sharpe 1998), and in other parts of the world such as California, U.S. (Erlandson et al. 1999), England (Mannino and Thomas 2001), Portugal (Cabral and da Silva 2003), and South Africa (Buchanan 1987, Henshilwood 2001, Shackleton 1973). Modern and archaeological evidence shows that where available, humans, whether habitually or erratically, make use of this resource as food. Additionally, *Patella* spp. is occasionally used as fishing bait in England (Deith 1985).

The taphonomy of limpet specimens in Troia was described and illustrated in Section 4.3.3. Furthermore, it was demonstrated in Section 4.4.3 that *Patella* spp. fragmentation in Yenibademli had identical characteristics to those described for the specimens in Troia. Based on their abundance and species-specific taphonomy, it is concluded that the great majority, if not all *Patella* spp. remains in Troia and Yenibademli consist of specimens that were collected live and subsequently consumed as human food. Some *Patella* spp. remains, which were fragmented in the same manner as described in Sections 4.3.3 and 4.4.3, were reported from the Chalcolithic Sivritepe and similarly interpreted as part of the food debris (Boessneck 1986). Another *Patella* specimen fragmented in a very similar manner as those demonstrated by Boessneck (1986) was recovered during more recent explorations at the same site. Despite Boessneck's (1986) observations, this specimen was published as an artifact (Korfmann 2000: 43, Figure 35c). Similarly, this type of limpet fragmentation has been considered to be artificial modification in other Bronze Age sites in the Aegean (Hood 1981: 677, Karali 1999: 112, Figure 24). Although it is difficult to make absolute judgments about the true nature of these fragmented remains based on published photographs and verbal descriptions, it is maintained that it is highly unlikely that the examples from Sivritepe and other sites from the Aegean have been exposed to deliberate modification and subsequently used as tools or ornaments. Like the entire *Patella* spp. assemblage from Yenibademli and those observed in Troia, their fragmentation also occurred most probably in the sediment once they were discarded as food refuse.

Hexaplex trunculus remains constitute ca. 1 % of the archaeomalacological assemblages of the Early Bronze Age in Troia and Yenibademli (cf. Sections 4.3.2 and 4.4.2, Tables 34 and 37). Additionally, there is a limited amount of *H. trunculus* remains originating from the Post-Bronze Age layers of Troia. Similar to the situations in Troia and Yenibademli, *H. trunculus* remains antedating the Middle Bronze Age in the Aegean are rarely encountered in the archaeomalacological assemblages (cf. Section 4.6.1, Table 11; also Karali 1999: 55, Table 1 and Çakırlar and Becks forthcoming). At least some of the *H. trunculus* specimens found in Early Bronze Age Aegean contexts represent beach-picked examples (Reese 1985). Others were interpreted as food refuse based on their fairly intact remains (Boessneck 1986). The few specimens from the early deposits of Yenibademli and Troia, and a few originating from the Post-Bronze Age Troia, either represent food refuse –probably by-catches of shellfish gathering activity which primarily targeted other species or, for Troia, also possible beach-picks. *H. trunculus* snails constitute a highly-rated part of the Mediterranean seafood menu today (Davidson 1972: 192-193). The amount, taphonomy, and stratigraphical occurrence of *H. trunculus* remains in Troia VI indicate that they are remnants of purple-dye manufacture (for a complete discussion cf. Section 5.7 below). Although it is possible that some of the *H. trunculus* specimens in the Troia VI and VIIa deposits are remnants of human dietary consumption, the general crushed state of these specimens do not imply that the this species was exploited primarily for dietary purposes. Crushed mollusks are not palatable because small sharp shell fragments can remain in the flesh, making the flesh, difficult, if not impossible, to digest.

Monodontinae and *Cerithium* spp. are also fairly common among the gastropod assemblages of Troia and Yenibademli. The quantity of Monodontinae in Troia is adequate enough to suggest that they represent food consumption. The Monodontinae were probably collected along with *Patella* spp. In Troia, some of the *Cerithium* spp. specimens may have been collected deliberately as a food resource, whereas others may represent incidental harvests as living animals or as empty shells. Still others may have been collected from the beach to be used as ornaments. The water-worn example from Troia (cf. Section 4.3.3., Figure 34c) attests that this specimen was empty when it was brought to the site. In Yenibademli, the fresh appearance of *Cerithium* spp. specimens indicates this species' role as human food at this settlement. The amount of both the Monodontinae and the *Cerithium* spp. found in Ulucak is too limited to imply food consumption of any significance.

The quantity and the taphonomic condition of the other gastropods (cf. Sections 4.3.3, 4.4.3, and 4.5.3) represented in the assemblages does not allow for assigning them as food species with any certainty. The fact that these species are infrequently represented in other archaeomalacological assemblages in the Aegean is affirmative of this consideration (cf. Section 4.6.1). When they are found in archaeological contexts, they often bear traces of natural or artificial post-mortem modification (Karali 1999, Reese 1982, 1983, 1987a, 1987b, 1992, 2006).

The most common bivalves represented in the studied mollusk assemblages are *M. galloprovincialis*, *O. edulis*, and *C. glaucum*, in phylogenetic order.

Valves of *M. galloprovincialis* and *O. edulis* are commonly found in fragmented form. The natural processes responsible for this type of fragmentation have been discussed in Section 4.3.3. Archaeological evidence for edible mussel (Mytilidae) consumption was found in many sites along the coasts of the world (e.g., Buchanan 1988, Erlandson et al. 1999). Today, mussels are used as human food commonly throughout the Mediterranean countries including Turkey (cf. Sections 3.8 and 5.1). They are also used in preparing fishing bait in Turkey and England (Deith 1985). Likewise, oysters (*O. edulis* or other members of the family Ostreidae) are also very common elements of the archaeomalacological assemblages across the world (e.g., see Custer and Doms 1990 for oyster middens of the eastern coast of the U.S. and Milner 2001 for North European examples) and a popular type of seafood today. The archaeological and modern analogies, their quantity, as well as the taphomomic nature of their remains in Troia and Yenibademli lead to the doubtless conclusion that *M. galloprovincialis* and *O. edulis* were harvested for human consumption as food.

Blegen et al. (1951: 66-67), who excavated in Troia between 1932 and 1938, found an analogy between the large quantities of *C. glaucum* shells in the archaeological deposits and the shellfishing behavior of the local workmen in their team and recalled that their "... workmen, who called them ἀλφάδες, since no equivalent Turkish name was known to them, said that these bivalves are regarded today as an especially succulent and dainty dish when properly cooked with pilaf.". They also reported that "[*C. glaucum*] still occur in quantities; it is reported, in the bed of Scamander, near its mouth." This account is the last written account of cockle gathering in the Troad. The species is still commonly consumed in the countries around the Mediterranean (Davidson 1972: 202). Their remains are among the most abundant features of the archaeomalacological assemblages from the Aegean coast and its immediate hinterland (cf. Section 4.6.1). If the soft body is kept undisturbed in its shell, *C. glaucum* can survive up to three days outside of its natural habitat. This biological fact helps to explain why mollusks were transported to coastal mound settlements and to the settlements further inland along with their shells, despite the burden of the inedible portion. Modern and archaeological evidence for the transportation of fresh shellfish to inland sites in the world has been discussed in various archaeological and ethnographic works (Buchanan 1988: 26, Moss 1993, Waselkov 1987).

The taphonomic state of the majority of the *C. glaucum* remains in Troia, Yenibademli, and Ulucak leaves no doubt as to their principle use as human food. Only a minor amount of specimens may have been used as ornaments or tools. Some specimens in Troia bear irregular holes on the upper section of the valve near the umbo. These were either artificially modified –presumably upon the consumption of the soft body– or the holes were formed by physical and chemical agents after the deposition of the valve. Factors in favor of the latter explanation have been described in Section 4.3.3. Similar *C. glaucum* specimens bearing holes have been regarded as remnants of bead making elsewhere in the archaeological contexts of the Aegean (e.g., Miller 1996, Shackleton 1988a), but before the Trojan examples can be related to a local bead production, the presence of their end-products in Troia, i.e., the small round shell beads (cf. Miller 1996), should be confirmed and both the "perforated" valves and the beads should be subject to experimental analysis. On the other hand, the holed valves themselves may have been used as pendants or beads. In fact, similar examples that were found in the Neolithic layers of Ulucak have been interpreted by the excavators as necklace beads (per. com. A. Çilingiroğlu and per. obv. at the İzmir Archaeological Museum). The true nature of these finds has not yet been thoroughly investigated. The hypothesis in favor of their artifactual nature should be verified by use-wear analysis and contextual evidence.

The nature of the abraded umbones and posterior margins of the 46 cockle specimens found in Yenibademli (cf. Section 4.4.3) is likewise problematic. Regardless of their nature and their purpose, these specimens make up a small part of the *C. glaucum* assemblage recovered from the archaeological deposits of Yenibademli.

Apart from cockles, mussels, and oysters, there are at least nine other bivalve species represented in the archaeomalacological assemblages. Some of these, such as the razorclams (*S. marginatus*) and scallops (Pectenidae), have been found as large accumulations. Their relatively high abundance in certain occupational periods in Troia indicates their significance as dietary resources, at least during those periods. The limited number of Pectenid remains from the Bronze Age layers of Troia cannot be considered to represent a significant food source in any way. Moreover, it has been detected that some of these Pectenidae specimens from the Bronze Age layers have been acquired as empty shells. The increasing abundance of the Pectenid shells among the Post-Bronze remains is considered to be evidence for their food value (For a further discussion on scallop consumption in Classical and Roman times, see Section 6.3 below).

While both razorclams and scallops are valued food items in the present day Mediterranean countries (Davidson 1972: 200-201, 207), it is important to note that the razorclams are only used as fishing bait in Turkey today (per. obv.). This modern practice brings up the question of whether or not the remains of *S. marginatus* in Troia can be related to fishing rather than shellfish consumption as human food. Although it can be argued that the refuse of shellfish used as bait would be located near the fishing grounds rather than on the settlement, it is also common practice today to prepare the shellfish for bait at the home base (per. obv.). One argument against considering the remains of *S. marginatus* in Troia or in Yenibademli as fishing bait is the virtual absence or rarity of the fish remains among the faunal remains of these sites (cf. Sections 2.7.1., 2.7.2., 2.7.3., where it has also been questioned whether the under-representation of fish remains at these sites is related to excavation techniques). Ethnographic studies have noted that the quantity of shellfish bait used in line fishing can almost equal the quantity of fish caught (Deith 1989). Directly, as human food, or indirectly, as fish bait, it is certain that the considerable amounts of razorclams at Troia and the small amounts found in Yenibademli were shellfished and contributed to the subsistence. The single specimen of *S. marginatus* in the Ulucak assemblage may have been brought to the site as an exotic curiosity or, more likely, accidentally along with the *C. glaucum* specimens.

It is quite uncertain whether other bivalve species occurring in the studied assemblages represent food species; if they do, their contribution to the human diet was certainly not of any great significance. The amount and the taphonomic condition of *A. noae* shells demonstrate that they have been either incidentally gathered along with other mollusk species or that their shells were deliberately collected to serve an unknown purpose. *A. noae* shells have been encountered in large quantities in some of the more southerly archaeological sites dating to the Bronze Age Aegean (cf. Section 4.6.1). The species is also among the commercial shellfish taxa of the Mediterranean today (Davidson 1972: 196, Peharda et al. 2002). It should be noted that they may also have been sought for the blister pearls they occasionally produce (Mienis 2000).

The Glycymerididae, the dog cockles, while not abundant at any of the Aegean Bronze Age sites, are common elements of the mollusk assemblages of the eastern Mediterranean. The dog cockles are consumed as human food today (Davidson 1972: 196). Interestingly, the majority of the Glycymerididae finds reported from the archaeological sites in the Aegean and inland Anatolia consists of shells which have been exposed to wave and/or sand action prior to final burial in the archaeological matrix (e.g., Becker 1996: 12, Figure 6c, Boessneck 1986, Reese 1986, 1987a, 1987b, 2006). Three of the six specimens found in Yenibademli are severely abraded examples (Figure 58). Since the taphonomy of the Troia specimens is unknown, it is not possible to ascertain whether the Glycymerididae recorded at this site represent shellfish consumption or the use of the empty shells in any other way.

While all the *Pinna* spp. (fan mussels) shells observed for this study are fragmentary, they do not bear any traces of work or artificial modification. In another work, it was suggested that the remains of the fan mussels found in the prehistoric deposits of the Aegean were used as spatulas (Karali 1999: 20). One of the *Pinna* species, namely *P. nobilis,* which may be present in the studied assemblages, produces small pearls. Today, diving for *P. nobilis* in the İzmir Bay with the intention of obtaining these pearls is one of the important factors threatening the survival of *P. nobilis* populations in this area (per. com. U. Önal). This modern practice brings to mind the possibility that they may also have been sought for this purpose in the past. A textile industry associated with the byssi of fan mussels is known from textual evidence post-dating 2nd century AD as well as ethnographic records in Italy (Laufer 1915, Davidson 1972: 199). Additionally, Davidson (1972: 199) provides a recipe for *P. nobilis.*

The majority of the *Spondylus* specimens in Yenibademli appear to have been long dead at the time of collection. These remains do not represent food items. The only archaeological context in the Aegean where *Spondylus* remains are numerous enough to represent food remains is the Bronze Age layers of Lerna (Gejvall 1969). The dubiosity of these finds has been discussed in Section 4.6.1.

The amount of *A. tuberculatum* and *Donax* remains in the archaeological deposits studied for this dissertation are meager. The remains of these mollusks are similarly scarce in other archaeological contexts in the Aegean. Since a number of *A. tuberculatum* remains in Yenibademli have been artificially modified, it should be questioned whether this species constitutes part of the diet at this Early Bronze Age settlement at all. Both *A. tuberculatum* and *Donax* spp. are consumed as seafood today in the Mediterranean (Davidson 1972: 202, 206 and per. obv. in Alexandria, Egypt).

While *T. decussatus* remains appear to be numerous in Troia, as mentioned earlier, their registered quantity at the site is questionable (cf. Sections 4.3.2 and 4.3.4). The taphonomy of a few saved examples attests that they have been collected live (Figure 21f).

The river clam, *Unio* spp., is not a commonly encountered species in Troia. It is represented by a single shell in Ulucak and it is absent in the assemblages of Yenibademli. With the exception of the Bronze and Iron Age layers of Kastanas (cf. Section 4.6.1; Becker 1986), *Unio* shells are neither a common component of the archaeological assemblages of the coastal Aegean nor of the inland sites of Anatolia and Greece (cf. Section 4.6.1). There is, however, a considerable amount of evidence from other parts of the world indicating that freshwater bivalves constituted a significant food resource for some prehistoric human groups (Parmalee and Klippel 1974). River clams are edible, but they have also found other uses in human culture: for example, their shells are used in para-medicine in Mexico and in the production of mother-of-pearl inlays in wooden handicraft products in Syria (per. obv.). Scanning electron microscope examinations on *Unio* finds in the Early Bronze Age contexts at Tell es-Sa'idiyeh in Jordan have shown that the specimens have been used as tools (Cartwright 2003). Although the presence of such tools or similar artifacts has not been detected among the specimens found in Troia, analytical research designed specifically to answer this question needs to be incorporated into the archaeomalacological study of this site, before the nature of the *Unio* spp. finds in Troia can be clarified.

To summarize, among the gastropods identified in the archaeological assemblages of Troia and Yenibademli, only *Patella* spp. represents a significant source of human food. Two additional groups of mollusk remains, specimens of Monodontinae and *Cerithium* spp., also represent mainly human food. *H. trunculus* was probably not very important as human food, but it was used chiefly as a raw material in local dye production in Late Bronze Age Troia. The presence of marine gastropods in Ulucak cannot be linked unambiguously to their purpose as human food. *M. galloprovincialis, O. edulis,* and *C. glaucum* finds unequivocally represent elements of human nutrition. Their abundance, taphonomy, and frequent occurrence in the other sites around the Aegean, and ethnographic parallels, substantiate this conclusion. Among these three species, in Yenibademli only *C. glaucum* and *O. edulis* were frequently eaten, while in Ulucak, only the *C. glaucum* were found in amounts large enough to characterize as a noteworthy food source. In Troia, two additional bivalves, namely Pectenidae and *S. marginatus*, may be listed among the mollusks that contributed to human nutrition. All other bivalves were either infrequently shellfished or consumed as human food or their presence at the archaeological sites signifies other types of human activity. Beach-collecting of dead specimens to be used in tool manufacture or for decorative purposes is manifested by the archaeological specimens that were abraded or otherwise modified by natural agents.

The possibility that any one of the mollusk taxa represented in the archaeological assemblages discussed here was favored, in addition to dietary items, as fishing bait –though presumably not very frequently– cannot be excluded. The shells of mollusks –in modified or unmodified form– may have served other purposes at the settlement subsequent to human consumption. Certain overlapping may occur between those species used as food or in the preparation of fishing bait and those used as raw materials in manufacturing tools and ornaments. The discussion above makes it manifest that detailed taphonomic descriptions of single archaeomalacological specimens are essential to unfold the different layers of past human activity represented by mollusk finds or to eradicate, unambiguously, the kinds of activities they do not represent.

5.2. Shellfishing grounds

Analyses of mollusk remains from the coastal sites of the world demonstrate that the most commonly exploited mollusk species are those living close to the shore (Waselkov 1987, also cf. Section 4.6.1). Peoples of the world have gathered and still gather shellfish from the most accessible locations close to their home base –be it seasonal camps or year-round settlements– in order to meet the time and energy costs of the procurement activity. Shellfishing at the inshore waters, apart from unusual and rare incidents, is conveniently profitable: coastal waters, including the lagoons, estuaries, and the littorals of the world seas are more productive than open seas (Barnes 1980: 49). Modern malacological knowledge allows for the reconstruction of the location and characteristics of the aquatic environments where the archaeomalacological mollusk specimens were harvested.

5.2.1. Shellfishing grounds in the Troad

In the Troad, a great range of aquatic habitats were already present within the Trojan Bay. The irregular deltaic coastlines with swamps, barriers and lagoons formed at the locations where the distributary channels of the Karamenderes and Dümrek Rivers reached the marine waters of the Trojan Bay were fairly close to the settlement of Troia throughout the Bronze Age (Figure 2; Kayan 1996, 1997a, Kayan et al. 2003, Kraft et al. 1980, 2003a, 2003b). In the lagoonal areas, the water was turbid and brackish. The bottom was variably muddy and sandy both at the rivers and at the lagoons. At the northern section of the Trojan Bay salinities fluctuated, though to a relatively limited extent, due to the seasonally changing influence of the salinity of the Marmara Sea (Aksu et al. 2002) penetrating into the bay and depending on evaporation and precipitation rates. The benthic types here varied from coarse sandy bottoms to muddy bottoms. Hard substrates may have been located both on the sides of the northern section of the bay, close to the shores of Kumkale and Sigeon ridges, and perhaps in the middle of the bay, at a distance from the shore. Along with the deltaic lagoons, the marine indentations at Kumtepe, Kesik, and Yeniköy provided environments with quiescent sheltered waters. To the north, in the Dardanelles, a fully marine environment with constant water circulation prevailed. To the west, the Beşik Bay embodied a lagoon and a shallow and quiescent littoral with a sandy bottom (Kayan 1991), while the western side of the Sigeon Ridge sloped sharply to the Aegean, forming a rocky shore exposed to high-energy waves. A series of pools with sandy bottoms, depending on sea-level changes and the strength of wave action, may have formed along this coast, but these were probably only temporary formations.

In this setting, the sheltered lagoonal environments closest to the settlement provided the most suitable habitats for *C. glaucum* beds. Additionally, the eastern indentations of the Sigeon lime stone ridge, namely the present Kumtepe, Kesik, and Yeniköy Plains, would make well-sheltered locations for *C. glaucum* populations to settle. The lagoonal area of the Beşik Bay was also an ideal habitat for them. Kayan (1991: 88) mentions that the shells of *O. edulis* and *C. glaucum* were abundant in the cores taken from the sand barrier that formed between the lagoon and the coastal part of the Beşik Bay. The abundance of *C. glaucum* and *O. edulis* in the archaeological record of the prehistoric settlements at Beşik Bay (Boessneck 1986, von den Driesch 1999) confirms that these mollusk beds were exploited by the former human populations in the Troad. However, the Beşik Bay was probably of secondary importance to the Trojans seeking *C. glaucum*, since a larger habitat with similar characteristics already existed virtually at the foot of the settlement during the Bronze Age.

The morphology of *C. glaucum* remains in Troia gives further insight to the possible range of habitats explored to obtain this species. The morphological variability (solidity and curve range) of the archaeological *C. glaucum* populations from Troia (cf. Section 4.3.5.1, Figures 42-44) suggests that the specimens brought to the settlement stemmed from a variety of populations that colonized different areas of the Trojan Bay, and probably of the Beşik Bay lagoon, with differential salinity, temperature, and substrate conditions. The contrastingly limited morphological variety in the modern populations living in the limited areas of the deltaic lagoons of Karamenderes (Figures 72-74) not only attests to the broad area of the Trojan Bay in the Bronze Age already well-established by geomorphological studies of the Trojan Plain, but it also demonstrates the use of its wide area by the Bronze Age inhabitants of Troia as a shellfishing ground.

Juvenile members of *C. glaucum* communities dwell among floating seaweed (cf. Section 4.1.2.9; Brock 1991, Ivell 1979b). The absence of specimens smaller than 10 mm in valve height in the studied *C. glaucum* populations from the Bronze Age site indicates that floating seaweed or marsh vegetation close to the shore were not explored to gather cockles still at the juvenile stage (0W), provided that the reason for the absence of small specimens is not a consequence of the hand-collecting mode of excavation.

Other common elements of the archaeomalacological assemblage of Troia, namely *M. galloprovincialis* and *O. edulis*, were collected from the hard substrates in shallow waters: rocks, pebbles or inundated and cemented remnants of old sand barriers, all possibly located in the middle and northern sections of the Trojan Bay. They have probably colonized those unsheltered areas of the Trojan Bay where water movement allowed sufficient flow of nutrients. Such areas would exist close to the opening of the bay. The littoral of the Aegean and the Dardanelles could also be exploited to gather these species.

Patella spp. and Monodontinae could be gathered from the rocky shores on both sides of the outer section of the bay where the influence of freshwater is negligible, allowing waters with marine salinity levels to prevail. They could also be obtained from the Aegean shore of the Sigeon Ridge or the Dardanelles shore of the Kumkale Ridge (Figure 2). Both of these coastal areas were rarely visited by the Trojans for shellfish gathering. *H. trunculus* could be collected from rocky and sandy substrates in the mid and northern sections of the bay, at a considerable distance from the lagoons but in shallow areas. *S. marginatus* and *T. decussatus* were gathered from the calm, but clear areas of the outer Trojan Bay with fine sandy bottoms, and also possibly from the Beşik Bay.

The inhabitants of Troia did not search for shellfish among seaweed. This is indicated by the low representation of species such as *Cerithium* spp. and *Pinna* spp. in the archaeological record and the absence of other seaweed dwelling species. However, this part of the depiction of shellfishing grounds may have been manipulated by excavation techniques or archaeozoological identification methods employed at the site. Shellfishing activity very rarely took place at the infralittoral and circumlittoral zones, i.e., at the open waters of the Aegean Sea or the Dardanelles. The rarity of species commonly living in the lower infralittoral and circumlittoral zones, such as *A. noae*, Glycymerididae, Pectenidae, and *S. gaederopous* –if they were at all alive when they were collected– constitutes the evidence for this interpretation. It should be considered, however, that these species are known to penetrate into the upper infralittoral zones, such as the northern section of the Trojan Bay, where their scarce remains in Troia may have originated from. The sandy and muddy river beds of the Karamenderes and Dümrek close to the site were exploited for *Unio* spp. only occasionally.

5.2.2. Shellfishing grounds of Yenibademli

Like Troia, Yenibademli was formerly located on a rocky promontory protruding towards an estuary. The main elements of the aquatic environments closest to the settlement included the Büyükdere River, the estuarine bay fed by this river, and the open sea to the north (Figure 3). This bay is surrounded by high ridges on both sides and is indented by the outreaches of these ridges. The ridges flanking the estuarine bay slope abruptly into the outer sea to the north, forming rocky shores. The geomorphological reconstruction of the formation of the Büyükdere Plain places the deltaic environment created by the distributary channels of the Büyükdere River immediately to the southwest of the Early Bronze Age settlement (Öner 2000, 2001). Accordingly, the northern slope of the promontory, on which the settlement was located, directly touched marine muds at the time of the Early Bronze Age occupation (Öner 2000, 2001).

If the reconstruction of the former location of the deltaic lagoons provided by geomorphological investigations at Yenibademli is accurate, then the main shellfishing grounds exploited by the inhabitants of Yenibademli should have been located ca. 1 km to the north of the settlement close to the opening of the estuary, or ca. 0.5 km further from the northeast and northwest of the estuary's opening, at the Aegean coast. According to the current results of the geomorphological investigations on the Büyükdere Plain, areas remaining to the south of these would not have provided the high-energy, saline and rocky environments necessary for *Patella* spp. colonization, which is the most abundant mollusk species represented in the archaeological deposits of Yenibademli. Instead, areas closest to the settlement would have hosted excellent beds of *C. glaucum*. Therefore, if it is considered that the geomorphological reconstructions reflect an accurate picture of the morphological features of the bay during the Early Bronze

Age, then it should be suggested that the deltaic lagoons, which were supposedly at the foot of the settlement, were only rarely exploited for shellfishing. *O. edulis*, theoretically colonizing the parts of bays where marine conditions prevail, could only be collected to the north of the settlement, within or outside the embayment. *O. edulis* remains are more common in Yenibademli than the *C. glaucum* remains, a fact which constitutes another point suggesting that the people of Yenibademli collected shellfish further away from the settlement and to a large extent deliberately chose to avoid exploiting the shellfish beds nearer to them, if the geomorphological descriptions are accurate. This depiction of the commonly exploited shellfishing grounds based on the current archaeomalacological and geomorphological evidence about the site demands further discussion. This issue will be re-visited towards the end of Section 5.2.

In Yenibademli, the relative abundance of the *P. caerulea,* which are more common organisms in the upper supralittoral zone than other *Patella* spp. living in the Aegean (cf. Section 4.1.1.1; Öztürk and Ergen 1999), implies that limpet gathering mainly involved the exploitation of upper supralittoral zones of the rocky shores. The low archaeological representation and the taphonomy of species other than *Patella* spp., *O. edulis*, and *C. glaucum* indicate that the infralittoral and the circumlittoral were very rarely visited to procure shellfish. Most of the species living in these environments, such as *S. gaederopous* and Glycymerididae, appear to have been collected on or close to the shore, when their shells were already empty. The Büyükdere River was either not exploited for riverine clams, *Unio* spp., or there were no *Unio* spp. populations present in this river (Balık and Ustaoğlu 1993 and per. com. M.R.A. Ustaoğlu).

5.2.3. The origin of mollusk assemblages at Ulucak

The only body of water in the area immediately surrounding Ulucak is the Nif Stream. It is most likely that the marine mollusks recovered from the excavations in Ulucak belong to populations living in the İzmir Bay situated to the west of the settlement. The coastal environment of the inner part of the İzmir Bay was defined by the rivers formerly flowing into the sea from the north, which created brackish environments with muddy substrate, beaches to the east, and shallow quiescent waters (cf. Section 2.2.3 and 2.4.3).

The archaeomalacological specimens in Ulucak belong to the mollusk species that live in habitats close to the shore in the supralittoral to infralittoral zones and lagoons, on substrates ranging from muddy to rocky. Among the mollusks represented in Ulucak, probably only *C. glaucum* was brought to the site to be used as human food (cf. Section 5.1). These *C. glaucum* may have originated from the eastern or northern coast of the inner part of the İzmir Bay. The easternmost littoral of the İzmir Bay, with sheltered waters and sandy bottoms, would make suitable habitats for this species. Likewise, the northern shore of the inner part of the İzmir Bay, where the deltaic lagoons of the Gediz and other rivers were formerly located, was also an ideal location to procure *C. glaucum.* It is not possible to tell with any certainty whether the *C. glaucum* habitats closest to the site or those located further northeast were chosen to gather *C. glaucum.*

The fact that only one specimen of *Unio* spp. was present in the assemblages from the Chalcolithic and Early Bronze Age layers of the site demonstrates that the exploitation of the Nif Stream for shellfishing was negligible during these periods – if this specimen is at all a member of the *Unio* populations of this particular river and if it had been collected while still living.

5.2.4. Summary

The mollusk species found in the three sites under investigation in this study come from coastal areas: lagoons, estuaries, and upper littorals of bays, and rocky shores. This pattern of coastal exploitation for shellfishing is in good agreement with archaeomalacological evidence from coastal shell-bearing sites from the Aegean and elsewhere in the world (cf. Section 4.6.1). The lower infralittoral and the circumlittoral zones were rarely exploited for shellfish. Mollusks were not sought among sea grass communities, underneath rocks, or in deep waters. In Troia and Ulucak, mollusk communities closest to the settlements were of prime interest for shellfishing. The geomorphological reconstructions of the nearby aquatic habitats fit well with the archaeomalacological reconstructions of the principle areas exploited for shellfish. Exploitation of rivers for mollusks constitutes an exception shellfish gathering took place at the rivers only rarely. The reasons for this may be related to the low density of mollusk communities inhabiting the rivers or to culinary preferences.

In Yenibademli, the comparison between geomorphological and archaeomalacological results does not correlate well with evidence from other sites where the correlation between geomorphological and archaeomalacological results leads to the conclusion that coastal inhabitants exploited the most accessible mollusk communities (cf. Section 4.6.1). A series of cultural factors that may have influenced foraging decisions in Yenibademli can be offered in an attempt to explain the present picture, i.e., that the primary shellfishing grounds of Early Bronze Age Yenibademli lay further away from the settlement. But such an attempt is unlikely to contribute to the discussion beyond speculation. This issue can be resolved only when the present chronology of the geomorphologic evolution of the Büyükdere Plain is revised with the aid of scientific dating methods, e.g., radiocarbon analysis.

Clearly, humans did not only collect shellfish from the areas closest to their home base. What caused them to seek mollusks at relatively distant habitats? During periods when shellfish consumption necessitated frequent activity, it is possible that people constantly moved between gathering locations. People could have changed their gathering location to obtain the targeted species, abandoning certain mollusk communities when their resources were exhausted, to seek recovered populations with normalized age profiles and new recruits. Hunter-gatherer-fisher societies had to relocate their camps frequently for the same reason (Waselkov 1987). Alternatively, they could have sought species other than those frequently targeted, moving the gathering locations to places with completely different environmental conditions. Alteration between gathering locations can also take place to enhance culinary variety. Seasons may also have affected the location of gathering activity –gatherers may have frequented different types of shellfishing grounds at different times of the year. Shellfish gathering societies of Northwest America, for example, avoided the outer beaches during the winter, when storms prevented access to these areas (Deith 1985).

It is also possible that the location of shellfishing grounds may have been altered in accordance with the character of the tasks that needed to be accomplished near the coast. The shellfishing grounds were essentially similar to fishing grounds. It is suggested that the available technical know-how in the Aegean greatly hindered the exploitation of pelagic migratory fish and fish of the continental shelf, binding their activity to inshore waters, lagoons and estuaries in the Bronze Age (Rose 1994: 188, 385). While the fish bone material available from the Bronze Age layers of Yenibademli and Ulucak are too limited to make any suggestions about the relationship between fishing and shellfishing grounds, the study of fish bones from Troia is in accord with Rose's (1994) interpretation of fishing grounds in the Bronze Age Aegean (Uerpmann and van Neer 2000), hinting at a possible socio-economic relationship between fishing and shellfishing activities at this site.

5.3. Techniques of mollusk procurement: gathering and cultivation

The possible techniques employed to obtain the mollusks in the archaeomalacological assemblages in question have been reconstructed based on ethnographic accounts and molluscan biology. The picture attained from this reconstruction diverges considerably from Homer's (*Iliad* XVI, 745-748) heroic depiction of oyster diving practiced in the Troad. Although diving for shellfish is not a practice completely avoided among the shellfish gatherers of the world (Waselkov 1987), in general, shellfish gathering does not require special techniques or skill. Most species can be gathered easily in terms of energy expenditure and strength requirement (Meehan 1983). Most of the mollusks represented in Troia, Yenibademli, and Ulucak were gathered using unelaborated techniques.

Patella species can be gathered by hand or using a lever made of wood, bone or metal (Forbes 1976a, Waselkov 1987).[33] Hand-picking them necessitates swiftness of movement, for once an individual is touched, it immediately secures a stronger grip to the hard surface it has been grazing upon and does not loosen this grip for some time. The large quantity of *P. caerulea* in Yenibademli implies that most of the shellfishing activity at this settlement could be done without even getting one's feet wet. The same is probably true about the way *Patella* spp. was collected around Troia. Monodontinae could be gathered from the same places where *Patella* spp. specimens were obtained. When Monodontinae populations were collected during their upward migration, it was also possible to obtain the individuals without going into the water. Gathering Monodontinae does not necessitate the use of any implements.

H. trunculus, which live on sandy and rocky substrate, can be gathered by hand. But the relatively low density of *H. trunculus* communities can turn the hand-collecting of this species into a time-consuming task. Instead, these carnivorous marine snails can be caught by baited baskets or ceramic pots (Ruscillo 2005). In order to scavenge the meaty bait, individuals gather up into vessels, from which they can be easily captured. Theoretically, other carnivorous marine gastropods, such as *P. granulatum* and *C. tritonis* can also be baited using the same technique.

M. galloprovincialis can be dislodged by lots from their dense beds by hand. Similarly, *O. edulis* can be obtained by dislodging the shells from the hard substrate they live on using a wooden stick or rock (Waselkow 1987). As illustrated in Figure 59b, specimens found still attached to the rock in Yenibademli attests that at least a portion of oysters were brought to the site without dislodging them from their natural benthos. Alternatively, the inhabitants of Yenibademli may have queried the oyster beds for building material, without bothering to remove the oyster shells from the stones of their selection. The technique of obtaining *O. edulis* from aquaculture represents a technical innovation dating to the Roman times. According to written sources, artificial oyster beds were built in Italy as early as 97 BC. At least since the 4th century AD oysters were cultivated using ropes. Roman glass vases from Italy depict oysters attached to solid crass poles of wood. (Günther 1897). The relationship between this innovation and the *O. edulis* remains from the Post-Bronze Age layers in Troia will be discussed more thoroughly in Section 6.3.

Individuals of *C. glaucum* can be gathered by hand or by using implements such as shovels and dredges that are made of wood or metal. Once a *C. glaucum* bed is located, which can be done by feeling for the individuals by foot, numerous individuals can be acquired in a small area and in a short time, without having to move around too much. The gathering can be done while semi-immersed in the lagoon or estuary water. The coeval presence of very small and very large individuals of *C. glaucum* in the archaeological deposits of Troia and Yenibademli (Figures 40, 63, and 64) indicates that no pre-selection took place during their harvest. The size range of the archaeological populations indicates that the individuals were collected randomly, without size discrimination.

[33] Numerous finds of bone, stone and metal tools have been found in the sites that are dealt with here, but none have been associated with shellfishing activity.

Retrieving *S. marginatus* individuals from their habitats constitutes a relatively difficult task. Securing sufficient yields requires certain experience, acquired either socially or by experimentation (Davidson 1972: 207, 2004). *S. marginatus,* a species which is very sensitive to vibration and light, can burrow faster in the sand than a human forager can dig to retrieve the individual. They rapidly retreat to the sand when approached. *S. marginatus* communities are not as dense or concentrated as, for example, *M. galloprovincialis* or *C. glaucum* communities. An acquaintance with the small hole-like trace they leave at the surface of the sand while burrowing aids the shellfisher to locate the individuals. In Brittany (France), *Solenidae* are collected by dropping salt into the suspected hole (Lacombe 2000), but individuals can also be reached using a spade or a similar implement.

Bivalves living in deeper waters can be caught using spears and pronged poles from boats (Waselkow 1987). Pectenidae, swimming bivalves, can be chased by hand in shallow water or haul trawled from a boat. Alternatively, they can be dived for.

Unio spp. can be easily gathered from the shallows of freshwaters or at times of the day when they crawl to the side of the riverbed, so that the shellfisher does not have to immerse him or herself in the water.

5.4. Shellfish processing

Modern examples and ethnographic studies show that shellfish gathered as human food is consumed in diverse ways: raw, boiled, steamed, pickled, dried, smoked, and fermented (Nash 2000). The archaeological traces of the activities involved in shellfish preparation are of differential nature and not very abundant. The taphonomic properties and the depositional patterns of the mollusk remains can lead to an understanding of the techniques used in shellfish processing prior to human consumption.

The most common way of consuming aquatic mollusks is to eat them fresh without any processing, often when the animal is still alive. All the marine mollusks representing food remains in the archaeomalacological assemblages investigated in this study can be consumed raw (Davidson 1972: 189-209). Sand-dwelling species, such as *C. glaucum, T. decussatus, S. marginatus*, et cetera, must be kept in salt water in order to remove the residual sand particles contained within the valves prior to consumption. More or less intact shells or shells bearing traces of deliberate breakage constitute the best evidence for human consumption of fresh shellfish flesh. It has been suggested that the ventral-posterior margins in *C. glaucum* and *Unio* spp. shells in Kastanas were deliberately broken off during meat extraction with a sharp implement (Becker 1986: 235-236). Gejvall (1938), too, reasoned that such breaks on *C. glaucum* valves in Troia were created with a knife-like implement (cf. Section 2.8). In the present study, it is maintained that this type of breakage, at least at the assemblages examined for this study, is not related to the processing activity, but occurs after the deposition of the valves (cf. Section 4.3.3).

While the consumption of raw shellfish flesh may be favored for its distinctive flavor, it is not the most effortless method of consumption. The muscles which enable locomotion are attached to the inner surface of the shell (Barnes 1987: 342). The removal of the soft-body by detaching it from these muscles and the shell is an energy consuming task. To avoid energy loss during consumption, the soft-body can be efficiently extracted from the shell by exposing whole animals to heat for a few minutes. Ethnographic studies recorded the practice of roasting limpets and other marine snails in the fire in the coastal villages of southern Greece (Forbes 1976a). Because the time necessary for the muscles to lose their function is very brief, the heating process leaves little or no trace on the archaeological shells. The abundance of unburnt, well-preserved *C. glaucum* specimens, as well as the presence of burnt specimens (Figure 35a) within the archaeological deposits recovered in Troia affirms that at least some of the *C. glaucum* individuals were prepared by boiling, steaming, or putting them directly on fire.

Accounts of intensification of gathering activity when mollusks are at their prime and then drying them for the lean period, as well as for trade inland in both hunter-gatherer-fisher and food producing societies, are known from ethnographic literature (Deith 1985, Henshilwood et al. 1994, Moss 1993, Waselkov 1987). Though not uncommon, shellfish drying is not a world-wide phenomenon. In some cultures, freshness of food is an absolute prerequisite for the edibility of resources (Meehan 1977). It can be empirically suggested that this is also partially true for Mediterranean cultures. Drying seafood for later consumption is a more common practice along the Atlantic coast of Europe than it is on the Mediterranean coast. While fresh seafood is a highly valued commercial product in the Aegean, an occasional sight in coastal villages is sun-drying fish and octopus (Davidson 2004, Forbes 1976a, and per. obv.). In Mediterranean climates, weather permits drying activities to take place during the dry period, usually towards the end of summer and before the beginning of the autumn rains. The archaeomalacological record can be investigated with the intention to trace the origins of such culinary tendencies in prehistoric times.

Results of experimental studies suggest that shellfish can be dried effectively for later consumption with relatively little effort and technology (Henshilwood et al. 1994). Henshilwood et al. (1994) suggest that the "mega-middens" of South Africa are remnants of mass shellfish drying activity. But unequivocal archaeological evidence for shellfish drying is scarce, because the process rarely leaves any trace on the mollusk shell. In Troia, Yenibademli, and Ulucak, such unequivocal evidence for shellfish drying is also not present.

Some of the clay-lined pits and similar structures such as water tanks recovered in cultural deposits have been interpreted as installations for shellfish storage (Bogoucki and Grygiel 1981, Homer 1955: 61, Özdoğan 1999). Özdoğan (1999) argues that the mollusk containing clay-lined pits found in the Late Neolithic layers of Hocaçeşme in Turkish Thrace represent specially prepared installations used to "store" mollusks. He sustains this interpretation by adding that the mollusks recovered from these pits were found sorted by size and species. Water tanks may have been used to enable surplus mollusks to maintain their fresh state temporarily. While contemporary aquaculture makes use of water tanks for mollusks and other aquatic animals, and keeping freshwater fish in tanks within settlements is known since the Sumerians (Nash 2000), the author is unaware of any ethnographic examples of the use of clay-lined pits used for mollusk storage. Clay-lined pits containing mollusk remains have been excavated in Troia from the Late Bronze Age phases of the site (Quadrate I8, Beh. 388: Unit 1016; Quadrate K8, Beh. 749: Unit 881; Quadrate A8, Beh. 1371 and 1389: Units 3464 and 3480 respectively). Such clay-lined pits may have been used as potable freshwater tanks and subsequently filled with rubbish, including mollusk shells, which also appear to have been discarded in ordinary pits subsequent to consumption. Unambiguous evidence hinting at mollusk storage is not present in the primary sites discussed in this stuy, while concomitant evidence, such as charred food remains found in pits and pithoi demonstrate that the concept and technology of food storage was well in practice.

On the other hand, Mesopotamian recipes written on cuneiform tablets suggest that the addition of shellfish flesh to fish sauce (the precursor of the Roman *garum* or similar) was a culinary innovation practiced as early as the 17th century BC (Bottéro 1996, Morales-Muniz 1993). Whether the acts of preserving shellfish in fish sauce or separately with salt or in salt water have been practiced in Troia, Yenibademli, or Ulucak cannot be proven with the current evidence. But, as indicated by textual evidence, the technique of preserving seafood in such ways was present within the contact realm of Troia in the Late Bronze Age (Becks 2006). Therefore the possibility exists that this preservation technique was known in Late Bronze Age Troia. It is, however, very questionable whether the environmental and climatic circumstances at Troia or Yenibademli would at all necessitate storing a food resource that was available year-round.

In addition to the activities pertaining to mollusk processing discussed above, ethnographic records and ethno-archaeological studies provide detailed information about off-site processing and consumption of mollusks. Ethnographers who worked among the hunter-gatherers and agricultural coastal dwellers report that not all shellfish are carried back to the home-base with the shell (Forbes 1976b, Meehan 1982). Some are de-shelled prior to transportation and occasionally shellfish are consumed during gathering (Buchanan 1988: 26). Bird et al.'s (2002) systematic ethno-archaeological studies show that processing and consumption of shellfish at or near the site of procurement manipulates contemporary household accumulations to the extent that the species composition of household shell refuse does not reflect the choice of prey. Off-site processing is clearly not an archaeologically visible activity.

Processing shellfish for purposes other than usage as human food may also leave traces in the archaeomalacological record. For example, cooking shellfish is not an effective way to produce fishing bait, because cooked flesh is soft and easily breaks away from the hook. Therefore, it can be suggested that archaeological mollusk shells bearing traces of cooking were not used for fishing bait. On the other hand, the most efficient way of removing the flesh for preparing bait is by smashing the shell. In today's Greece, this method is used to extract the meat off of *C. vulgatum* shells to be used in fishing bait (Petersen 2004). In Franchthi Cave, broken shells of this species occur throughout the Mesolithic deposits (Shackleton 1988a and 1988b). If it can be confirmed that the shattering of the archaeological mollusk shells at this site resulted from deliberate action and not from post-depositional processes, it can be stated with a considerable amount of certainty that the Franchthi specimens were not used as human food, based on the fact that shattering causes shell fragments to cleave to the meat, disabling human digestion.

In summary, while ethnographic studies of hunter-gatherer-fisher societies and agricultural communities show that an array of activities are involved in shellfish consumption following shellfish gathering, in contrast to the high visibility of archaeological mollusk remains, the activities concerning their consumption are difficult to detect from the archaeomalacological record. Contemplating on these processes contributes two very important considerations towards more accurate interpretations of the archaeomalacological records of Troia, Yenibademli, and Ulucak, and sure enough, at other archaeological sites: 1) the mollusk remains may represent individuals that were consumed months after they had been gathered; and, 2) the mollusk remains found on the sites might represent only part of the shellfish consumed by the former inhabitants.

5.5. Nutritional value of mollusks

One of the common quantitative methods used to assess the dietary and economic value of mollusks in a past society is to calculate the nutritional value (i.e., calorific and protein contents) represented by an archaeomalacological assemblage. While such an analytical approach is not suitable for the analyses of the archaeomalacological assemblages studied here (cf. Section 3.3), the nutritional value of molluscan flesh should not be overseen as a factor that may have influenced decision making in mollusk foraging. Mollusk flesh has a chemical composition of lipids, inorganic material (minerals such as iron, phosphor, calcium, magnesium, potassium, zinc, et cetera), protein, and carbohydrates (Prieur 2005). Moreover, the amino acids it contains are proportionally comparable to eggs, meat, and milk (Nash 2000). It is also rich in vitamins, especially in vitamins A and B (Souci et al. 2000). The Mytilidae also include vitamin C in small amounts (Souci et al. 2000: 482). Based on similar data pertaining to the nutritional value of mollusks, some scholars have argued that nutrition secured from mollusks can easily compete with other food items in the daily menu, if not augment or replace them when necessity arose (e.g., Erlandson 1988).

Table 12. Nutritional values of commonly represented mollusk taxa.

Species	Water	Lipid	Protein	Carbohydrate	Study Area	Source
P. caerulea	75.4	7.3	9.9	4.1	Antalya Bay, Turkey	Küçük et al. 1996
M. articulata	66.1	0.4	12.8	17.4	Antalya Bay, Turkey	Küçük et al. 1996
M. galloprovincialis	85.1	14.0	8.7	13.6	Dardanelles, Turkey	İlhan and Gülyavuz 2004
P. jacobeus	77.5	23.6	10.7	17.6	Dardanelles, Turkey	İlhan and Gülyavuz 2004
O. edulis	76.7	15.8	15.5	13.0	Dardanelles, Turkey	İlhan and Gülyavuz 2004
C. glaucum	83.2	10.0	-	15.4	Lago Lungo, Italy	Ivell 1979a
T. decussatus	83.9	13.8	9.8	12.6	Dadanelles, Turkey	İlhan and Gülyavuz 2004
U. crassus	81.7	4.0	5.6	37.3	Antalya, Turkey	İlhan et al. 2004

All values are expressed in (%) percentages.

Furthermore, some properties of the mollusk flesh differ considerably from other animal flesh. Mollusk flesh is among the few foods originating from animals, which, aside from protein and fat, also contain carbohydrates (Griffiths 1977). It also holds a different line of proteins, acids and fats that can contribute to the human diet in ways not comparable to the properties of meat yielded from mammals and their secondary products (Meehan 1977 and 1983). The ratios of nutritional components vary depending on species and on populations, but independent of flesh and shell weights. Species-specific nutritional studies of mollusk flesh are scarce and scattered in the scientific literature. This is especially true for the Mediterranean populations.

Today nutritional values of different food items play an important role in the consumption behavior of western societies. While it is true that the chemical components of different species could not have been known in the past, the unconscious 'craving' for protein, lipids, and certain minerals, et cetera, might have played a role in the choice of mollusk prey. Certain mollusk species may have been favored over others because they were fatty, sour or saturating. Likewise, some species may have been avoided because nutritional yields do not meet the energy spent to obtain them. The desire for certain species with different tastes, thus with different nutritional properties, may have stimulated movement between different gathering locations.

The nutritional values of eight common mollusk taxa (or their close relatives)[34] represented in the assemblages are reviewed in Table 12. Studies that present results from the analyses of populations obtained from locations closest to the archaeological sites were chosen for the comparative review of nutritional values. Commonly presented nutritional values in the literature are the average lipid, protein and carbohydrate contents. From the figures provided in Table 12, it can be deduced that there are marked nutritional differences between the species. *P. jacobeus* and *O. edulis* are the richest species in lipids, proteins and carbohydrates, whereas other species are variably rich in only one or two of these elements (Table 12). It can be inferred that these two species are also the richest in kcal g $^{-1}$. The lipid content of *M. articulata* is very depressed in relationship to its carbohydrate content. *U. crassus* flesh is depleted in lipids and proteins, but rich in carbohydrates. *P. caerulea* is poor in carbohydrates. The nutritional contents of *M. galloprovincialis, C. glaucum*, and *T. decussatus* are similar.

Additional nutritional data has been provided only for the *C. glaucum* populations at Lago Lungo, Italy (Ivell 1979a). Ivell (1979a) translates the lipid and carbohydrate contents of *C. glaucum* into average kcal g $^{-1}$ as 4.5, where the estimated amount of dry flesh weight per individual is 300 mg. Seasonal variations in nutritional values have also been provided only for *C. glaucum.* Ivell (1979a) observed that the condition index (i.e., dry flesh weight) of the *C. glaucum* at Lago Lungo was poorest during June and July in 1973. The calorific content of these populations was lowest between May and July. High summer temperature and accompanying dystrophy are seen as factors influencing these decreased nutritional values (Ivell 1979a).

Because of the general and seasonal variability in the nutritional value of aquatic mollusk species, it is possible to suggest that the differences in the nutritional values between the mollusk species represented in the archaeomalacological record of Troia, Yenibademli and Ulucak may have influenced seasonal or general preferences for consuming certain mollusk species. But it is not plausible to consider that aquatic mollusks have substituted or challenged the nutrition secured from other available food resources in the context of these Bronze Age settlements practicing advanced Mediterranean polyculture –gathering of wild plants, animal husbandry, hunting, and fishing. In the context of Mediterranean cultures with agriculture, even the domesticated cereals have high protein content: ca. 8-14 %

[34] *Pecten jacobeus* has been included in this table as a representative of Pectenidae. Information about the nutritional values of other Mediterranean species of this family is not present. Likewise, *U. crassus* is included as an analogue for *Unio* spp. and *M. articulata* represents Monodontinae.

(Diamond 1997: 138). The protein contents of the commonly represented mollusks (Table 12) are in the same range with the protein contents of the common cereals recovered at these sites, namely einkorn, emmer and barley. Moreover, ethnographic studies revealed that even in societies who are capable of creating shell-middens consisting almost only of mollusk remains, shellfish provided ca. 10-15 % of the protein and ca. 5-9 % of the calories obtained from the animal sector of the diet of the whole population (Meehan 1977).

5.6. Harvest time

The wide-ranging role of shellfish within the annual economic cycle of hunter-gatherer-fisher groups is demonstrated by ethnographic studies (Deith 1985). The variety of these roles show that determining the harvest time of archaeological mollusk assemblages constitutes valuable information for revealing the role shellfish played in the subsistence strategies of the past societies under investigation. Naturally, the data about the harvest time of archaeological mollusk assemblages is valuable in this regard, only when the factors affecting preference for one season or seasons over others have been laid out.

Ethnographic accounts of hunter-gatherer-fisher groups, as well as (although scarce) agricultural coastal societies are helpful in deducing the factors affecting the time of shellfish gathering. Malacological information on the behavior of modern mollusk populations presents further possibilities to interpret the gathering time of archaeomalacological populations. Ancient texts offer minute information on the seasonality of shellfish gathering in the Aegean. Factors influencing the choice of harvest time may vary from biological factors affecting the condition of the prey populations such as upwelling and spawning events; to environmental phenomena such as wind patterns and tidal rhythms; to economic factors such as the schedule of other subsistence activities, such as vertical transhumance, agricultural work, wild plant gathering, and the season of expected lean periods; to cultural activities such as religious ceremonies or festivities.

In his *Historia Animalium* (Book 8, p. 30) Aristotle writes that "mollusks are best in the breeding time, as the calamari, the sepia and the octopus." This excerpt implies that at least as early as the 5th century BC, a notion related to spawning times of aquatic invertebrates and in relation with it, a preference for certain times to consume marine mollusks existed in the Aegean societies. Most mollusk species have higher meat weight to shell weight ratios prior to and following spawning (Waselkov 1987). The periodic change in mollusk meat yield per individual, rather than an awareness of spawning times, may be influential in shaping the schedule of shellfishing activities. Ethnobiological studies among contemporary Brazilian shellfishers have shown that modern shellfishers who empirically gather shellfish at cyclical patterns when the shellfish is "fatty" (in their own words), do indeed collect shellfish at times when the mollusks are relatively more nutritional than at other times (Nishida et al. 2006). On the other hand, while spawning time is regarded as a good time to eat shellfish by some, for others the same period is time to avoid shellfish consumption, because the flesh is fatty, and its appearance is not particularly palatable (Davidson 1972: 197). Seasonal upwelling events also enhance the nutritional value of the mollusk flesh by bringing up waters that are enriched in nutrients and trace elements from deep waters (Claassen 1998: 25-26).

Modern studies among shellfish gatherers in Brittany indicate that unfavorable wind is one of the most important factors negatively influencing shellfish gathering on the tidal flats (Lacombe 2000). On the other hand, tidal rhythms appear to be a more important factor affecting shellfish gathering strategies among coastal foragers living at places where tides constitute a significant co-efficient in the coastal environment (Bird and Bliege-Bird 1997 and Thomas 2002). In such environments, shellfishing activity increases during daily and seasonal low tides.

The seasonal availability of resources is a very important aspect in the annual rhythm of the subsistence activities pursued by societies populating temperate climatic zones. Thus, the timing of activities concerning resources that are available year-round is dependent on the timing of other subsistence activities. It can be expected, for instance, that sedentary agricultural coastal dwellers do not spare time for shellfishing, fishing, or procuring raw material for specialized industries when they are busy with agricultural activities (Deith 1986). While ethnographic studies focusing on shellfish gathering strategies among agricultural peoples are generally meager, two accounts from the southern Aegean report on the role of shellfish in coastal villages. Leake (1830, cited in Purcell 1996) observed that during the summer, the coastal villagers of the Peloponnese Peninsula supplemented the inadequate agricultural reserves by eating shellfish from the rocks. According to Leake's (1830) description, the nutrition gained from mollusk gathering supplements food shortage during expected lean periods. The mollusks that are referred to in this account must be *Patella* spp. or Monodontinae. In the 20th century, in her observations on the gathering practices in the Argolid, Forbes (1976a) notes that marine mollusks were gathered throughout the year, but the shellfish gathering activity became intensified during the summer. According to Forbes (1976a: 131), shellfishing activity intensifies during the summer, because it is the "most pleasant and practical time for it." Although there is no indication in Forbes' (1976a, 1976b) accounts that shellfish gathering intensified during the summer to supplement the shortage of wild plants or any other wild or domestic food resource –domestic or wild –, Deith (1988a) linked Forbes' account of the intensification of shellfish gathering in the summer to the shortage of wild plants during the dry season. This situation demonstrates how much care should be taken when interpreting the ethnographic accounts for the purpose of archaeological explanation.

In this study, the stable isotope geochemistry and the incremental growth pattern of only one species, namely *C.*

glaucum, from the Bronze Age layers of Troia, has been investigated with the aim of determining the schedule of cockle harvesting at this site. The results of these investigations indicate that the Bronze Age cockle populations were collected year-round, with increasing activity during warmer and drier months (cf. Section 4.7.4). When factors affecting decisions concerning harvest time are evaluated in view of environmental phenomena, such as the results of isotopic analyses, modern *C. glaucum* behavior (i.e., growth patterns), and the Bronze Age subsistence economy of Troia, the motives behind the emphasis on cockle gathering during warmer months can be construed and the role of this resource in the annual economy can be interpreted with greater accuracy.

In temperate regions, spawning in marine mollusks generally occur seasonally when water temperatures reach appropriate levels. Although this time is the summer for most marine species of the Mediterranean (Atay 1984), at least three different spawning intervals have been recorded for *C. glaucum* populations living in the Mediterranean (Ivell 1979a and Zaouali 1980). While it is not correct to appropriate these observations for the *C. glaucum* populations in the Troad, the variability in the valve heights of modern and archaeological individuals that were collected at Stage 1 implies that the spawning behavior of the Bronze Age *C. glaucum* populations in the Troad also followed a polycyclic pattern (cf. Section 4.7.4). If the implication that spawning occurred year around can be vindicated by malacological studies in the Troad, then it can be suggested that the condition of the cockle populations before or after spawning did not play a decisive role in the scheduling of cockle gathering in Bronze Age Troia. Further reflections on the possible relationship between spawning and cockle gathering times should be avoided until malacological studies on the spawning pattern of the *C. glaucum* populations in the Troad are pursued.

In Section 2.4, it was reasoned that the upper Marmara current in the Dardanelles must have penetrated the Trojan Bay, influencing the water conditions there. Upwelling at the southern end of the Dardanelles occur between late Summer and Autumn, but the normally positive effect of upwelling events is depleted due to the lack of nutrients in deep Black Sea water (Jönsson 2002). These upwelling events must have either a negligible or a negative effect on the *C. glaucum* populations living in the lagoons of the Troad. Summer dystrophy, on the other hand, may have been an important factor influencing the condition index of *C. glaucum* populations in the Troad. It is unknown whether the upwelling cycle in the Dardanelles was identical to the present. It has been recorded that the flesh and calorific content (i.e., the condition index) of the *C. glaucum* populations living in Lago Laguno (Italy) was very low during late spring and early summer months due to the effect of summer dystrophy (cf. Section 5.5; Ivell 1979a). It is probable that summer dystrophy negatively affected the nutritional properties of the Bronze Age *C. glaucum* populations in the Troad as well. If that was the case, since the archaeomalacological evidence points at increasing cockle gathering activity during the warmer months in the Bronze Age Troia, it cannot be suggested that changes in the nutritional values of cockle populations was a significant factor affecting the time of cockle gathering.

The tidal factor can be altogether omitted as a possible condition affecting shellfish gathering in the Aegean. The tidal effects in these waters are negligible. The strong and persistent northerly winds prevailing in the Dardanelles area (See Neumann 1991) can, however, be considered as an environmental factor affecting shellfishing activity. These winds have the potential to make outdoor activities rather unpleasant year-round, rarely stopping during summer days (Neumann 1991). However, despite the negative effects of the winds, necessary outdoor activities are pursued throughout the year in the Troad. While cockle gathering may have been occasionally cancelled due to strong winds, the archaeomalacological evidence does not offer any hints suggesting that the prevailing winds during the summer or any other time of the year were a major factor influencing cockle gathering. It is much more plausible to consider the wind as a factor hindering pelagic activities (e.g., tuna fishing) in the Bronze Age Troia.

In the absence of written records or iconographic depictions, the archaeological record in Troia is devoid of any evidence of cultural activities such as ceremonies, festivities, or fasting that may have affected the timing of certain subsistence activities. Since transhumance was probably not a necessity of subsistence in Troia (Riehl 1999: 58), the impact of such activity on the schedule of shellfish gathering is, then, also not a subject that needs to be addressed.

In light of the current evidence, it is most plausible to consider that the schedule and necessities of other subsistence activities determined the annual schedule of *C. glaucum* collecting at Troia. *C. glaucum* was available year around; resource availability was not a factor affecting the decisions regarding when to gather cockles. However, the seasonal availability of resources predetermined the schedule of other subsistence activities in the Troad.

In the Mediterranean, most household and agricultural activities are conducted during the dry season. Cereals are harvested in the summer, while olives are harvested in late autumn. Grapes become ripe in early autumn, shortly before the olives. Legumes are harvested in late spring if they are going to be consumed fresh and in August if they are to be stored. Mudbrick is produced during summer months. Firewood is provisioned before the beginning of cold months. Overall, the rainy months of the winter can be considered a rest time for the agricultural communities of the Mediterranean and today's Troad is no exception.

On the other hand, subsistence activities pertaining to the exploitation of wild resources follow a different annual pattern. For example, tuna fish (*Thunnus thynnus*) can be caught between April and December in the Dardanelles, with a short dead season between January and April during which the tuna remain sedentary in the Black Sea (Uerpmann and van Neer 2000). But the lagoons and estuaries, where most of the fishing activity in Bronze Age Troia took place, remain productive fishing grounds year around (Whitehead et al. 1987). Gathering the green leaves of a variety of wild plants, such as sweet fennel (*Foeniculum vulgare*) and common mallow (*Malva sylvestris*) in Greece and Turkey is done during the rainy months (Forbes 1976a and 1976b and per. obv.). Likewise, hunting can be pursued year around.

The emphasis placed on cockle gathering activity during the warmer months may have arisen from the need to supplement the diet when agricultural reserves were becoming smaller and wild plants were scarce due to low precipitation rates, in a manner similar to Leake's (1830) account of shellfish gathering as practiced during historical times in southern Greece (cf. above, this section). Unequivocal archaeological evidence indicating short or prolonged periods of food shortage crises is absent in the archeological record of Troia. If the motive of cockle gathering during the dry periods was to supplement the shortage of other food resources, then these were probably recurrent lean periods, not completely unexpected single events. There is no evidence indicating that *C. glaucum* was a resource that was consumed solely during food shortages or at times of crisis. On the other hand, the emphasis on *C. glaucum* gathering during the warmer months may be indicative of the leisurely nature of the activity, which was conducted more frequently during the dry months simply because warm water temperatures and dry air make the gathering activity more pleasant and easier to pursue, as reported by Forbes (1976a). Both of these explanations are plausible. Although these two explanations ascribe different roles to *C. glaucum* gathering in the subsistence economy of Troia, the procurement of this year-round available resource with intensified activity during the warmer months shows that *C. glaucum* exploitation was an integrated part of the subsistence strategies and the place of *C. glaucum* was well-established as an augmentation to the diet.

Finally, it should be acknowledged that seasonality data can inform on the time of year when aquatic mollusks were collected, but for agricultural societies who are experienced in food storage, the season of shellfish collecting does not have to be synonymous with the time of shellfish consumption. As Deith (1985) states, "the only indisputable information [inferred from the seasonality data of mollusk remains] about the people who discarded the shells is that they were on the shore at the time(s) of the year indicated by the seasonality studies."

5.7. Dye production in Troia

Some species of carnivorous marine snails, such as the members of the Family Muricidae and Thaiidae, secrete a purple substance through their hypobranchial gland (Reese 1980, Sabelli 1987). This purple substance can be extracted using a variety of techniques to produce a color-fast dye (Baker 1974). The tradition of producing dye from these marine snails is wide-spread along the coasts of the world, from South America to Japan (Reese 1980). In studies of Mediterranean archaeology and history, this tradition has received particular scholarly attention because the phenomenon offers one of the rare inter-disciplinary opportunities to verify or appraise ancient texts with archaeological evidence (Becker 2001). Ancient authors, from Homer (*Odyssey* 6.53: 306) to Pliny (*Natural History* 9.62.135), mentioned the dye in association with trade-related prosperity and political prestige and provided detailed descriptions of production methods (For a review of textual evidence see: e.g., Blum 1998, Reese 1980, Ruscillo 2005).

In 1879, von Martens (and later Virchow in 1880) suggested that the Muricidae remains from Troia constituted the archaeological evidence for the dye industry mentioned in ancient texts. This was the first time in Mediterranean archaeology that mollusk remains were associated with this dye industry. Discoveries of several other sites with Muricidae remains and a great deal of scientific and popular literature written on the subject followed. Many scholars (e.g., Ackermann 1996, Baker 1974, Ruscillo 2005, per. com. A. Aksoy) experimented with ancient descriptions of production methods to reproduce the dye and investigated its chemistry, whereas others reported the remains of Muricidae and Thaiidae in the archaeological contexts of the Mediterranean as evidence for local dye production (e.g., Becker 1996, 2001, Çakırlar 2003, Gates 1999, Karali 1999, Minniti 2005, Reese 1980, 1987b, 2000, 2005, Stieglitz 1994, Zidermann 1990). The vast literature concerning the archaeological evidence for 'murex' dye production cannot be reviewed here in detail. The purpose of this section is to evaluate *H. trunculus* remains in Troia and to assess their relationship with the local dye production at this Bronze Age settlement.

There are three species in the Mediterranean that secrete the purple substance used to manufacture this dye: *Hexaplex trunculus, Bolinus brandaris,* and *Purpura haemastoma* (Reese 1980). The presence of only one of them, *H. trunculus,* has been securely attested in the assemblages from the New Excavations in Troia (cf. Section 4.3.2). Reese, who studied shell samples from Blegen's excavations in Minnesota and Stockholm, also recorded only *H. trunculus* remains (Reese 1987b and per. com. D.S. Reese). Reese's studies confirm Gejvall's (1938: 4) account, which describes the *H. trunculus* remains coming from the Troia VI levels uncovered during Blegen's excavations. Although von Martens includes *P. haemastoma* in his list of mollusk remains from Schliemann's excavations (1879, cf. Section 4.2.1), based on the observations on the archaeomalacological assemblages from the New Excavations and Reese's reports, it can be argued with confidence that *H. trunculus* predominates among the species of marine snails represented at the site and suitable for the production of this organic dye.

It might be appropriate to mention once more that *H. trunculus* is edible and is eaten in the Aegean (cf. Section 5.1; Davidson 1972: 192). Its flesh is also suitable to be used as fishing bait. After the flesh is eaten, used as bait, or collected for dye manufacture, its shells can be used as building material (Çakırlar 2003, Gates 1999, Karali 1999: 45) or burnt for lime production (Gates 1999, Reese 1980). Current archaeological evidence from Troia does not imply that their shells have been recycled as building material or lime production at this settlement.

Since *H. trunculus* flesh is edible and suitable to be used as fishing bait, a re-evaluation of the *a priori* associations of *H. trunculus* remains with dye production in Troia is necessary. In this way, erroneous and exaggerated interpretations of the archaeological record can be avoided. Careful examination of the amalgamated evidence of *H. trunculus* remains in Troia with regard to the current understanding of 'murex' dye production can aid a more accurate assessment of the nature of dye production at this site.

A number of conditions need to be met for the production of the 'murex' dye. Obtaining sufficient amounts of raw material at low production costs is the first pre-requisite. The Late Bronze Age settlement of Troia was overlooking a shallow estuary. This estuary provided an ideal habitat for *H. trunculus* colonies. The presence of *H. trunculus* remains at all periods of occupation in Troia attests that the inhabitants of the site exploited this habitat for variable amounts of *H. trunculus* catches from the beginnings of occupation in Troia until Byzantine times (Tables 25-29).

The 'murex' dye production has almost always been associated with local textile industries (e.g., Becker 1996, 2001, Korfmann 1998). Large amounts of snails are required to manufacture even the smallest amount of dye necessary to affect the tiniest piece of textile (Baker 1974, Doumet 1980, Reese 1980 and 1982, Ruscillo 2005, Stieglitz 1994). Ruscillo (2005) reports that her experiments with the murex dye required 160 individuals to produce 590 ml dye, which were used to dye four swatches of 15 cm x 20 cm of cloths. This result confirms the opinion that only large quantities of Muricidae remains can be shown as evidence for dye production and its relation to textile industry. Relatively recently, Becker (1996) introduced the idea that small quantities of Muricidae remains can also be considered as remnants of dye production, albeit at a house-hold level (1996). Becker's (1996) suggestion is one way of explaining small patches of crushed *H. trunculus* remains found in the archaeological contexts of the Mediterranean, including those isolated finds in Troia. But there remain two other possibilities to explain small amounts of archaeological *H. trunculus* remains: (1) they may have been the shells of those individuals used for fishing bait; and, (2) although they may have been used in dye production, they were not used to color textile, but to produce temporary tattoos in a manner similar to henna. Ruscillo (2005) notes that the stain of the 'murex' dye remains on the hands for about a month after the production process. The small relative amounts of *H. trunculus* remains uncovered in layers pre- and antedating the Late Bronze Age in Troia are similar to the relative abundance of Muricidae remains from contemporary sites elsewhere in the Aegean (cf. Section 4.6.1). This situation suggests that the specimens from the Early Bronze Age, Middle Bronze Age, and Post-Bronze Age layers were consumed as food or fishing bait, and possibly to tattoo the body.

H. trunculus remains in sufficient amounts, indicating the presence of a local dye industry comes only from the Late Bronze Age layers of Troia. Results from old and recent excavations show that most of the *H. trunculus* remains dating to this chronological period originate from the southeastern sector of the excavations around the Troia VI citadel wall (Schliemann 1881: 357 and Blegen 1938; also cf. Section 2.8 and for a much more detailed discussion of the 'murex' contexts uncovered during early excavations, see Çakırlar and Becks in print). Earlier excavators of Troia report that large accumulations of *H.trunculus* remains covered vast areas immediately inside and outside the Troia VI citadel wall (Blegen 1938, Schliemann 1881: 359). The deposits of *H.trunculus* found in Quadrates I/K8 immediately outside the city wall and tower VIh during the New Excavations should be connected to the deposits reported by Blegen between House VIG and the fortification wall, corresponding to Quadrates J7-8, and Schliemann's excavations (1881: 359). Unless the Late Bronze Age inhabitants of Troia were culturally immune to dangerously bad smells, considering the absence of evidence for a secondary use of the shell refuse for lime production or building use, the presence of large amounts of *H. trunculus* remains in the residential areas poses a perplexity. The stench the manufacturing process produces is notorious (Ruscillo 2005, Steiglitz 1994, per. com. A. Aksoy). The *H. trunculus* accumulations associated with the Late Bronze Age Troia citadel must be secondary deposits associated with the disuse periods of the areas in which they are recovered. Their spatial distribution cannot be directly related to the actual location of dye-production, as it was suggested by Blegen (1937).

But besides the remains in and around Quadrates IK8, other accumulations of *H. trunculus* shells have been found in Troia during the New Excavations until 2005. These concentrate in Quadrates K4 and KL16/17. Deposits of *H. trunculus* shells in Square KL16/17 are associated with hearth installations; probably used to boil the shellfish to reveal the color from the glands (cf. Çakırlar and Becks in print). These are in proximity with two other unusual shell deposits, which separately contain *S. marginatus* and *Patella* spp. shells. These species are not associated with any known ancient industry. As discussed before (in Section 5.1), they are edible. The amount of *H. trunculus* remains recovered in Quadrate KL16/17 do not suffice to indicate substantial dye manufacture for a large textile industry, but that small-scale production took place in this area at the fringes of the settlement during Troia VI Early times. Korfmann's (1998) association of the "10 kg of *H. trunculus* remains" discovered in Quadrate KL16/17 during the 1997 excavation season in Troia with "a well-developed, sophisticated textile industry" has little ground. It is likely that these finds are related to ornamental painting of the human body.

It should also not be ignored that, all other archaeological units that contain a large number of *H. trunculus* remains include the shells of food species. Neither the deposits uncovered by Blegen's team nor the accumulations found in the New Excavations consist solely of *H. trunculus* remains. This situation is in contrast with Virchow's reported claim that the *H. trunculus* deposits in Troia solely consist of the remains of this species and no others (von Martens 1879, cf. Section 2.8).

The second requirement for the production of dye is the technological know-how. As described in Section 5.3, the knowledge of baiting the carnivorous Muricidae snails can improve the time and energy consumed to acquire the individuals of these species. This knowledge may represent a Middle Bronze Age innovation in the Aegean, linked to the development of dye industries in the region. Textual, pictorial, and archaeomalacological evidence suggests that manufacture of this dye from *H.trunculus* remains was already known in Middle Bronze Age Aegean (ca. 1750) by the Minoans in Crete (Reese 1987b, Stieglitz 1994). Troia was part of a maritime Aegean trade and exchange network already in the Early Bronze Age and maintained and developed cultural contact with other settlements in the region in the eras that followed (Becks 2006, Blum 2006, Korfmann 1996, Ünlüsoy 2006). If dye production was not developed in Troia independently, it is highly probable that Trojans of the Late Bronze Age learned about the technique of producing dye from marine snails from their southern neighbors.

The first step of the basic method of dye production technology is to extract the hypobranchial gland from the shell, which will only precipitate color when exposed to air (Doumet 1980). Although the hypobranchial gland can be removed by hand or by piercing the shell, the most efficient method is by crushing the shell (Reese 1980). As described before (Section 4.3.3), the majority of *H.trunculus* remains in Troia, except for small individuals, consist of crushed shells. The fragmented condition of the *H. trunculus* remains in Troia show deliberate breakage. Due to the compact and solid nature of the *H. trunculus* shell, this type of fragmentation, *en masse*, could not have taken place after the deposition of the shells as a consequence of trampling, re-depositional activity or soil contraction. Shattering of the mollusk to extract the animal for human consumption is not a plausible explanation for the origin of this kind of breakage pattern. Almost all *H. trunculus* remains post-dating the Middle Bronze Age in the Mediterranean were found in crushed form at varying degrees (e.g., Becker 2001, Karali 1999, Reese 1987, and Forres 1987). Earlier finds of dye producing snails in archaeological sites around the Aegean have been found largely intact, at best in small accumulations, or already beach-worn, indicating that they were used as food or that they were collected dead (Karali 2005a and per. obv. at the Neolithic habitation site of Yeşilova in Bornova, İzmir).

The technique of producing dye from marine snails additionally involves various equipments and installations, such as vats and basins (Becker 2001). The exposed hypobranchial glands had to be oxidized by air for three days prior to a long cooking phase (Stieglitz 1994). Remains of such installations associated with Muricidae remains were found elsewhere in the Mediterranean (Gates 1999, Reese 1980, Steiglitz 1994). A basin in Sarepta (Lebanon) still contained dye-residue at its bottom at the time of recovery (Reese and Rose 2002). Such major installations associated with *H. Trunculus* accumulations were not traced in Troia. A possible location for these installations is the lower plain, near the shellfishing grounds exploited for the procurement of the raw material. Such an arrangement not only reduces the transportation costs, but also keeps the stench produced during production (Stieglitz 1994, Reese 2000, per. com. A. Aksoy) away from the settlement proper. However, it is also possible that in Troia, instead of vats and shallow basins, pithoi or other vessels were used to oxidize the glands, by way of occasionally stirring the decaying mixture once in a while. This is, however, a relatively labor-intensive process.

In view of the quality and quantity of the *H. trunculus* remains in Troia and their association with other archaeological and archaeomalacological finds, it appears that the dye production at this settlement was not substantial enough to create shell heaps as large as in Sidon, but also not so unimportant as to leave only scarce and isolated patches of remains in the archaeological deposits. While it should be acknowledged that although the Late Bronze Age *H. trunculus* remains in Troia are abundant and there is good reason to consider that they represent only the tip of the iceberg, the extant evidence does not suffice to elaborate further on a major local dye or textile industry when evaluated in comparison with very large coastal heaps from Iron Age sites in the Levant, such as the one near Sidon (Lebanon). There, the remains consisted of a 120 m long and 7-8 m high midden of broken Muricidae snails only, representing archaeological evidence adequate for the mass production of the infamous ancient purple dye (Forres 1987: 119). Such unambiguous evidence is clearly not represented by the archaeological remains of *H. trunculus* in Troia.

Still it must be clearly stated that in the context of Late Bronze Age Eastern Mediterranean, Troia provides the most abundant evidence for this dye production, which probably served the local textile industry (For further discussion and results relating to this phenomenon from latest excavations up until 2008, see Çakırlar and Becks in print). The end-products of these two industries must have played an important role in the economy of Late Bronze Age Troia, securing the settlement's significance in the regional trade network.

In this chapter, the different facets of the practice of mollusk procurement and utilization as represented by the mollusk remains in Troia, Yenibademli, and Ulucak have been discussed. In Chapter 6, variation and continuity in the archaeomalacological record, the patterns of which have been previously presented in Chapter 4, will be discussed in a wider context, seeking the relationships between archaeomalacological patterns and cultural, economic, environmental, and climatic phenomena.

Chapter 6. Variation and continuity in the archaeomalacological record

The quantitative patterns of variation and continuity in the archaeomalacological record have been described in Chapter 4. In the present chapter, several factors will be considered in an attempt to explain these apparent patterns. Because change and continuation in each feature of the archaeomalacological record can be interpreted in a variety of ways, these features will be discussed individually in separate sections of this chapter. Patterns in the archaeomalacological record are interpreted in terms of spatial and temporal variability in the environment, changes in the social context, subsistence, and culture of the individual settlements. Both in pre-agricultural and more advanced societies, foraging behavior is determined by a combination of these factors (e.g., Deith 1985, Meehan 1983). Excavation techniques, taphonomic processes and the nature of the excavated areas also have an impact on the patterns of the archaeomalacological record. The aim of this chapter is to isolate the environmental and economic factors that influence the archaeomalacological record, in order to reach an understanding of shellfishing and shellfish consumption in Troia, Yenibademli, and Ulucak.

6.1. Species diversity

Species diversity, i.e., the range of taxa in the archaeomalacological assemblages, shows variation between different sites (Tables 17, 30, 34, and 37). Such variation can be both a reflection of differential methodology, such as excavation techniques, identification methods and the size of the studied sample, and a reflection of behavioral patterns adapted to the exploitation of resources available in the environment.

The difference in the species diversity of the assemblages from Troia and Yenibademli results from the differential methodology involved in the compilation of primary archaeomalacological data. The taxonomic list of the mollusk remains from Troia should actually include more species than those that are listed in the present study (cf. Table 24). Mollusk remains that are identified to the family level usually enclose two or more species. The taxonomic list produced from the mollusks recovered from the New Excavations in Troia can be expanded by including the species identified among the archaeomalacological remains from the early excavations at the site (cf. Section 2.8; Boettger 1935, Gejvall 1938, Virchow 1881: 134, von Martens 1879). However, such an attempt to expand the diversity of the mollusk species will not greatly alter the palaeoenvironmental and palaeoeconomic interpretations based on the archaeomalacological record of Troia. An important contribution to the mollusk species represented in Troia is made by Gejvall (1938) by the addition of *Sepia* spp. to the mollusks found in Troia I layers. There are three commercially important species of *Sepia* (cuttlefish) in the Mediterranean: *S. elegans, S. officinalis*, and *S. orbignyana* (Mangold and Boletzky 1987). Although all of them have an internal shell, only *Sepia officinalis* (the common cuttlefish) grows large enough (up to ca. 35 cm in length) to be preserved represented in the archaeological contexts of the Eastern Mediterranean. Like other Cephalapods, the common cuttlefish is a free-swimming mollusk. It is demersal and neritic, living on sandy bottoms in depths ranging between 0-150 m (Mangold and Boletzky 1987). Since individuals can be caught using nets, harpoons, or spears, the exploitation of *S. officinalis* in Troia should be related to the fishing activities at this settlement, rather than shellfish gathering.

The variety of the mollusk species identified in Ulucak is relatively smaller. While this is related to factors such as relatively small sample size and the settlement's distance from the sea, it is also a reflection of selective behavior. This can best be demonstrated by contrasting it to the mollusk assemblages originating from prehistoric sites located further inland. For example, the mollusk assemblage from the prehistoric layers of Aphrodisias in western Anatolia shows a broader species diversity with 13 marine species (Reese 1986) than that represented by the Ulucak assemblage. Worked and unworked marine shells in Aphrodisias were transported from the coast as exotic goods converted into personal or household ornaments, whereas the majority of the Ulucak assemblage was transported to this site for its nutritional value. When mollusks are valued strictly for their contribution to the subsistence economy, emphasis is placed on quantity rather than variety. The role mollusks played in the subsistence of the Chalcolithic and Early Bronze Age settlements of Ulucak is thus reflected in the molluscan species diversity. The species diversity represented by the Ulucak assemblage can of course be expanded by studying a larger sample, which is only possible by future excavations in the Early Bronze Age layers of this site. But it is highly improbable that other qualities of the assemblage, such as the relative abundance of taxa –also a reflection of the mode of consumption behavior– will resemble those inland mollusk assemblages with a predominantly artifactual nature.

The contrast between the number of shelled mollusk species reported from the Aegean and the variety of the mollusk species represented in the archaeological record also demonstrates how species diversity of the archaeomalacological assemblages is largely a reflection of the relationship human societies built with the environment. Only about 5 % of the shelled mollusk species present in the Aegean have been recorded from the archaeological settlements of Troia, Yenibademli and Ulucak (cf. Section 2.5.2, 4.3.2, 4.4.2, 4.5.2 and 4.6.1). The situation is not different for the other coastal archaeological sites in the basin. The narrow diversity of the archaeological mollusk species reflects the limitation of mollusk gathering activity primarily to the relatively more nutritional species that could be acquired with ease in the coastal area between the lagoons and the infralittoral. Despite high production levels, the littorals have biomasses with lower diversity (Barnes 1990: 55). From this low diversity then only a few mollusk species are targeted by human predators. Selective gathering is common practice in shellfish gathering societies, as demonstrated by archaeological and ethnographic evidence elsewhere (e.g., Meehan 1983). The absence or low representation of small mollusk species in the archaeological deposits of the Aegean, such as *C. neritea*, *Abra* spp. or *Donax* spp., which are common elements of the lagoons and bays of the Aegean coast today, indicates that the nutritional value offered by the individuals of a species was an additional factor affecting species diversity represented by the archaeological deposits.

6.2. Proportion of mollusk remains in the faunal assemblages

The differences between the proportions of the mollusk remains in the faunal assemblages of Troia, Yenibademli and Ulucak, and between cumulative chronological units were demonstrated in Section 4.7.1. Although it is very difficult to distinguish the archaeological factors, i.e., sampling techniques employed during excavation, differential preservation, and the nature of the excavated areas, from the ecological and cultural factors affecting the proportion of mollusk remains in faunal assemblages, the quantitative relationship between the mollusk remains and vertebrate remains can form a good basis to assess the contribution of mollusks to the subsistence of temporally or spatially defined cultural groups (cf. Section 3.3; Glassow and Wilcoxon 1988). The obvious differences between the proportional representation of mollusk remains in the shell middens created by cultures without domestic animals, "shell-bearing sites" of the agricultural societies, and the amount of mollusk remains found in inland sites well illustrate the point that ecological and behavioral factors are at play in creating this feature of the archaeomalacological record.

In the Troad, there is a marked decrease in the amount of the mollusk remains in relationship to the vertebrate remains throughout the known occupational history of the region. In Sivritepe (ca. 4600-4500 BC), the mollusk remains outnumber the faunal remains, by making ca. 97 % of the total of the faunal remains (5533 in NIS; Boessneck 1986). Although Boessneck (1986) was of the opinion that the badly preserved state of vertebrate bones, an argument he supported by the absence of fish remains, resulted in the high proportion of mollusk remains, he also states that Sivritepe represents a site where the primary food item consisted of marine mollusks. Such evidence does not have parallels in the Aegean archaeology, probably reflecting the limited amount of studies comparing the proportions of vertebrate and invertebrate reimans in archaeozoological assemblages of the Aegean, rather than actuality. The archaeomalacological evidence from Sivritepe is more akin to the archaeological phenomenon represented by the Mesolithic shell middens along the coast of Europe, South Africa, or Japan. If the proportion of the mollusk remains in Sivritepe is not heavily altered by taphonomic conditions, it must be reflecting the specific orientation of the animal subsistence strategy towards the sea. It is important to note that the limited number of vertebrate remains reflect an animal economy based on cattle, sheep, and goats (Boessneck 1986). The joint evidence from mollusk and vertebrate remains implies that this site represents a subsidiary community or the site of a camp of a contemporary inland settlement, used specifically for the procurement and processing of marine mollusks. During the time when the site was in use, Sivritepe was situated at a favorable spot that provided easy access to both the Beşik Bay and the Trojan Bay, which were at their widest extent during the 5th millennium BC, as well as the littoral of the Aegean. Shellfish consumption at Sivritepe may have taken place at the spot or shellfish may have been dried or otherwise handled for later consumption elsewhere inland. Both types of activities would have left such a large heap of mollusk shells near the coast, analogous examples of which are present in the ethnographic literature (Deith 1985).

Although not in quantified form, the archaeomalacological evidence from other sites in the Troad antedating Troia I resembles the situation in Sivritepe. The investigators of the 5th millennium site of Alacalıgöl highlighted the overwhelming abundance of mollusks at the disturbed surface of the site (Gabriel et al. 2004 and per. obv.). The archaeological deposits of the lower layers of Kumtepe (Kumtepe A and B) are similarly characterized by abundant mollusk remains, the amount of which decreases in the later layers contemporary to Troia I (per. com. M. Uerpmann and per. obv.). Kumtepe A dates to the 5th millennium BC, whereas Kumtepe B is the sole representative of 4th millennium BC settlement in the Troad.

Regardless of differential preservation conditions and the inconsistency in the type of archaeomalacological data available for each site, these three 5th millennium settlements appear to have enjoyed marine mollusks at a relatively greater degree than the human groups that occupied the Troad in the subsequent millennia. Along with the high proportion of mollusk remains, the locations of these sites suggest a coastal settlement pattern that enables easy access to both mollusk resources and fertile hinterland sufficiently sizable to allow for successful agriculture and animal husbandry.

At the time of Troia I and Troia II, during the 3rd millennium BC, the faunal assemblages studied from settlements in the Troad –namely the faunal assemblage of the Yassıtepe layers contemporary with Troia I, the assemblage of Kumtepe C, also contemporary with Troia I, and the assemblage from Troia itself– appear to contain proportionally lower amounts of mollusk remains than the 5th millennium sites in the Troad. The mollusk remains make up more than half of the faunal remains from Troia I phases of Yassıtepe (von den Driesch 1999). This proportion is essentially comparable to the figures provided for the Troia I and Troia II layers of Troia (cf. Section 4.3.1). Across the sea, Yenibademli represents a settlement culturally affiliated with Troia I is well known. The impression drawn by the proportion of mollusk remains from this site also demonstrates similarities with the assemblages of the Troad.

Three factors may have been effective in forming these faunal assemblages, which, by comparison to the 5th and 4th millennium BC assemblages, appear to be depleted in mollusk remains:

By the 3rd millennium BC, substantial changes had taken place in the area and nature of the resource bases in the Troad. By this time, the Beşik Bay lagoon had already completed its formation and was at its largest extent (Figure 2; Kayan 1991). The sediments of the Karamenderes had filled the area between the northern edge of the Troia Ridge and the southern end of the valley, where the delta was located at the time of maximum marine transgression, ca. 7000-6000 BP (Figure 2, Kayan et al. 2003). The marine indentation at Kesik, at the southern side of which Kumtepe was located, had become marshland (Figure 2; Kayan et al. 2003). The woodland was starting to open (Riehl 1999: 83, Uerpmann 2003). Intensive, small-scale agriculture was being practiced around Troia (Riehl 1999: 83). The role of sheep, goat, and cattle had increased in the animal sector of the subsistence economy (Uerpmann 2003). These changes indicate degradation of both the terrestrial and the marine sections of the natural realm from the 5th millennium BC to the 3rd millennium BC. The possible role of anthropogenic activities in transforming the relatively pristine landscape of the 5th millennium has been recognized both by archaeobotanical and archaeozoological studies (Riehl 1999, Uerpmann 2003). Greater dependence on agricultural products and domestic animals, together with the reduction of the size of the Trojan and Beşik Bays during the course of time from the 5th millennium BC to the 3rd millennium BC may have led the 3rd millennium BC inhabitants of the Troad to place less emphasis on marine mollusks. It is most likely that the alteration in the mode of marine mollusk exploitation, induced by changes in the landscape and in subsistence strategies, is the type of human behavior reflected in the reduced mollusk proportion of the faunal assemblages dating to the 3rd millennium BC in the Troad. The emphasis given to mollusk exploitation in the subsistence system of the Early Bronze Age settlement in Yenibademli, inferred from the proportion of mollusk remains, is essentially similar to that in Yassıtepe and Troia I. This situation must be a consequence of the similarities between the environments and the subsistence systems of these settlements.

It is here appropriate to consider the mollusk remains of the Early Bronze Age settlement at Ulucak in terms of their proportional representation in the faunal assemblage and integrate the archaeomalacological evidence from this site to the discussion about the ecological and behavioral meanings of these figures. The proportion of the mollusk remains (Table 19 and Figure 65) in the Early Bronze Age in Ulucak was formed by the excavation of ca. 95 % of the mollusk remains from a single context (cf. Section 4.5.4). It is highly probable that the mollusk content of this deposit represents a single event. Whether this deposit represents a unique event or is representative of the structure of subsistence patterns at the Early Bronze Age settlement in Ulucak cannot be confirmed without the recovery of further archaeological features dating to the Early Bronze Age. At this stage of research, it is not possible to view the proportion of mollusk remains in the faunal assemblage as representative of the general exploitive patterns in Early Bronze Age Ulucak. What this deposit clearly indicates is that marine mollusks have been brought to the settlement for their dietary value.

The presence of marine mollusk accumulations at the site represents the relationship of this settlement with the coast, which should be explained in terms of subsistence economy. The mollusks found in Ulucak may have been obtained directly from the source by the people of Ulucak who traveled to coastal regions. While it is known from ethnographic and archaeological studies that hunter-gatherer-fisher societies make occasional trips to the coast and transport shellfish to the camp sites from grounds as far as 16 km (Buchanan 1988: 88, Deith 1985, Moss 1993), it is also possible that the relationship between the settlement of Ulucak and coast was an active relationship between the coastal dwellers (perhaps transhumant groups traveling seasonally between the coast and the hinterland) and the people of this inland settlement, involving economic and social interaction. Ethnographic studies record that coastal foragers take gifts of shellfish to inland people in exchange for meat procured from wild terrestrial mammals (Meehan 1977). A comparable mechanism could have operated in the formation of the archaeomalacological assemblages of Ulucak. The mollusks may have been brought to the site by the residents of nearby coastal areas and exchanged for other organic or inorganic products that are not available at the coast. From archaeological records yielded from elsewhere in the world it is known that raw or processed shellfish was an exchange item between the coast and the hinterland along with other goods such as salt and obsidian (Mitchell and Foster 2000). Shells may have been brought to this inland site dried, salted or fresh. Shelled aquatic mollusks can remain alive in a cool environment for two to three days, and 4-6 hours or longer in direct sun (Henshilwood et al. 1994 and per. obv.). The presence and the nature of an exchange between the Early Bronze Age coastal sites around the Inner Bay of İzmir and Ulucak can be better defined by comparative studies of artifact assemblages in the future.

The abrupt decline in the proportion of mollusk remains in the faunal assemblage originating from the Troia III and Troia IV layers is mainly attributable to the unsatisfactory sampling techniques employed during excavation (cf. Section 4.3.4). Likewise, although there is an increase in the proportion of mollusk remains in the Troia V layers, it is clear that this figure has also been negatively influenced by insufficient sampling. It is unfortunate that such an inconsistency exists in the archaeomalacological record of the Troia III through Troia V periods, for the settlement at Troia probably went through substantial demographic and economic changes during the time period covered by these layers (cf. Section 2.6.1 and 2.7.1; Blum 2006, Korfmann 1995, 1996, Riehl 1999). If the new elements in the material culture appearing in the Troia IV and V periods are representing new demographic elements –for example, a social group originating from inland Anatolia– it is highly probable that this sector of the society was traditionally unacquainted with marine resources and perhaps reluctant to exploit the coastal waters. If this was the case, the proportion of mollusk remains in the faunal assemblages dating to the period may be reflecting these developments.

Moreover, these changes in the archaeological record coincide with a cool episode in the global climate that showed its effect as a dry period in the lowlands of the eastern Mediterranean and cooler SST values in the Aegean (cf. Section 2.3.2). The time lapse in the formation of the calcite that grew on the walls of the water reservoirs of Troia also overlaps with these developments (Frank et al. 2002). If this break in the calcite formation was triggered by low groundwater levels induced by dry climatic conditions, the combined effects of low groundwater and low sea-level must have caused a further reduction in the area of the Trojan and Beşik Bays and the formation of seasonally wet areas. This would connote that the environments which were formerly hosting mollusk populations were no longer available for the exploitation of mollusk populations. Therefore, when these environmental and economic developments are considered together, a decrease in the proportion of mollusk remains as a reflection of a change in the mode of mollusk exploitation would be expected. But the current archaeomalacological evidence the Middle Bronze Age Troia is ambiguous and limited to be securely correlated with these suggested environmental and demographic changes.

In the Late Bronze Age layers of Troia, the amount of mollusk remains in relation to the amount of vertebrate remains increases (cf. Section 4.3.1, Table 21, Figure 32). The NIS counts are probably inflated due to factors related to the taphonomy of the mollusk finds, such as the crushed state of *H. trunculus* shells (cf. Section 4.3.3). Despite that, the amount of mollusk remains is proportionally lower in the Late Bronze Age than it is in the Early Bronze Age. This situation is also explicable by environmental and economic developments. During the Late Bronze Age, the areas of the Trojan and Beşik Bays continued to lessen in size as the sediments of the Karamenderes and the Dümrek Rivers pushed the lagoonal coastline further away from Troia (Figure 2). This meant that the most important shellfishing grounds went through further reduction of area. Moreover, they were further away from the settlement than they ever were before during the occupational history of the site. For the first time in the history of occupation at Troia, distance from the mollusk beds (despite the introduction of donkey and horse; Uerpmann 2003) may have significantly influenced the frequency of mollusk foraging. Distance is an important determining factor in the selection of animal resources exploited by prehistoric human populations (McKillop 1984). The ca. two-kilometer distance that formed between the deltaic coastline and the settlement since the Early Bronze Age required more time and effort to be devoted to obtain the same amount of mollusks gathered with considerably less time and effort during the Early Bronze Age. Additionally, the increase in the meat gain secured from cattle (Uerpmann 2003), intensification of agriculture (Riehl 1999), strong centralization of the social organization and the increase in the population (Becks 2006) hint at the development of subsistence strategies that put emphasis on surplus-producing and nutritionally more rewarding domestic resources, as well as terrestrial game animals (Uerpmann 2003, 2006). Additionally, ethnographic examples demonstrate that badly planned urbanization and migration to cities might increase the demand for cheap resources, such as easily gathered shellfish (Thomas 2002). The proportion of mollusk remains recovered from Late Bronze Age Troia does not suggest that urbanization resulted in an increase in the demand for aquatic mollusks. But it should be remembered that there is always an unknown amount of shellfish consumed at the gathering site and an unknown amount of mollusk flesh brought to the site detached from the shell (cf. Section 5. 4). In fact, the archaeological finds ("remains related to activities involving fire" and ceramic shards dating to the Troia VI and Troia VII periods) found at the foot of the northern slope of the Trojan Ridge may be associated with off-site activities related to the exploitation of lagoonal resources, such as the mollusks. It is probable that the amount of mollusk flesh consumed already at the gathering site and brought to the settlement without the shells was larger during the Late Bronze Age than in the Early Bronze Age, because the increased distance between the site and the primary gathering locations required relatively more energy input during the transport of intact mollusks to the settlement. If off-site processing or consumption of mollusks took place at substantial levels in order to meet the energy-costs of the foraging activity, then the picture attained from the proportion of the mollusk remains in the archaeological deposits of the settlement proper cannot reflect the whole amount that was consumed.

The faunal assemblage originating from the Troia VIIb layers of the settlement are marked by a further decline in the proportion of mollusk remains (cf. 4.3.1, Figure 32). This depletion in the mollusk content of the faunal assemblage of Troia VIIb is well in line with the continuity of the decrease in mollusk remains in subsequent cultural layers, but divergent from the Troia VI and Troia VIIa layers. A number of factors may have affected this picture, including demographic and cultural changes, and associated organizational changes, as well as environmental changes. As it has been suggested for the Middle Bronze Age, if foreign elements in the material culture indicate the arrival of new groups and if these groups are peoples unfamiliar with marine mollusk procurement, the decrease in the proportion of mollusk remains might be reflecting this situation. Changes that occur in the archaeozoological and archaeobotanical record of the site point at the increased importance of domestic food animals in this period (cf. Section 2.7.1). The decrease in mollusk remains might be related to this factor, which indicates that animal protein and fat were well secured by products obtained from domestic animals.

On the other hand, further reduction in the size of primary gathering grounds can be seen as a reason for the depletion in the proportion of the mollusk remains dating to this period. The low proportion of the mollusk remains in the faunal assemblages dating to the Post-Bronze Age (Figure 32) must also be related to the formulation of new subsistence strategies determined by ecological, economic, and cultural factors. During the Post-Bronze Age, the deltaic coastlines were ca. 3-4 km away from Troia and the area of the Trojan

Bay was at its smallest extent during the occupational history of the site (Kayan et al. 2003, Kraft et al. 2003a). The site was practically land-locked. The distant location of the coastline necessitated shellfishing expeditions. In addition, a great extent of the land between the settlement and the deltaic lagoons and the marine bay was covered with marshland. The distance and the uninviting environment covering it probably turned the shellfishing activity into an economically unfavorable activity. Furthermore, during the historical periods, the exploitation of the wild realm appears to have focused on terrestrial resources, particularly on fallow deer (Fabiš 2003). The analysis of Post-Bronze Age fish bones show that fishing took place at the Karamenderes and Dümrek Rivers relatively more frequently during these periods (Uerpmann and van Neer 2000). Based on the analysis of macrobotanical remains, Riehl (1999) suggests that the use of the marshland remained minimal during the Post-Bronze Age. The relatively low representation of the mollusk remains in the faunal assemblages fit well with the environmental depiction of the lower plain and the location of the coastline and complements the picture drawn by other archaeobiological studies indicating the terrestrial orientation of resource exploitation in the Post-Bronze Age Troia. It cannot be ruled out, however, that the proportion of the mollusk remains, as well as other features of the archaeomalacological assemblages of the Post-Bronze Age Troia, is largely an outcome of the recovery techniques employed during excavation. Just how sensitive the Post-Bronze Age sampling techniques were to the mollusk remains cannot be judged by archaeozoological analyses.

In summary, although the proportional representation of mollusk remains in the faunal assemblages must be approached with caution and analyzed in view of the effects of factors including sampling techniques and the contexts of mollusk remains, the value of this type of evidence as a proxy for the changes in the contribution of shellfishing to the subsistence economy, as well as to explain the relationship of shellfishing patterns with environmental changes is clear. Based on the evidence presented by the proportion of mollusk remains in the archaeological deposits dating to the 5th millennium BC, it could be postulated that the greater area of the littorals and the proximity of the seashore allowed the mollusks to substantially contribute to the diet of the settlements in the Troad. Lower levels of coastal and terrestrial environmental degradation permitted a larger level of dependence on the easily accessible wild resources such as the marine mollusks. In the Early Bronze Age, the proportions of marine mollusks in Troia, Yenibademli, and Yassıtepe in the archaeozoological assemblages are very large, indicating the orientation of the coastal foragers towards subsistence strategies emphasizing marine mollusks. The high proportion of the mollusk remains in Early Bronze Age Ulucak is explained by the context of these finds. They probably represent a single event. The decrease in the proportion of mollusk remains in the faunal assemblages through time demonstrates that the increasing distance between the settlement and the resource base in the Troad had a negative impact on the role shellfishing played in the subsistence economy or, alternatively, less mollusks were brought to the site intact, because processing or consumption took place near the resource base. The resistance of these resource bases to frequent exploitation and large yields had probably diminished as the resource base became significantly smaller and the population in Troia became larger during the Late Bronze Age. Shellfishing was probably no longer regarded a reliable food procurement activity as early as the beginning of the Late Bronze Age.

6.3. Relative abundance of mollusk taxa

It has been shown in Chapter 4 that the relative abundances of the mollusk taxa in the archaeomalacological assemblages of Troia, Yenibademli, Ulucak, and other sites located in the Aegean demonstrate variability across time and space. Changes in the availability and abundance of mollusk species in the environment, harvesting ease, handling time, nutritional gain, culinary value, and technological innovations introduced through acculturation processes are considered to be the factors influencing this variability in the archaeomalacological record. Similar to the discussion in Section 6.2, integrating the available archaeomalacological evidence from the Troad antedating the time when occupation in Troia began will enable a more comprehensive understanding of the relationship between shellfishing patterns and the environmental changes. Furthermore, evidence from other sites in the Aegean (already reviewed in Section 4.6.1) will be added.

The differences in the relative abundance of species records are primarily determined by the differences in the availability of mollusk resources in the environments surrounding the sites. Altering and stable coastal environmental conditions play a major role in shaping relative taxonomic proportions in Aegean sites. This can be easily inferred from the review of the species spectra from the Aegean sites presented in Section 4.6.1 (also cf. Table 11). The inhabitants of the settlements located on rocky promontories or close to rocky shores mainly procured those species occupying rocky littorals such as *Patella* spp. and, to a lesser extent, Monodontinae. At those sites, which are located close to estuarine or lagoonal environments, the most commonly procured species were those which dwell in sandy or muddy environments and tolerate salinity fluctuations, such as *C. glaucum*. At sites located at the coasts of semi-enclosed bays where the influence of freshwater was small and low sedimentation rates allowed the presence of some hard substrates within the bays, those species that characteristically occupy these habitats, such as *O. edulis*, *M. galloprovincialis*, *A. noae* and Muricidae were collected most frequently. At sites located further inland, the species spectra of the mollusk remains reflect the environmental characteristics of the nearest shoreline.

The mollusk spectra represented at the archaeological sites in the Troad complement the reconstruction of the coastal environments in the area. The predominant abundance of *O. edulis* and *M. galloprovincialis* remains in Sivritepe, Kumtepe, and Alacalıgöl indicates the presence of shallow waters with hard substrates and little freshwater influence in close proximity to the sites. The Trojan and the Beşik Bays must have provided a vast environment for these species during the time period between the maximum marine transgression (ca. 7000-6000 BP) and the invasion of the large areas of the bays with alluvial sediments, when lagoons and marshlands were formed near the sites (ca. 5000 BP). The mollusk shells recovered in the geomorphological samples from the Trojan and Beşik Bays support this observation. An abundance of oyster shells have been reported from the marine deposits that lie directly above the bedrock near Pınarbaşı at the southern end of the Trojan Bay (Kayan et al. 2003). Oyster shells were also found in the coarse sandy and gravelly deposits that built the coastal barrier forming the lagoon at Beşik Bay (Kayan 1991). Abundant remains of *Ostrea* spp. shells reported from Hanaytepe may be contemporary with this early phase of sedimentation in the Trojan Bay (Calvert 1881, also see Section 4.2.1)

The gradual silting of the Trojan Bay and the formation of deltaic and coastal lagoons are apparent in the shift from the abundance of *O. edulis* and *M. galloprovincialis* in the archaeomalacological assemblages in the Troad, to a predominantly *C. glaucum* assemblage in Troia I and II. This change appears to take place sometime between the Troia I and Troia III periods. Although the proportion of *A. noae* remains is very limited in the archaeomalacological assemblages from Troia I layers during the New Excavations, from an environmental point of view, Gejvall's (1938) account reporting an abundance of *A. noae* in the Troia I layers recovered during Blegen's excavations does not contrast with the picture drawn for Troia I in the present study. *A. noae*, *O. edulis*, and *M. galloprovincialis* are known to share the same environments. It may be *A. noae*'s tendency to populate depths between 3-5 m more densely than shallower waters (cf. Section 4.1.2.1) that hindered the human exploitation of this species in the Troad. Although Gejvall's (1938) reflections about the abundance of *A. noae* might have some truth, they also appear to be biased. The infrequent presence of rocky shore species and species dwelling in sandy shallow littorals indicates that because such environments were not very close to the settlement, shellfishing rarely took place in these areas during the Early Bronze Age.

The abundance of *Patella* spp. and *C. glaucum* remains in the Troia I layers of Yassıtepe is directly related to the location of this site on a high rocky promontory next to the Beşik Bay. Likewise, the abundance of *C. glaucum* in Ulucak is a representation of the connection between this site and the coastal areas of the İzmir Bay.

In view of these strong correlations between the character of the coastal environments and the relative abundance of mollusk taxa found in the archaeological sites near the coast, the reconstruction of geomorphological features around Yenibademli requires reconsideration. Based on the archaeomalacological evidence, the following reconstruction of the coastal environment is suggested: the deltaic lagoons of the Büyükdere were located towards the bottom of the valley, in a more southerly location than it has been suggested by the geomorphological studies of the site's environs (Öner 2000, 2001). *C. glaucum* lived in this southern part of the estuary. Towards the north and to the immediate south and southeast of the settlement, marine waters prevailed. Here the bottom deposits were not entirely covered with alluvial silt during the Early Bronze Age, between 2900-2600 BC. *O. edulis* populations dwelled at the bottom of this marine embayment, while the *Patella* spp. colonized the shore rocks flanking it, especially to the north of the settlement where waves provided a more suitable habitat for *Patella* spp. colonization. The close proximity of *O. edulis* beds to the settlement is attested by the presence of specimens still attached to rocks found in the archaeological deposits of the settlement proper (cf. Section 4.4.3; Figure 59b). The Early Bronze Age settlement of Yenibademli was located on a rocky promontory, at the bottom of which sufficient water circulation and the relatively stable salinity permitted *Patella* spp. populations to live just at the foot of the settlement. There are only two prerequisites for such an environment to be present during the first half of the 3rd millennium BC in the Büyükdere Valley: a low sedimentation rate and a deep basin. The past velocity of the Büyükdere River has not been reconstructed. There is presently no information on the natural vegetation that covered the hills of the valley during the Early Bronze Age, but it can be guessed that they were not barren like they are today. The hills of the valley are very steep; steeper than, for example, the lower Karamenderes Valley. There is no reason why the slopes should suddenly become gentler towards the bottom of the basin. Geomorphological cores dug 35 meters below the present level of the plain did not reach the bedrock (Öner 2001, 2002). The similarity between the species abundance records of the earlier and later layers of Yenibademli suggests that the coastal environment and the shellfishing practices were largely stable during the course of occupation at the site, although a slight increase in the abundance of *C. glaucum* remains in the archaeomalacological assemblages may have resulted from the changes in the substrate of the bay from detrital to sandy-muddy conditions due the effect of sedimentation and delta-progradation.

To come back to Troia, for the same reasons related to the sampling techniques described above (cf. Section 6.2), it is very difficult to delineate the factors that were influential in shaping the changes in the relative abundance of species in the archaeomalacological assemblages from the Troia III, IV, and V levels. The similarity of the relative abundance of species in Troia III assemblages with those assessed for the earlier Troia I and Troia II layers provides affirmative evidence that the change in the proportion of mollusk remains is attributable to sampling strategies and is not indicative of environmental changes or changes in subsistence strategies. The continuing trend in the relative abundance of species in Troia IV in contrast with the low representation of mollusk remains in the faunal assemblage can be explained along the same lines of argument.

Similarly, the figures depicting the relative abundance of species in Troia V assemblages have been negatively affected by insufficient sampling techniques employed during excavation (cf. Section 4.3.4). The increased abundance of *S. marginatus* remains in Troia V is linked to the recovery of a large accumulation containing specimens of this species. It is not possible to attribute this accumulation securely to changes in prey choice or shellfishing strategies due to cultural or environmental changes. However, the absence of such *S. marginatus* accumulations in the layers dating to the earlier periods of the settlement and the increasing presence of these in the archaeological deposits that formed during the subsequent Troia VI period may be signifying a marked change in the coastal environment or in prey choice during the Middle and Late Bronze Ages. As already discussed, the *S. marginatus* remains might be related to fishing rather than shellfish consumption as human food (cf. Section 5.1). Besides, retrieving razorclams from fine sandy bottoms with clear waters requires relatively more physical effort and experience (cf. Section 5.2.1 and 5.3). The requirement of experience to collect *S. marginatus* eliminates the possibility of linking these finds with the new demographic elements with their origins in inland Anatolia. Instead, they may be representing new *S. marginatus* populations that are partially invading the northern section of the Trojan Bay, gradually replacing the *O. edulis* and *M. galloprovincialis* beds there. The decrease in the relative abundance of *O. edulis* and *M. galloprovincialis* remains attests to this. A drop in the eustatic sea-level and accelerated sedimentation in the Trojan Bay may have resulted in the formation of shallow fine sandy bottoms to the north of the deltaic coastline, where *S. marginatus* could be acquired with relative ease. Alternatively, a drop in the sea-level may have caused the formation of sandy bottomed pools to the west of the Sigeon Ridge or shallow areas in the marine waters of the Beşik Bay, where *S. marginatus* could also have been procured. In addition, a variety of cultural explanations can be offered for the increasing presence of *S. marginatus* in Troia. The difficulty involved in harvesting these species may have increased its culinary value. For this reason, *S. marginatus* may have become a type of food that was demanded on certain occasions to enhance culinary variety. If the species was used as fishing bait, its increase in the archaeomalacological record may be indicating changes in fishing practices. But studies of fish remains in Troia do not signify a technological change involving fishing practices in the Middle Bronze Age or Late Bronze Age. Isolated *S. marginatus* accumulations may also be traces of the effects of short-time events (e.g., social conflict, affluence, scarcity, revelry, et cetera) in the archaeomalacological record; however, it is very difficult to distinguish these securely from the effects of general changes in the environment and subsistence strategies.

Until the beginning of Troia VI, the relative abundance of species in the archaeomalacological assemblages in Troia and in other sites in the region suggests that shellfishing was practiced primarily to fulfill dietary needs using the most optimal methods. Although the relative abundance of species represented in the archaeomalacological assemblages largely reflects the continuation of this straightforward trend, the increase in the proportions of *Patella* spp., Monodontinae, *H. trunculus,* and *S. marginatus* points to firm changes in the role of shellfishing in the economy of Troia. The increasing presence of these species in the archaeological record from Troia VI shows that more mollusks were gathered for their culinary value and as raw material in dye manufacturing. Culinary value can play an important role in procuring those resources, which do not have a prime dietary role but complement the nutrition of a society. The increasing abundance of these species also suggests that the areas, which were previously visited ephemerally, were more regularly used for shellfishing activity during the Late Bronze Age. Although the focal location of the activity was still the deltaic lagoons and the nearby sheltered areas close to the settlement, shellfish gathering in the areas further to the north, to the west of the Sigeon Ridge on the coast of the Aegean and to the northeast on the coast of the Dardanelles became more frequent during Troia VI. Frequenting mollusk beds other than the most convenient areas may also indicate the exhaustion of mollusk populations due to prolonged intense foraging. Ethnographic examples demonstrate that over-exploitation of near-shore mollusk beds decreases the size and abundance of exploited mollusk populations, compelling foragers to gather shellfish in deeper water (Thomas 2002). As it will also be discussed in Section 6.4, exhaustion of the primary mollusk resource, namely *C. glaucum* beds in the deltaic lagoons, is unlikely to be the reason for this apparent shift. Most probably, interrelated changes in the subsistence economy and the surrounding environment (cf. Section 6.2) led culinary value, desire for variety and obtaining raw material for dye manufacture to become important factors affecting shellfishing practices during this period. Greater nutritional gain could certainly not have constituted a reason for the search for *Patella* spp. and Monodontinae (cf. Section 5.5; Table 12). It is also very unlikely that the nutrition acquired from species like *H. trunculus* and *S. marginatus* would recompense the time and effort spent to gather these species in an optimal way. *H. trunculus,* a species that was not preferred previously probably due to its relatively patchy and insignificant occurrence in the northern section of the Troia Bay, was now demanded as raw material for dye manufacture. The practice of manufacturing dye from Muricidae snails probably dispersed from the southern Aegean to the entire region sometime around ca. 1750 BC.

Small changes that occur in the relative abundances of species among the different phases of Troia VI can be to a large extent attributed to the differential nature of excavated areas and the problems associated with assigning different stratigraphical units to separate occupational periods. As mentioned earlier (cf. Section 2.6.1), the successive periods of Troia VI and VIIa represent the rebuildings of the citadel walls in the course of the Late Bronze Age, which do not necessarily indicate social, demographic, or cultural changes in the settlement.

The similarity between the relative abundance of species in the assemblages dating to the Troia VI and VIIa periods affirms the suggestions that environmentally, culturally and economically, this period was a continuation of the Troia VI period. For example, the presence of crushed *H. trunculus* remains in the deposits of this occupational period demonstrates that the custom of dye manufacture continued from Troia VI (Table 27).

The relative abundance of food species is not markedly different in the subsequent Troia VIIb period (Table 28), but the decrease in the abundance of *H. trunculus* remains in the excavated areas point to the decline of the dye manufacture. While the species spectrum of the mammalian record (per. com. H.-P. Uerpmann) and changes in the set of cultivated plant crops (Riehl 1999: 83) of this period are attributed to social and environmental changes, the relative abundance of species in the archaeomalacological assemblages originating from the Troia VIIa and VIIb phases does not have any strong indications about environmental changes or changes in the use of the coastal environment for shellfish procurement.

Post-Bronze Age shellfishing in Troia assumes an altogether different character when compared with the previous periods. During the Post-Bronze Age, the shellfish gathering activity becomes selective. The significant shift that occurs in the relative abundance of species in the assemblages dating to the Troia VIII, IX and X periods (cf. Section 4.3.2, Table 29, Figure 33) is related to both environmental and economic changes. It should be noted once again, however, that the appearance of this shift may be an outcome of selective sampling during the excavations. The increase in the abundance of *O. edulis* and Pectenidae, already occurring in the Troia VIII assemblages, indicates that the main area of shellfishing has moved from the deltaic lagoons to the bay and possibly to the open sea. The deltaic lagoons are located closer to the Dardanelles at this point in time than to the settlement, whereas the Trojan Bay is significantly reduced in area. The reduction of the lagoonal areas was probably accelerated due to the negative effects of hydraulic engineering programs carried out in the region, such as the building of the aqueduct on the Kemerdere River and possibly by the irrigation works conducted on the lower Trojan Plain. The freshwater influx to the Trojan Bay thus reduced; the salinity of the Trojan Bay must have increased, consequently creating favorable environments for Pectenidae and *O. edulis* to penetrate. The Trojans probably turned to collect these readily available, more nutritious species (cf. Section 5. 5) also to meet the energy cost of covering the long distance between the shellfishing grounds and the settlement. Additionally, the Post-Bronze Age inhabitants of Troia may also have tried to limit the time they spent at the *C. glaucum* beds due to health reasons. Based on her studies on the macrobotanical remains from Troia, Riehl (1999) suggests that the Post-Bronze Age inhabitants of the site avoided the marsh areas in the lower plain, which were adjacent to *C. glaucum* beds, because they wanted to preclude the possibility of catching malaria. The significance of acute malaria in shaping the relationship between coastal dwellers and the environment has been discussed elsewhere before, as a threat inherent to the wetland areas of the Troad both during prehistoric and historical periods (Aslan 2003, Witter-Backofen 2003). It is also known from the ethnographic record elsewhere that the presence of mosquitoes in wetland areas influenced decisions concerning the timing and place of shellfishing activity (Meehan 1983). Such are the environmental conditions that may have affected the changes in the relative abundance of species in the Post-Bronze Age assemblages of Troia.

Based on archaeological evidence, Hitcher (1999) reports that in the course of the "Romanization" of Provence, oysters replaced mussels, and especially "from the 1st c. C.E. [Common Era, i.e., AD], the consumption of oysters, and to a lesser degree, scallops (*Proteopecten glaber*) became a province-wide phenomenon…involving all population classes."[35] The popularity suggested to have been brought about by the Romanization of a coastal province in the Western Mediterranean has a resemblance to the changes occurring in the archaeomalacological assemblages in the Post-Bronze Age Troia. Hitchner (2000) also argues that unlike in Rome, in Provence mussels and oysters were not prestige foods, but staples. This may simply be due to the nature of prestige items; they are rarely obtainable but highly demanded objects. The availability of shellfish at a coastal site is obviously not comparable to that in Rome during the Roman times. Despite these parallelisms showing that acculturation processes might be an important factor influencing patterns of shellfishing and consumption in the Post-Bronze Age Troia, current archaeomalacological evidence does not suffice to make a detailed and accurate evaluation of the culinary or social value of shellfish during Period IX. Additionally, Hitchner (2000) reports that mussels returned as an important part of the diet during the Byzantine times replacing oysters, as a consequence of growing demand on cultivated young oysters. Archaeomalacological evidence from Medieval Troia (Period X) does not imply such a change. But this picture may be falsified due to the restricted sample size dating to Period X.

While an Archaic to Hellenistic date (Troia VIII) is earlier than the first written accounts of oyster cultivation as well as the dispersion of the oyster trend through Romanization, it is surely no coincidence that the relative abundance of *O. edulis* in the archaeological deposits of Troia IX is roughly contemporary with the beginnings of contact with Rome and the first written accounts of oyster cultivation. If such a cultural mechanism enhanced the consumption of *O. edulis* and Pectenidae in Roman and Byzantine Troia, then the archaeomalacological evidence dating to these periods cannot be explained solely in environmental or economic terms. Given the frequency of the relationship between Roman Troia and Rome (Rose 2006) and the fact that Troia was a tourist site already in the Classical times increases the possibility that the high abundance of *O. edulis* and Pectenidae in the Post-Bronze Age site is the archaeological representation of the local or foreign demand for these food species, the former of which may have been procured through aquaculture. But do the limited remains of *O. edulis* in Roman and Byzantine Troia, although proportionally dominant in the mollusk assemblage, represent oyster culture? If morphological descriptions of the *O. edulis* remains from the Post-Bronze Age layers were available, this question could be answered by plotting the size profiles of the specimens and by analyzing the negative imprints left by the implements used in aquaculture, such as ropes, wooden rods or even pottery.

[35] *Proteopecten glaber* is synonymous with *Flexopecten glaber*, a Pectenid species whose existence in the archaeological deposits of Troia III and Troia VI has been confirmed (cf. Section 4.1.2.5).

Unio spp., as mentioned before, must have occurred in the freshwater sources around Troia and Ulucak in fairly large amounts. They are not represented in the assemblages in Yenibademli and it is not clear whether *Unio* spp. existed at all on Gökçeada (cf. Section 5.2.2). They are represented in Troia and Ulucak in very low numbers. The increased exploitation of freshwater fish in the Post-Bronze Age layers of Troia (Uerpmann and van Neer 2000) is not accompanied by increased exploitation of freshwater clams. The increasing presence of freshwater fish taxa in the archaeozoological assemblages has been attributed to amplified sedimentation of the Trojan Plain and coeval shift of the river mouth to the north (Uerpmann and van Neer 2000). If the slight turn from lagoonal fisheries to freshwater fisheries during the Post-Bronze Age was caused by accelerated sedimentation rates, the shellfishing sector of the economy appears to have responded to these environmental developments by renewed emphasis on open sea and bay shellfishing rather than riverine exploitation. Similar dualities resulting in the differential palaeoenvironmental interpretations of the archaeomalacological and itchthyoarchaeological evidence have been recorded elsewhere (Rose 1994: 246). There may be three reasons for this duality in the Trojan recod: (1) relative emphasis on freshwater fisheries is not a consequence of palaeoenvironmental changes, but a cultural phenomenon; (2) the choice for exploiting nearby freshwater fisheries did not affect decisions in shellfishing because *Unio* spp. are not as gregarious as the marine species that were available; or, (3) freshwater clams are not considered palatable. Based on present evidence, culinary preference is the most likely explanation for the scarcity of *Unio* spp. in the archaeomalacological assemblages. *Unio crassus* contains relatively less flesh and protein than other shelled aquatic mollusks. The dietary need for animal fat and protein, transformed into an unconscious sense of taste, may have led to the avoidance of this species.

Factors such as (1) changes in the availability of mollusk populations in the environment due to environmental developments; (2) species-specific harvesting methods and handling time; (3) nutritional content of mollusk flesh; (4) changes in the culinary value given to certain species; (5) the influence of technological innovations acquired from other societies; and, (6) changes in the other sectors of subsistence, were all linked to explain the variability and continuity observed in the species compositions of the archaeomalacological assemblages presented in this study. In more than one case, it was shown that differential sampling during excavations was the most likely factor that influenced the formulation of the archaeomalacological record. The effects of the differential nature of the excavated areas were also discussed.

It was concluded that similar to the observations made for other sites in the Aegean, changes in species spectra through time and differences between the sites were related to environmental conditions of the coastlines. The gradual shift from the *O. edulis* and *M. galloprovincialis* dominated assemblages of the 5th millennium BC to the *C. glaucum* dominated assemblages of the 3rd millennium BC was explained by geomorphic changes in the coastal areas. An alternative reconstruction of the geomorphic evolution of the terrain surrounding Yenibademli was offered based on the species spectrum of this site. This alternative reconstruction places the Büyükdere Delta to a more southerly location in the Early Bronze Age. The species spectrum of Yenibademli also shows that coastal environment near the site did not go through substantial changes throughout the period of occupation at this site. The increasing proportion of *S. marginatus* remains in Troia starting with the Troia V period is attributed to changes in the habitats provided by the Trojan Bay and alternatively, to changes in culinary practices.

In Troia, habitats far away from the settlement were visited more frequently in the Late Bronze Age, indicating the change in the role of mollusks in the diet and economy. Value given to mollusks as resources fulfilling basic nutritional needs decreased. The similarity between Troia VI and VIIa species spectra attests the unity of these periods. The species spectrum of Troia VIIb does not suggest marked differences in shellfishing practices or in the environment. 'Murex' dye industry, however, seems to have gone through a major decline during this period.

Gathering practices became more selective during the Post-Bronze Age. Both cultural, most possibly the influence of Romanization, and environmental changes appear to be at play in this change. *O. edulis* may have been cultivated in Roman and Byzantine Troia. The relative emphasis placed on river fishing is not accompanied by a similar trend in the shellfishing practices of the period.

6.4. Size and age

Shellfish gatherers have probably always selected larger specimens over smaller individuals (Claassen 1988). Intentional selection of smaller and younger individuals is viewed as an unlikely choice and has no ethnographic examples (Reitz and Wing 1999: 315). When larger specimens are available and easily accessible, preying on the largest possible group of individuals in a mollusk community is the most pragmatic way of shellfish gathering in order to meet the energy costs of the activity. On the other hand, selective gathering of large specimens is not an advantageous method, since it is very likely that the energy devoted to selection for size during gathering will not be returned by the calorific gain obtained from the yield.

Intense foraging activity leads to decreased individual size in the prey population (e.g., Jackson et al. 2001). A modern example of decreased individual size in mollusk populations caused by intense gathering activity was reported from Micronesia, mollusks became higher in demand as a cheap food resource as a consequence of rapid and unsustainable urbanization (Thomas 2002). Similarly, increased human population is thought to have caused the over-exploitation of wild resources, resulting in diminishing shell sizes of archaeological mollusks (Jerardino 1997, Prummel 2005, Yesner 1984, 1988). When stress caused by foraging is decreased, individuals reach larger sizes again (Marcello and Thomas 2001, 2002). But, as discussed earlier (cf. Section

3.4), size is not always a function of age. In mollusk populations, size profiles may also be indicative of differential environmental conditions. Lagoonal invertebrate populations can easily attain the carrying capacity of the restricted habitat they live in due to a general lack of predation pressure. Populations increase in number; competition for food and other resources intensifies (Barnes 1980: 65). This situation may result in reduced individual sizes. In marine mollusks size is also known to increase with decreasing latitude (Graus 1979, Ivell 1979a, Vermeij 1978, 1980).

In this study, the size (VH) profiles of *C. glaucum* populations from Troia, Yenibademli, and Ulucak and the modern habitats in the Troad were investigated (cf. Sections 4.2.2.1, 4.3.5.1, 4.4.6, 4.5.5, 4.7.3). In Troia, the positive relationship between size and age of *C. glaucum* populations have been disclosed by plotting the demographic profiles of the archaeological and modern populations using increment counts. Additionally, the size profiles (L, H, W) of *P. caerulea* were plotted (cf. Section 4.4.5). These size profiles of the archaeomalacological populations are considered to represent the size of prey that was available in the nature. In the following paragraphs, possible explanations for the trends in the archaeomalacological size profiles obtained from the study of archaeomalacological assemblages will be discussed. Moreover, questions will be raised as to whether changes in the size of prey mollusks influenced human gathering behavior.

In Troia, the increase in the shell size of *C. glaucum* populations neatly coincides with the decrease in the proportion of mollusk remains in the faunal assemblages and in the relative abundance of *C. glaucum* in the archaeomalacological assemblages from the Early Bronze Age to the Late Bronze Age (cf. Section 4.3.5.1). This situation attests to the idea that the shell size of archaeological *C. glaucum* assemblages is good evidence for the intensity of the mollusk gathering activity in Troia. The abundance of larger individuals among the extant *C. glaucum* populations in the Troad, which represent growth under no human foraging pressure, confirms this observation.

In the Early Bronze Age, the frequency of *C. glaucum* gathering and large yields probably led to the decrease in individual size in the mollusk populations. Most individuals were gathered before they could reach sizes larger than ca. 27 mm. Human impact on the Trojan Bay was not only imposed by the Trojans. Other settlements scattered around the Trojan Bay also had their share of the coastal resources in the area, augmenting the exploitation pressure placed on these resources both before and during the Early Bronze Age. Even so, it appears that very small individuals were rarely procured –if the threshold of ca. 10 mm VH is not considered to be a consequence of excavation techniques or differential preservation. The continuity in the abundance of *C. glaucum* remains from Troia I to Troia II indicates that the intensity of gathering activity persisted despite the lowered amount of flesh gained from each lot. This type of activity may be denoted as an opportunistic approach to foraging, showing a high level of dependence on aquatic mollusks.

In the Late Bronze Age, the mean shell size of archaeological *C. glaucum* populations becomes larger in comparison to the Early Bronze Age populations. The increase in the mean size of archaeological specimens is a reflection of the increase in the abundance of older individuals. This situation shows that despite lesser dependency on *C. glaucum* gathering –as indicated by the decrease in the relative abundance of *C. glaucum* remains in the faunal assemblage– cockle gathering activity yielded more flesh per lot gathered. The increase in the mean amount of nutritional gain obtained per individual may have, however, been evened out by the energy spent to cover the distance between the settlement and the *C. glaucum* beds. If the energy output during transport is considered negligible, the picture drawn by the relative abundance of species, and the size and age profiles of *C. glaucum* in the Late Bronze Age should indicate that although the gathering activity took on a different form and the dependence on aquatic mollusks was lessened, the contribution of mollusks to the diet may remained the same or even became larger for those who consumed shellfish flesh. In a complex society such as the one represented by the Late Bronze Age Troia, it would be erroneous to presume that mollusks or any other food resource contributed to the diet of all sectors of the society equally. Ethnographic data provided by observations among native Northwest Americans indicate that shellfish gathering and consumption are associated with certain groups in a society (Moss 1993).

Moreover, numerous archaeological sites including finds dating to the Troia VI and VIIa periods were detected in the Troad, indicating dense settlement and population (Aslan and Bieg 2003, 199: Figure 33, 201: Table 3). Presumably these settlements, like the pre-Bronze Age and Early Bronze Age settlements in the Troad, also made use of the aquatic resources of the Trojan Bay. But the archaeomalacological evidence suggests that the human impact on the *C. glaucum* populations was relatively lower in the Troad during the Late Bronze Age. The suggested increase in the population size of the Late Bronze Age settlement and the urbanization process that seems to have taken place during this period do not seem to have had a negative impact on the *C. glaucum* populations by imposing intensified foraging pressure on them. Additionally, the increase in the mean size of *C. glaucum* shells during the Late Bronze Age eliminates the possibility that a considerable amount of shellfish processing and consumption took place off-site, near the mollusk beds.

The relatively larger size of the *C. glaucum* populations from Yenibademli can be attributed both to the low level of exploitation these populations were exposed to and to environmental factors. In order to distinguish which factor was influential on the size profile of *C. glaucum* shells from this site, it is necessary to investigate the growth rates of geographically coherent modern populations and compare the results with those of the archaeological populations. It is, however, uncertain if a habitat similar to that of the former Büyükdere estuary is extant on the coasts of Gökçeada.

The *C. glaucum* shells from Ulucak constitute the smallest assemblage among the archaeological and modern populations that were investigated for this study. If the assemblage in Ulucak is representative of the overall size range of the past populations that lived in the inner part of the İzmir Bay, their small size implies that the populations at İzmir Bay did not grow to become as large as their contemporary northerly counterparts, either due to adverse environmental factors or due to predation pressure, the latter possibly imposed by humans. If the latter is true, this is suggestive of the existence of a considerable shellfishing community living on the coast of the Inner Bay. On the other hand, the relatively narrow size range represented by this assemblage may suggest a size selection during or after the procurement of the individuals from the nature. It was mentioned above that size selection during gathering, especially in favor of smaller individuals, is a very unlikely mode of foraging (cf. this section, above).

Another possible explanation for the small size of the Ulucak assemblages can be construed with aid from ethnographic accounts. Ethnographic records show that in some cultures, specific shellfish beds are owned by corporate groups (Moss 1993). An intriguing suggestion would be then the possibility that the Early Bronze Age people of Ulucak, if it was indeed they who procured the *C. glaucum* from the coast, were able to gather cockles only from limited locations where environmental conditions were not very favorable for *C. glaucum* growth and therefore not exploited by coastal foragers. Humans may have competed for shellfishing grounds just as they did for hunting grounds. Gathering grounds cannot be *a priori* conceived as communal or natural space. If the *C. glaucum* were transported to Ulucak as gifts or exchange items, then smaller individuals may have been deliberately selected to be bartered. Which of these processes were at play in the formation of the size profile of the Ulucak assemblage may be revealed by future research in the İzmir Region.

A significant reduction in the mean size (length, height, and width) of *P. caerulea* populations in Yenibademli has been observed from the earlier occupational phases to the later occupational phases (cf. Section 4.4.5). If the delta of the Büyükdere shifted considerably northwards during the period of occupation at Yenibademli, the diminishing size of the consumed portion of *P. caerulea* populations might be a reflection of the effect of the reducing habitat of *P. caerulea* populations settled in the bay. Such a suggestion can be tested when the geomorphological studies investigating the development of the Büyükdere Valley delineate the rate of the northward shift of the delta area between 2900-2600 BC. Biological factors, for example a decrease in algae populations, which the limpets graze on, may also have caused the mean size decrease in local *P. caerulea* populations. But such a change in algae communities is not detectable by archaeobiological methods. In light of other archaeomalacological evidence, such as the persistent but slightly decreasing abundance of mollusk remains in the archaeological deposits of Yenibademli throughout its relatively short occupational history and the predominating presence of *P. caerulea* shells, the most probable explanation for the size reduction in the archaeological *P. caerulea* populations is the negative effect of human foraging activity.

There is no strong evidence indicating that the diminution in the shell size of the *P. caerulea* populations at Yenibademli was accompanied by depletion in the density and abundance of *P. caerulea* populations caused by severing environmental conditions. Although it is well known that islands are delicate ecological systems easily affected by the impact of resource exploitation (Keegan and Diamond 1987) and this factor should be kept in mind as a possible explanation for the abandonment of the Early Bronze Age settlement at Yenibademli, the size reduction in the mollusk populations that were available to the inhabitants of the site cannot be convincingly considered as part of a process that led to a general depletion of resources on the island. While smaller sizes of the procured *P. caerulea* individuals certainly meant decreased nutritional yields for the same amount of effort devoted to the gathering activity, there is no clear evidence suggesting that this development led to an important change in role of the gathering activity in the subsistence economy or was part of the process that eventually led to the abandonment of the settlement. Whether past human populations were aware of the negative impacts of over-exploitation on the yields of the shellfish gathering activity and took precluding measures in order to secure successful foraging is arguable.

6.5. Other morphological attributes

The solidity and curve of three different archaeological populations and one modern population of *C. glaucum* have been revealed by morphometric investigations (cf. Sections 4.2.2.1, 4.3.5.1, 4.4.6, 4.5.5, 4.7.3). Additionally, the relationship between the height and length, and length and width of *P. caerulea* populations in Yenibademli were also explored to detect the chronological changes in the morphological features of these populations.

Although the shell curve and shell solidity of the *C. glaucum* populations from Troia, Yenibademli, and Ulucak differ significantly from each other and between different chronological periods, these apparent differences are considered to be a co-dependent of the differential size range of these archaeological populations. The valves of older bivalve individuals are more convex than those of younger, smaller, and faster growing individuals (Vermeij 1993: 29), whereas older bivalves tend to be heavier than longer (i.e., more solid) due to augmented crowding of growth increments with increased age. Therefore, because of the significant differences in the size range of archaeological *C. glaucum* populations investigated for this study, indices of other morphological attributes of the specimens, i.e.; solidity and curve are not suitable proxies to detect changes in the environmental conditions such as SST and salinity, building analogy with Mariani et al.'s (2002) observations on the correlation between the shell shape of *C. glaucum* and the host environment.

Figure 78. Morphological differences between *C. glaucum* (top) and *C. edule* (bottom) (After Brock 1991: 8-9, Figure 2).

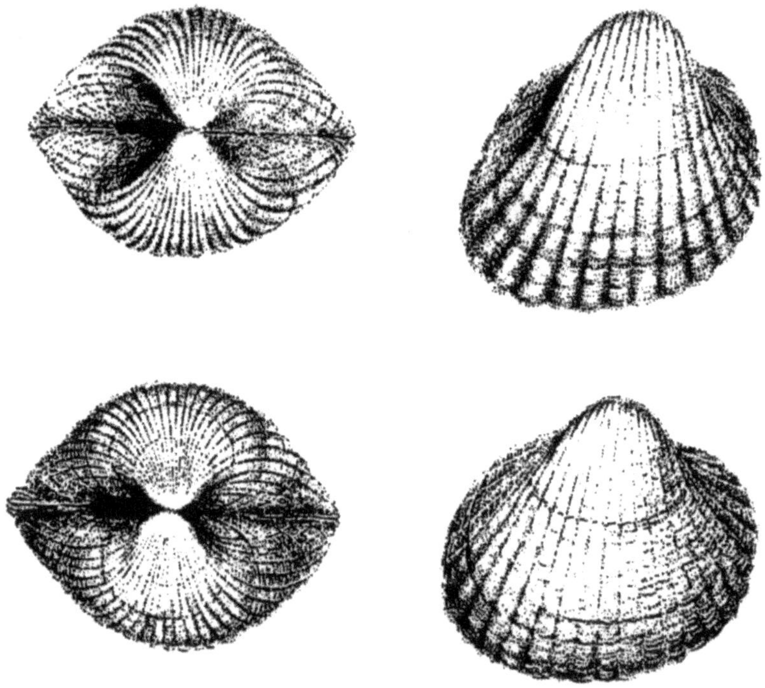

On the other hand, statistical results indicate a significant difference in the shell curve and solidity relationship between the archaeological and modern *C. glaucum* populations (cf. Section 4.7.3, Figures 72, 73, and 74). This difference can also be observed visually in the curve and thickness of the sections of *C. glaucum* valves demonstrated in Figures 30-31 and 48-49. The "lange Form" recorded in Kastanas (Becker 1986) and the "thick shells" of New Halos (Prummel 2005) appear to be analogous to the solid and curved valves represented in Troia, Yenibademli, and Ulucak. The curved *C. glaucum* shells from the archaeological deposits of Troia were commented on before, already in the 19th century by von Martens (1879). *C. glaucum* populations with this shell form seem to have died out in the Karamenderes Delta. Modern *C. glaucum* in the Troad has assumed a rather *C. edule*-like shell, where valves are less convex, but also where the difference between the valve length and valve height is reduced (Figures 71, 72, 73; Brock 1991: 8-9, Figure 2, also cf. Brock 1991: 21, Figure 21). According to our present knowledge, the change in shell form occurred sometime between the end of the Bronze Age and the present day. Barnes (1980: 56) reports that such a change in shell morphology has been observed in *C. glaucum* populations that settled estuarine habitats as opposed to their usual lagoon habitat (Barnes 1980: 56). Although it is not possible to speak of the presence of an estuarine bay in the Troad today, the deltaic lagoons of the Karamenderes, while still isolated from the influence of the marine realm in terms of salinity and wave action, are located at their closest position to the Dardanelles. This shift in the position of the deltaic lagoons and the consequent increase in the influence of turbulent and saline marine waters may be the cause of shell form change in *C. glaucum* populations in the Troad. On the other hand, instead of gradual changes in the geomorphology and environment of the Trojan Bay, recent processes may have caused the relatively abrupt occurrence of this morphological change. Anthropogenic control of the Karamenderes flood regime and water pollution both in the Karamenderes and the Dardanelles may have led to decreased levels of dissolved oxygen, carbon and nutritional sources in the host environment of the extant *C. glaucum* populations in the Troad, depleting shell carbonization rates and therefore shell solidity. The scope of the present study did not allow for detailed investigations of these ecological relationships. Corroborative malacological and archaeomalacological data from the Aegean, which could allow further interpretation of this type of morphological data in terms of palaeoenvironmental conditions, is lacking.

There are no significant differences between the relationships of height, length and widths of *P. caerulea* populations from the earlier and later phases of settlement in Yenibademli. This result is affirmative of the suggestion that the size reduction observed in the *P. caerulea* shells from Yenibademli is a consequence of prolonged human exploitation rather than changes in environmental conditions (cf. Section 6.4).

6.6. Incremental growth pattern in *C. glaucum* populations

Only the incremental growth patterns of modern and archaeological *C. glaucum* populations in the Troad were determined. Changing and persistent trends observed in the incremental growth of these populations have implications about behavioral and environmental patterns in the Bronze Age.

The gray incremental zones observed on the individuals in the Late Bronze Age assemblage are mainly broader than their Early Bronze Age counterparts (cf. Section 4.3.5.1). Two possible interpretations are considered to explain the palaeoenvironmental conditions responsible for this change in the incremental growth pattern. If the gray incremental zones of the Late Bronze Age are broader due to an increase in the amount of shell material secreted, then the broadening of gray incremental zones may be attributed to improved conditions in the host environment of the *C. glaucum* populations, such as depletion in summer dystrophy, decreased evaporation, or increased precipitation due to milder summer months. If, on the other hand, the gray incremental zones of the Late Bronze Age are broader despite stability in the amount of shell material secreted per microscopic increment within the incremental zone, then broader gray incremental zones should be indicative of a longer period during which environmental conditions are consistently not favorable for *C. glaucum* growth. The fact that there are no clear differences between the breadths of the white increments of these *C. glaucum* populations suggests that the second interpretation is more suitable to explain the change between the growth pattern of the Early and Late Bronze Age populations, because improved environmental conditions during the Late Bronze Age should also have been manifest in the breadths of the white increments. For such a change to occur in the growth pattern of *C. glaucum* populations, the amplitude of the seasonal differences in the aquatic conditions in the lagoonal areas of the Trojan Bay must have been relatively greater during the Late Bronze Age. Global climatic proxies (cf. Section 2.3.2), changes in the vegetation cover suggested by regional palynological research, as well as archaeobotanical studies on the macrobotanical remains from Troia (cf. Sections 2.5.1 and 2.7.1) and changes that occurred in the geomorphology of the Trojan Bay all point at a slow shift towards drier conditions in the Troad. These drier conditions must have increased the amplitude of seasonal differences in the aquatic environment, increasing the impact of summer evaporation and winter precipitation, resulting in the formation of broader gray incremental zones in the Late Bronze Age *C. glaucum* shells.

The variation in the breadths of both complete and incomplete incremental zones is great among the individuals of both modern and archaeological populations (Figures 26, 27, and 45). While this situation hampers expected growth calculations and a more refined assessment of harvest times, it also suggests that *C. glaucum* gathering took place year-round. While *C. glaucum* gathering took place year-round during the Bronze Age in Troia, gathering activity intensified during the months when SST and air temperatures were higher, and the precipitation rate lower, roughly between June and November according to the results of seasonal observations on the extant *C. glaucum* populations in the Troad.

The results of the examinations of the marginal growth increments have suggested that the proportion of the individuals with gray incremental zones at their margins is larger in the Late Bronze Age assemblages than it is in the Early Bronze Age assemblages (cf. Section 4.3.5.3). In view of the observation that the gray incremental zones are generally broader in the Early Bronze Age specimens than in the Late Bronze Age specimens, it is not possible to attribute the differences observed at the marginal growth zone securely to changes in harvest time. Instead, the increase in the amount of marginal gray zones in relationship to the amount of white incremental zones suggests that *C. glaucum* gathering took place year-round, with a concentration during warmer months in the Late Bronze Age, as it had in the Early Bronze Age.

Chapter 7. Summary, conclusions, and prospects

7.1. Summary

The aim of this study was to understand the environmental (including climatic) factors and cultural and economic stimuli affecting change and continuity in the role and nature of shellfishing as an economic activity and the role of shellfish as an element of human nutrition at Troia, Yenibademli, and Ulucak based on the study of the mollusk remains yielded from excavations at these sites. While issues such as the location and techniques of shellfish procurement, the nutritional value and processing of aquatic mollusks, dye production, harvest time, as well as the impact of prolonged human exploitation on mollusk populations and its consequence for the gathering activity were all undertaken in detail, discussions as to whether aquatic mollusks constituted a staple or supplemental food resource and estimates of the percentage contribution of aquatic mollusks to the diet have also been purposefully avoided. Further, questions dealing with the gender, ethnicity, and class of shellfishers and shellfish consumers have been deliberately left aside.

Research methods included common archaeozoological quantification techniques using NIS (number of identified specimens) and WIS (weight of identified specimens) counts, morphometric analyses, stable isotopic ($\delta\ ^{18}O$ and $\delta\ ^{13}C$) analyses, and incremental shell growth analyses of archaeological and modern mollusk populations. Modern ecological phenomena and ethnographic accounts were utilized to interpret the analytical results.

Differences in the proportions of mollusk remains in faunal assemblages and in the relative abundance of mollusk taxa in the archaeomalacological assemblages were calculated. The taphonomy and the differential contextual origins of the archaeomalacological assemblages were delineated to the greatest possible extent. The morphological characteristics of the *C. glaucum* populations from the Bronze Age layers of all three sites and the extant *C. glaucum* populations in the Troad were determined. Additionally, the morphological characteristics of the *P. caerulea* populations in Yenibademli were investigated. Based on observations of the extant *C. glaucum* populations in the Troad, the demography and the harvest time of the *C. glaucum* populations from the Bronze Age Troia were surmised. The archaeomalacological record from other sites in the Aegean basin –with a chronological focus on the Bronze Age and a geographical focus on Northwest Anatolia– was reviewed and re-evaluated with respect to their environmental context.

Variability and continuity in the archaeomalacological record were evaluated with regard to; (1) the altering cultural demography of human populations; (2) the requirements of local economic strategies; (3) environmental changes; (4) climatic fluctuations; (5) seasonal rhythms affecting subsistence economies; and, (6) human foraging pressure.
The effects of differential preservation, the differential nature of excavated areas, excavation techniques, and archaeozoological methodology on the formation of the archaeomalacological results were also acknowledged.

The primary sites that are dealt with in this study represent sedentary human populations which were already pursuing well-established, multi-pronged subsistence economies at the beginning of the Early Bronze Age, exploiting domesticated plants and animals, gathering wild plants, fishing, and hunting. They were also involved in long-distance trade and subject to acculturation processes.

The results of the study shows that the differences between the surrounding environments of the sites and the necessities of the subsistence systems, as well as acculturation processes, were the main factors shaping the main differential characteristics of the archaeomalacological assemblages of these three sites.

The archaeomalacological assemblages of these three sites mainly represent food refuse. Some may have been used to prepare fishing bait, whereas most specimens of rarely encountered species were brought to the settlements empty. The crushed *H. trunculus* shells from the Late Bronze Age layers of Troia are thought to be remnants of a flourishing dye industry that can be linked to a growing local textile industry, both serving regional trade under the cultural influence of the Minoans.

In Troia, shellfish gathering mainly took place between the lagoonal and estuarine environments of the Trojan Bay and the mesolittoral zone of the coastal areas. Confirming the results of archaeobotanical and archaeozoological investigations at the site, this situation shows that even if the Trojan Bay never served as a harbor for the Bronze Age Troia (Kayan et al. 2003), it was of foremost importance to the settlement as a rich resource base. The infralittoral and the circumlittoral zones of the sea and the rivers were seldom exploited for shellfish. While gathering could generally be done by employing easy techniques that rarely required implements, knowledge of mollusk behavior was prerequisite to obtaining species such as razorclams (*S.* marginatus) and the 'murex' snails (*H. trunculus*). Oysters may have been cultivated in Troia during the Post-Bronze Age. Despite the existence of food storage technology in these advanced societies, the processing of shellfish for later consumption was not attested by the archaeomalacological evidence. The year-round availability of aquatic mollusks probably made preservation techniques superfluous.

The role of shellfishing in the subsistence economies of the settlements in the Troad appears to have declined gradually from the 5th millennium BC to the end of the Byzantine settlement in Troia. The large proportion of marine mollusk remains from the settlements dating to the Early Bronze Age and immediately before implies that the exploitation of marine mollusks constituted an important part of the subsistence strategies. Settlements made use of the mollusk beds that were closest to them. The Trojan Bay and the deltaic lagoons of the Karamenderes and the Dümrek Rivers were the primary shellfishing grounds of Troia, while the Beşik Bay and the shores of the Aegean and the Dardanelles were rarely visited to gather shellfish. Cockle (*C. glaucum*) gathering in Troia was an integrated part of the subsistence since the beginning of the settlement. At that time, the proximity of cockle beds to the settlement made this activity a profitable pursuit in terms of nutritional gain. The abundance, demographic structure, size profiles, and growth increments of cockle remains in Troia indicate that the Early Bronze Age Trojans were opportunistically exploiting cockles and other shellfish year-round to the extent that both the mollusk populations and the gathering activity were negatively affected by the consequences of this intensive foraging activity.

The mode of mollusk exploitation changed to become slightly more discriminatory during the Late Bronze Age than it was during the Early Bronze Age. As the distance between Troia and the mollusk beds increased, the benthic structure of the Trojan Bay became devoid of any hard features, and the area of the continental shelf became smaller, the mode of mollusk exploitation became less intensive. Simultaneously, changes took place in other sectors of economic activities in the settlement. As a consequence, the culinary value, the desire for variety, and the need to obtain raw material for industrial work became important motives for gathering mollusks during the Late Bronze Age. Shellfish exploitation was now not only a part of the subsistence economy but also part of commerce and probably of long-distance trade. *C. glaucum* beds close to the settlement were still the most frequented environments for shellfishing, but visits to the outer bay and to the rocky shores became more frequent. The decrease in cockle consumption caused a significant increase in the mean valve size and the amount of older individuals in the cockle populations gathered during the Late Bronze Age. This situation positively influenced the amount of flesh acquired by the activity.

During the Bronze Age, cockles were exploited year-round, with increased activity during the drier and warmer months. Increased cockle gathering activity during warmer months is considered to be a part of the life rhythm of increased household and agricultural activities during these months. This situation is affected both by the necessity to supplement food resources and favorable weather conditions allowing for the exploitation of aquatic resources with relative ease. Incremental growth analysis shows that there is, (1) an increase in the proportion of individuals demonstrating gray marginal incremental zones at the time of collection in the Late Bronze Age samples; and, (2) an increase in the breadths of gray incremental zones. These changes in incremental growth patterns are linked to a gradual shift to drier environmental conditions in the Troad.

During the Post-Bronze Age periods, shellfish procurement assumed a rather selective mode. Oysters (*O. edulis*) and scallops (Pectenidae) started to dominate the archaeomalacological assemblages of Troia in the Post-Bronze Age. Shellfishing grounds moved from the closest areas to the settlement, such as the lagoons, to the Trojan Bay and probably to fully marine waters, in search for larger and more nutritious species. Acculturation processes, innovations in aquaculture, as well as the terrestrial orientation of subsistence strategies are thought to have been influential in this process.

Several problems occurred when testing the implications of the archaeomalacological evidence in view of proposed climatic, environmental, cultural, and economic changes. For instance, the archaeomalacological evidence from the Troia II period could not be evaluated to test archaeologists' postulations about "insecure times" associated with the increase in the amount of pits in the archaeological deposits (cf. Section 2.6.1) because none of the studied assemblages dating to this period originate from these pit contexts. Changes observed in the archaeomalacological assemblages from Troia III, Troia IV, Troia V could be more securely attributed to failures in the sampling strategies employed during the excavations than to environmental or economic changes. For this reason, the risk-buffering mode of plant cultivation suggested for the Middle Bronze Age Troia (cf. Section 2.7.1) could not be related to a similar trend in mollusk exploitation patterns. While changes between the assemblages of the Troia VIIa and VIIb periods might be indicative of the decline of the 'murex' dye industry, there is nothing that explicitly suggests environmental, economic, or cultural changes.

The bulk of the *H. trunculus* remains from the Late Bronze Age Troia represent an archaeological case that demonstrates the effects of technological and cultural innovations on the archaeomalacological record. The present study evaluated *H. trunculus* remains from Late Bronze Age Troia by incorporating the results from older excavations at the site and by comparing the combined evidence with other cases in the eastern Mediterranean. It was concluded that purple dye production at Troia began already during the Troia VIa period in the lower town, near the edge of the settlement and continued until Troia VIIa. The present evidence of *H. trunculus* accumulations at Troia represents the largest in scale among its contemporaries in the Eastern Mediterranean. In addition there is a possibility that the amount of on-site *H. trunculus* remains might lead to an underestimate of the capacity of the local dye industry, which must have constituted a major component in Troia's role in the regional Late Bronze Age trade network.

Limpets (Patellidae), oysters (*O. edulis*), and cockles (*C. glaucum*) were the most exploited mollusk species in the Early Bronze Age Yenibademli. Rocky supralittoral and mesolittoral zones, the estuary of the Büyükdere, and the deltaic lagoons were the most important shellfishing grounds. In light of the correlation between coastal environmental parameters and archaeomalacological evidence from contemporary sites in the Aegean, the relative abundance of species in the archaeomalacological assemblage of Yenibademli either reflects an unusually selective mode of shellfish gathering or calls for a revised chronology for the evolution of the Büyükdere Plain. The intensity of limpet (*P. caerulea*) gathering at Yenibademli affected the local limpet populations causing depletion in the mean shell size of the populations. Although this situation may have hindered shellfish gathering activity, it is unlikely that this constituted one of the factors that led to the abandonment of the site ca. 2600 BC. The relatively larger size of the cockle shells in Yenibademli can either be attributed to the low degree of foraging pressure on the populations or it can be linked to differential environmental conditions.

The majority of the cockle remains in Ulucak probably reflects a single event, not exploitive patterns. The remains might be remnants of a social interaction that was part of an exchange system between the coastal dwellers and the Early Bronze Age inhabitants of Ulucak. Alternatively, the cockles might have been gathered from the coast by the residents of Ulucak themselves. The latter interpretation enlarges the catchment area of this inland settlement towards the coastal areas. It is most likely that the marine mollusk remains in Ulucak originate from the easternmost or northeast coast of the inner section of the İzmir Bay.

While the peoples of the Bronze Age Aegean relied on land rather than on marine animals for subsistence, the persistence and ubiquity of aquatic mollusk remains in all Bronze Age sites along the coasts of the Aegean indicate that aquatic mollusks were part of the diet and shellfishing was a characteristic part of the subsistence activities. Marine mollusk assemblages from prehistoric contexts in the Aegean do not vary significantly in terms of species diversity and are limited to inshore species living in shallow waters. Changes in the relative abundance of species are mainly attributable to the characteristics of the coast line at different sites and geomorphological changes that affected them through time. In Troia, changes in the species compositions of the archaeomalacological assemblages neatly correlate with the evolution of the coastal plains in the Troad. Gathering techniques were similar across the region and comparable to those employed at other coastally adapted societies. The spectra of exploited species do not appear to have been shaped by cultural differences. This does not mean, however, that the types of shellfish consumed did not become part of group identities gradually.

The morphological differences (solidity/curve relationship) between various archaeological *C. glaucum* populations are considered to be a function of the differential size and age profiles of the respective populations. At this stage of archaeomalacological research in the Aegean, the size and curve indices of the *C. glaucum* populations cannot be linked unambiguously to the differential characteristics of ancient coastal environments. The morphology of the *C. glaucum* populations from Troia, Yenibademli, and Ulucak is essentially similar. A significant morphological difference has been observed between the archaeological and modern populations. This observation has important implications for the effect of the ecological changes on benthic organisms in lagoonal environments, not only in the Troad but potentially throughout the Aegean.

7.2. Concluding remarks

The relevance of an accurate understanding of the interaction between advanced societies and their environment for illuminating various questions regarding the past is often neglected in Aegean archaeology. However, as Fagan (2004:103) powerfully points out, "even if a community grows prosperous off long-distance trade, it must generate large local food surpluses to support public works. That implies ample crops, good weather and a carefully fostered relationship with the environment." The relationship between environment and subsistence constitute the foundation of advanced societies. The present study contributed to the understanding of the interrelationship between human settlements and the coastal environments in the Aegean by delineating the different aspects of the man (as predator) and mollusk (as prey) relationship in the dynamic cultural, economic, and environmental context of three cases: Troia; Yenibademli; and, Ulucak. The study explicitly demonstrated that the prehistoric and early historic peoples of the Aegean had a far more varied diet and subsistence system than it is commonly acknowledged.

It may appear that two interrelated questions remain unanswered: how much shellfish did the peoples of the prehistoric and early historic Aegean consume and how important was shellfishing? In the context of agricultural societies that are involved in regional and long-distance trade (probably also of the domesticated and wild surplus), it is not possible to answer this type of questions with any degree of accuracy. Nevertheless, it is possible to reach a relative understanding of the role of shellfish and shellfishing in the prehistoric and early historic Aegean.

There is no recorded traditional society that has relied on shellfish for subsistence (Meehan 1977, 1982, 1983, Noll 2002: 16, Waselkov 1987). There is currently no general consensus about the meaning of large shell middens in terms of the contribution of shellfish to prehistoric diets. While most scholars estimate that shellfish does not constitute more than 5-10 % of the calorific intake of coastal hunter-gatherer-fishers (Noll 2002: 16-17), others think that the contribution of shellfish to the diet of coastal foragers was greater (Buchanan 1988). Human groups who rely heavily on shellfish and various other marine resources are usually those that live on coastal environments that lacked rich and suitable terrestrial biotopes supporting a wide assortment of edible plants, herded or domesticable wild animals, and arable land. Shell midden builders of the Americas and Australia exemplify these peoples (See Diamond 1997 and Santoro et al. 2005). Around all three of the primary sites discussed in this dissertation, arable land was abundant in supply. In addition, at all the Middle and Late Holocene settlements discussed in this study, a vast variety of crops were seasonally available to the human populations and storage techniques were long known. Shellfish was available year-round, but so were the domestic animals. Even so, the archaeological deposits of the Aegean sites are full of mollusk remains. Therefore, in a context where the possible variety of economic activities is so vast, looking for answers to the two aforementioned questions necessitates a different approach.

Although a general impression about marine resources is that their yield is difficult to predict and therefore their consumption is difficult to regulate, as pointed out by Glassow and Wilcoxon (1988) and Yesner (1980), the dietary and economic advantage of shellfish gathering stems from, (1) the ease of shellfishing activity; (2) the predictability of the location of shellfishing beds; and, (3) high yields. These advantages compensate for the limitations of the low nutritional value offered by the single mollusk individual. They also explain how the exploitation of marine mollusks insured the successful subsistence of agricultural-dependent human populations living along the coasts of the Aegean.

Moreover, although shellfish is often regarded as a trivial contributor to the diet or as emergency food resource that is consumed during periods of famine or low terrestrial productivity (Deith 1988a, Forbes 1976a, Glassow and Wilcoxon 1988), the archaeomalacological record of the Aegean does not indicate usage solely during times of hunger. On the contrary, shellfish gathering probably consolidated the stability of economic and cultural systems of coastal societies in the Aegean by presenting an additional and perennially available supplement to the dietary items provided by agriculture, animal husbandry, hunting, fishing, and the gathering of wild plants. Thus, resistance against lean or crisis periods imposed either by social instability or unexpected climatic changes could be reinforced by the exploitation of the mollusk beds when it was necessary. A general lack of evidence for nutritional deficiency among the human populations of the Bronze Age Troad supports this view. The palaeoanthropological investigations of the human remains found in Troia and in the Beşiktepe Cemetery confirms the picture of plenty and variety drawn by archaeozoological and archaeobotanical investigations (Riehl 1999, Uerpmann 2003, Uerpmann et al. 1992, Uerpmann and van Neer 2000, von den Driesch 1999, Witter-Backofen 2003). Certainly, shellfish exploitation contributed to this picture by ensuring essential proteins, vitamins, and minerals. In fact, the availability of good shellfishing grounds may have played a decisive role in the shaping of the prehistoric settlement patterns along the coastlines of the Aegean. Furthermore, the exploitation of aquatic mollusks enhanced the diet by giving ancient populations the freedom to exploit and experiment on the resources that are more likely to fail. Such was the importance of shellfish and shellfishing to the agriculturally inclined coastal dwellers of the Aegean during the Middle and Late Holocene.

7.3. Prospects for future research

In cases when the palaeoenvironmental and palaeoeconomic questions necessitating the evaluation of spatial and temporal variations of the archaeomalacological record could not be pursued effectively in this study, this was largely due to the inconsistency in the archaeomalacological record caused mainly by the application of dissatisfactory sampling techniques during excavation and discarding of excavated mollusk remains. This situation calls for the urgent standardization of the recovery techniques employed in Aegean archaeology. There is no excuse for the insufficient sampling of organic remains that constitute key evidence for the exploitation of marine resources and the reconstruction of the environment. If coastal economic systems of the prehistoric settlements in the Aegean are to be understood in greater detail, the employment of more sophisticated recovery techniques is of the essence. Mollusk remains only represent one sector of coastal subsistence economies. The full value of archaeomalacological studies in Aegean archaeology will be realized and appreciated only when the objectives of archaeological projects place emphasis on questions regarding palaeoenvironmental conditions and the development of subsistence strategies, and when excavators start designing research strategies suited to answer these questions. Scientific progress in this direction will surely alter the current impression of the role of marine resources in prehistoric and historic Aegean radically.

Further work on the biometry and seasonality of diverse archaeomalacological assemblages in the region, covering a wider chronological period, will certainly advance the understanding of the ecological relationships between the shell morphology of aquatic mollusks, the effects of human exploitation on marine resources, and the annual scheduling of subsistence activities in Mediterranean economies. Standardization of analytical methods, e.g., morphometric measurements, is of utmost importance to create a compatible archaeomalacological database for the region. When sufficient funding is available, archaeomalacological assemblages from all sites in the Troad should be re-evaluated. Their archaeomalacological analyses should be improved to include morphometric and growth pattern analyses of all commonly encountered mollusk species. Post-Bronze Age remains from Troia should also be investigated in order to bridge the gap of knowledge between the Bronze Age and modern mollusk populations. There was not sufficient time within the scope of the present study to sample modern populations more frequently; for example at monthly intervals. If the monitoring of extant mollusk populations in the Troad continue by increasing the periodicity of the observations and collections, then the harvest time of extant mollusk populations can be more precisely determined and the ecological and economic connotations of morphological features will be better resolved. Malacological research on the growth rates and isotopic contents of the *P. caerulea* populations in the Mediterranean has showed that the shells of this species are accreted in equilibrium with the seasonal changes in the host water (Bianucci and Longinelli 1982). These observations on modern populations indicate that incremental growth and isotopic analyses of archaeological *P. caerulea* shells from the Aegean have the potential to answer questions regarding the palaeoeconomy and palaeoenvironment at those sites where *P. caerulea* shells were found in large amounts, including Yenibademli.

The lack of detailed archaeomalacological and malacological studies in the Aegean basin –particularly of comparative morphological indices, incremental growth patterns, as well as stable isotopic compositions– greatly hindered the verification of the results of the present study. It is hoped that the results presented in this study will be inspirational and help trigger a wider consideration of the potential of mollusk remains in Aegean archaeology and cooperation with the field of malacology. Further research requires a broader database compiled from a variety of archaeological sites and modern collections from geographically coherent locations.

Additionally, thorough excavation and analysis of the archaeozoological material from the historic village garbage heaps antedating the population exchange between Greece and Turkey in the Troad has great potential to elucidate the palaeoeconomical and palaeoenvironmental meaning of the organic assemblages from the archaeological sites in the region. The results of such a study can serve as a substitute for the lack of ethnoarchaeological studies concerning the subsistence economy of the region, which needs to be filled without delay.

Furthermore, the role of different marine resources in the diet of an archaeological population can also be revealed by applying stable isotopic ($\delta^{13}C$ and $\delta^{15}N$) analyses on human skeletal remains of the population under investigation (e.g., Harrison and Katzenberg 2003, Papathanassiou et al. 2000, Richards and Hedges 1999). Stable isotope values of carbon and nitrogen in human tissue –such as bone collagen and tooth enamel– are used to infer the ratio of seafood consumption versus terrestrial resources. Future geochemical investigations of the human skeletal finds recovered in and around Troia and Ulucak will certainly contribute to the understanding of the role of marine resources in the diet of the Bronze Age peoples who populated these sites.

Finally, it is hoped that this documentation of the present and past condition of Holocene mollusk populations in the Trojan Bay can serve as a source of historical perspective for present and future studies concerning the restoration and management of coastal ecosystems in the Troad in particular and in the Aegean coasts in general. Ecological science has recently started to realize the "myth of the pristine nature" prior to the Industrial Revolution and to appreciate the potential of archaeological data as witnesses of past marine life (Jackson et al. 2001). The results of the present study clearly demonstrate that human populations of unknown density, which exploited terrestrial and coastal resources by keeping herds, practicing agriculture, hunting, fishing, and shellfishing, have affected their environment. This is evident in the shifting size and age structures of archaeomalacological populations. Morphological and incremental growth analyses imply that anthopogenically and climatically induced environmental changes in the coastal shelf and lagoonal habitats in the Troad have affected the *C. glaucum* populations. These environmental changes may have included changes in precipitation rates, pluvial regimes, temperatures, and coastal geomorphology. All these environmental phenomena have directly influenced traditional subsistence systems in the Mediterranean and they still do. Further work on archaeomalacological assemblages in the Mediterranean may help clarify the impact of regional climatic changes and human exploitation pressure on aquatic mollusk populations, coastal marine life, as well as on patterns of human adaptations to coastal environments.

References

ACKERMANN B. 1996. Das Sekret. *Jetzt: Das Jugendmagazin der süddeutschen Zeitung* 25: 22-25.

AKSU A.E., R.N. HISCOTT, M.A. KAMINSKI, P.J. MUDIE, H. GILLESPIE, T. ABRAJANO, and D. YAŞAR. 2002. Last glacial-Holocene paleoceanography of the Black Sea and the Marmara Sea: stable isotopic, foraminiferal and coccolith evidence. *Marine Geology* 190: 119-149.

ALBARÈDE F. 2003. *Geochemistry: An Introduction*. Cambridge, Cambridge University Press.

ANDREASEN N.H. 1998. *The Exploitation of a Brackish Lagoon Bivalve,* Cerastoderma glaucum, *Bruguiere, at the Neolithic Settlement of Makriyalos, Central Macedonia, Greece*. Unpublished M.Sc. Thesis. University of Sheffield.

ANONYMOUS. 1982. *Yurt Ansiklopedisi*. Ankara, Anadolu Yayıncılık.

ANSELL A.D. 1968. The rate of growth of the hard clam *Mercenaria mercenaria* (L.) throughout the geographical range. *Journal du Conseil, Permanent International pour l'Exploration de la Mer* 31: 364-409.

ARİSTOTLE. *Historia Animalium.* Translated by D. W. Thompson. 2004. ebooks@Adelaide.

ARNERI E., G. GIANNETTI, and B. ANTOLINI. 1998. Age determination and growth of *Venus verrucosa* L. (Bivalvia: Veneridae) in the southern Adriatic and the Aegean Sea. *Fisheries Research* 38: 193-198.

ARTHUR M.A., D.F. WILLIAMS, and D.S. JONES. 1983. Seasonal temperature-salinity changes and thermocline development in the mid-Atlantic Bight as recorded by the isotopic composition of bivalves. *Geology* 11: 655-659.

ARTÜZ L. 2002. Marmara ve Boğazlar'ın ekolojisi ve değişimler. Paper delivered at the *Boğaziçi Üniversitesi Sualtı Bilim ve Teknoloji Toplantısı 2002.*

ASLAN R. 2003. "The relationship between man and landscape in the Troad during the Ottoman Period." In: G.A. Wagner, E. Pernicka, and H.-P. Uerpmann (Eds.), *Troia and the Troad: Scientific Approaches*. Berlin, Springer Verlag: 31-41.

ASLAN R. and G. BIEG (with contributions from P. JABLONKA and P. KRÖNNECK). 2003. Die mittel- bis spätbronzezeitliche Besiedlung (Troia VI and Troia VIIa) der Troas and der Gelibolu Halbinsel. *Studia Troica* 13: 165-214.

ATAY D. 1984. *Kabuklu Su Ürünleri ve Üretim Tekniği*. Ders Kitabı 257. Ankara, Ankara Üniversitesi Ziraat Fakültesi Yayınları 914.

AUGIERRE M.L., S. RICHIANO, and N.Y. SIRCH. 2005. Palaeoenvironments and palaeoclimates of the Quaternary molluscan faunas from the coastal area of Bahia Vera-Camarones (Chubut, Patagonia). *Palaeogeography, Palaeoclimatology, Palaeoecology* 229: 251-286.

BAILEY G.N. and J. PARKINGTON. 1988. "The archaeology of prehistoric coastlines: an introduction." In: G. Bailey and J. Parkington (Eds.), *The Archaeology of Prehistoric Coastlines: New Directions in Archaeology.* Cambridge, Cambridge University Press: 1-10.

BAILEY G.N., M.R. DEITH, and N.J. SHACKLETON. 1983. Oxygen isotope analysis and seasonality determinations: limits and potential of a new technique. *American Antiquity* 48: 390-398.

BAKER J.T. 1974. Tyrischer Purpur ein antiker Farbstoff, ein modernes Problem. *Endeavour* 33/118: 11-17.

BALAKRISHNAN M., C.J. YAPP, D.J. MELTZER, and J.L. THELER. 2005. Palaeoenvironment of the Folsom archaeological site, New Mexico, USA, approximately 10,500 14C yr BP as inferred from the stable isotope composition of fossil land snail shells. *Quaternary Research* 63: 31-44.

BALIK S. and M.R.A. USTAOĞLU. 1993. Preliminary investigation of the freshwater fauna of Gökçeada (Imbroz) Island. *Biologia Gallo-Hellenica* 20/1: 299-303.

BAR-MATTHEWS M. and A. AYALON. 2005. Evidence from speleothem for abrupt climatic changes during the Holocene and their impact on human settlements in the Eastern Mediterranean region: dating methods and stable isotope systematics. *Zeitschrift fur Geomorphologie*. Supplement-Vol. 137: 45-59.

BAR-MATTHEWS M., A. AYALON, A. KAUFMAN, and G. WASSERBURG. 1999. The eastern Mediterrenean palaeoclimate as a reflection of regional events: Soreq Cave, Israel. *Earth and Planetary Science Letters* 166: 85-95.

BARNES R. 1980. *Coastal Lagoons: The Natural History of a Neglected Habitat.* Cambridge Studies in Modern Biology 1. Cambridge, Cambridge University Press.

----. 1987. *Invertebrate Zoology*. 5th ed. Philedelphia, Sounders College Publishing.

BAR-YOSEF D.E. 1991. "Changes in the selection of marine shells from the Natufian to the Neolithic." In: O. Bar-Yosef and F.R. Valla (Eds.), *The Natufian Culture in the Levant*. Ann Arbor, Michigan, International Monographs in Prehistory: 629-636.

----. 1999. *The Role of Shells in the Reconstruction of Socio-Economic Aspects of Neolithic through Early Bronze Age Societies in Sinai.* Unpublished PhD thesis. Hebrew University of Jerusalem.

----. 2005a. (Ed.), *Archaeomalacology: Mollusks in former environments of human behavior. Proceedings of the 9th ICAZ Conference in Durham, 2002.* Oxford, Oxbow Books.

----. 2005b. "An introduction to archaeomalacology." In: D. E. Bar-Yosef (Ed.), *Archaeomalacology: Mollusks in former environments of human behavior.*

Proceedings of the 9th ICAZ Conference in Durham, 2002. Oxford, Oxbow Books: 1-4.

BASEDOW M. 2000. *Beşiktepe –Das spätbronzezeitliche Gräberfeld.* Studia Troica Monographien 1. Mainz: Philipp von Zabern.

BECKER C. 1986. *Kastanas: Die Tierknochenfunde.* Wissenschaftsverlag Volker Spiess, Berlin.

----. 1996. Nourriture, cuilléres, ornaments…les témoignages d'une exploitation variée des mollusques marins à Ayios Mamas (Chalcidique, Grèce). *Anthropozoologica* 24: 3-17.

----. 2001. "Did the people in Ayios Mamas produce purple-dye during the Middle Bronze Age? Considerations on the prehistoric production of purple-dye in the Mediterranean." In: H. Buitenhuis and W. Prummel (Eds.), *Animals and Man in the Past: Essays in Honour of Dr. A.T. Clason, Emeritus Professor of Archaeozoology, Rijksuniversiteit, Groningen.* Groningen, ARC-Publicatie 41: 122-133.

BECKS R. 2006. "Troia in der späten Bronzezeit –Troia VI und Troia VIIa." In: M.O. Korfmann (Ed.), *Troia: Archäologie eines Siedlungshügels und seiner Landschaft.* Mainz, Philipp von Zabern: 155-166.

BECKS R., P. HNILA, and M. PIENIĄŻEK-SIKORA. 2006. "Troia in der frühen Eisenzeit –Troia VIIb1-VIIb3." In: M.O. Korfmann (Ed.), *Troia: Archäologie eines Siedlungshügels und seiner Landschaft.* Mainz, Philipp von Zabern: 181-188.

BELKNAP D.F. and Y. MART. 1999. Sea-level low stand in the eastern Mediterranean: Late Pleistocene coastal terraces offshore northern Israel. *Journal of Coastal Research* 15: 399-412.

BIANUCCI G. and A. LONGINELLI. 1982. Biological behavior and accretion rates of *Patella caerulea* L. as indicated by oxygen isotope measurements. *Palaeogeography, Palaeoclimatology, Palaeoecology* 37: 313-318.

BİLECENOĞLU M. and E. TAŞKAVAK. 1999. General characteristics of the Turkish marine ichthyofauna. *Zoology in the Middle East* 18: 41-56.

BINTLIFF J.C. 1977. *Natural Environment and Human Settlement in Prehistoric Greece.* B.A.R. Supplementary Series 28. Oxford, Oxbow Books.

BINTLIFF J.C. and W. VAN ZEIST. (Eds.). 1982. *Palaeoclimates, Palaeoenvironments and Human Communities in the Eastern Mediterranean Region in Later Prehistory.* B.A.R. International Series 133 (i). Oxford, Oxbow Books.

BIRD D.W. and R.L. BLIEGE-BIRD. 1997. Contemporary shellfish gathering strategies among the Meriam of the Torres Strait Islands, Australia: testing predictions of a central place foraging model. *Journal of Archaeological Science* 24: 39-63.

BIRD D.W., J.L. RICHARDSON, P.M. VETH, and A.J. BARHAM. 2002. Explaining shellfish variability in middens on the Meriam Islands, Torres Strait, Australia. *Journal of Archaeological Science* 29: 457-469.

BLEGEN C. 1937. Excavations at Troy. *American Journal of Archaeology* 41/4: 553-597.

----. 1963. *Troy and the Trojans.* London, Thames and Hudson.

BLEGEN C., C. BOULTER, J. CASKEY, and M. RAWSON. 1958. *Troy IV: Settlements VIIa, VIIb and VIII.* Princeton, Princeton University Press.

BLEGEN C., J. CASKEY, and M. RAWSON. 1951. *Troy II: The Third, Fourth and Fifth Settlements.* Princeton, Princeton University Press.

----. 1953. *Troy III: The Sixth Settlement.* Princeton, Princeton University Press.

BLEGEN C., J. CASKEY, M. RAWSON, and M. SPERLING. 1950. *Troy I: The First and Second Settlements. Troy I-IV.* Princeton, Princeton University Press.

BLUM H. 1998. *Purpur als Statussymbol in der griechischen Welt.* Bonn, Antiquitas Reihe 1. Band 47.

BLUM S.W.E. 2006. "Troia an der Wende von der frühen zur mittleren Bronzezeit –Troia IV und Troia V." In: M.O. Korfmann (Ed.), *Troia: Archäologie eines Siedlungshügels und seiner Landschaft.* Mainz, Philipp von Zabern: 145-154.

BODUR T., Y. KAYA, and M. GÖKOĞLU. 2004. Antalya Körfezi'nde Nuh'un Gemisi (*Arca noae* Linnaeus, 1758)'nin dağılımı ve bazı biometrik ölçümleri. *Türk Sucul Yaşam Dergisi* 2: 243-246.

BOESSNECK J. 1986. Weichtierresser vom Beşik-Sivritepe. *Archäologischer Anzeiger*: 329-338.

BOESSNECK J. and J. SCHÄFFER. 1986. Tierknochenfunde aus Didyma II. *Archaeologischer Anzeiger*: 251-301.

BOETTGER C.R. 1935. Über Weichtierschalen in den Kulturschichten von Troja. *Sitzungsberichte der Gesellschaft Naturforschender Freunde zu Berlin* 1934: 297-303.

BOGUCKI P.I. and R. GRYGIEL. 1981. The household cluster at Bresc Kujawski 3: small-site methodology in the Polish Lowlands. *World Archaeology* 13: 59-72.

BOLLE H.-J. 2003. "Climate, climate variability, and impacts in the Mediterranean area: an overview." In: H.-J. Bolle (Ed.), *Mediterranean Climate: Variability and Trends.* Berlin, Springer Verlag: 5-86.

BOND G., B. KROMER, J. BEER, R. MUSCHELER, M.N. EVANS, W. SHOWERS, S. HOFFMANN, R. LOTTI-BOND, I. HAJDAS, and G. BONANI. 2001. Persistent solar influence on North Atlantic Climate during the Holocene. *Science* 294: 2130-2136.

BOND G., W. SHOWERS, M. CHESEBY, R. LOTTI, P. ALMASI, P. DE MENOCAL, P. PRIORE, H. CULLEN, I. HAJDAS, and G. BONANI. 1997. A pervasive millennial-scale cycle in North Atlantic Holocene and glacial cycles. *Science* 278: 1257-1266.

BOND G., W. SHOWERS, M. ELLIOT, M. EVANS, R. LOTTI, I. HAJDAS, G. BONANI, and S. JOHNSON. 1999. "The North Atlantic's 1-2 kyr climate rhythm: relation to Heinrich events, Dansgaard/Oeschger cycles and the Little Ice Age."

In: P.U. Clark, R.S. Webb, and L.D. Keigwin (Eds.), *Mechanisms of Global Climate Change at Millennial Time Scales Geophysical Monograph* 112. Washington D.C., AGU: 35-58.

BOOKSTEIN F.L. 1982. Foundation of morphometrics. *Annual Review of Ecology and Systematics* 13: 451-470.

BOTTEMA S. 1979. Pollen analytical investigations in Thessaly (Greece). *Palaeohistoria* 21: 19-40.

BOTTÉRO J. 1996. "The most ancient recipes of all." In: J. Wilkins, D. Harvey, and M. Dobson (Eds.), *Food in Antiquity.* Exeter, University of Exeter Press: 248-255.

BOYDEN R. 1972a. Behaviour, survival and respiration of the cockles *Cerastoderma edule* and *C. glaucum* in air. *Journal of the Marine Biological Association of the United Kingdom* 52: 661-680.

----. 1972b. Relationship of size to age in the cockles *Cerastoderma edule* and *C. glaucum* from the River Crouch Estuary, Essex. *Journal of Conchology* 27: 475-489.

BOYDEN R. and P.J.C. RUSSELl. 1972. The distribution and habitat range of the brackish water cockle (*Cardium* (*Cerastoderma*) *glaucum*) in the British Isles. *Journal of Animal Ecology* 41: 719-734.

BRAUDEL F. 1972. *The Mediterranean and the Mediterranean World in the Age of Philip II*, Vol. 1. Translated by Siân Reynolds. New York, Evanston, San Francisco, London, Harper and Row Publishers.

BROCK V. 1991. *An interdisciplinary study of the evolution in the cockles* Cardium (Cerastoderma) edule, C. glaucum, C. lamarcki. Vinderup, Vestjydsk Forlag.

BRÜCKNER H., A. VÖTT, A. SCHRIEVER, and M. HANDL. 2005. Holocene delta progradation in the Eastern Mediterranean: case studies in their historical context. *Méditerranée* 1/2: 95-106.

BUCHANAN W.F. 1988. *Shellfish in Prehistoric Diet. Elands Bay, S.W. Cape Coast, South Africa.* Cambridge Monographs in African Archaoelogy 31. B.A.R. International Series 455. Oxford, Oxbow Books.

BUITENHUIS H. 1995. "The faunal remains (Chapter 9)." In: J. Roodenberg (Ed.), *The Ilıpınar Excavations* 1. Istanbul, Nederlands historisch-archaeologisch Institut: 151-156.

BUTZER K.W. 2005. Environmental history in the Mediterranean world: cross-disciplinary investigation of cause-and-effect for degradation and soil erosion. *Journal of Archaeological Science* 32: 1773-1800.

CABRAL J.P. and A.C.F. DA SILVA. 2003. Morphometric analysis of limpets from an Iron-Age shell midden found in West Portugal. *Journal of Archaeological Science* 30/7: 817-829.

ÇAKIRLAR C. 2003. *Animal Exploitation at Kinet Höyük during the Late Iron Age.* Unpublished M.A. thesis. American University of Beirut.

ÇAKIRLAR C. in print. To the shore, back and again: archaeomalacology of Troia. *Studia Troica* 18.

ÇAKIRLAR C. and R. Becks. in print. Murex dye production in Troia: assessment of archaeomalacological data from old and new excavations. *Studia Troica* 18.

CALVERT F. 1881. "Thymbra, Hanai Tepeh." In: H. Schliemann. *Ilios: Stadt und Land der Trojaner.* Leipzig, Brockhaus: 706-720.

CARPENTER R. 1966. *Discontinuity in Greek Civilization.* Cambridge, Cambridge University Press.

CARRÉ M., I. BENTALEB, D. BLAMART, N. OGLE, F. CARDENAS, S. ZEVALLOS, R.M. KALIN, L. ORTLIEB, AND M. FONTUGNE. 2005. Stable isotopes and sclechronology of the bivalve *Mesodesma donacium*: potential application to Peruvian paleoceanographic reconstructions. *Palaeogeography, Palaeoclimatology, Palaeoecology* 228: 4-25.

CARTWRIGHT C. 2003. Unusual use of freshwater mussel (*Unio* sp.) shells during the Early Bronze Age at Tell es-Sa'idiyeh, Jordan. *Environmental Archaeology* 8/1: 85-89.

CHARISI S.D., B. SCHMITZ, and C. PERISSORATIS. 1998. Palaeoceanography of the Aegean Sea during formation of sapropel S_1: $\delta\ ^{18}O$ and $\delta\ ^{13}C$ records from foraminiferal and pteropod shells. *Earth Sciences Centre. Göteborg University. Series A, 32*: 1-24.

CHEMELLO R. and M. MILAZZO. 2002. Effect of algal architecture on associated fauna: some evidence from phytal molluscs. *Marine Biology* 140/5: 981-990.

CHILARDI S., L. GUZZARDI, M.R. IOVINO, and A. RIVOLI. 2005. "The evidence of *Spondylus* ornamental objects in the central Mediterranean Sea: two case studies: Sicily and Malta." In: D. E. Bar-Yosef (Ed.), *Archaeomalacology: Mollusks in former environments of human behavior. Proceedings of the 9th ICAZ Conference in Durham, 2002.* Oxford, Oxbow Books: 82-90.

ÇİLİNGİROĞLU A. and E. ABAY. 2005. Ulucak Höyük Excavations: new results. *Mediterranean Archaeology and Archaeometry* 5/3:5-21.

ÇİLİNGİROĞLU A., Z. DERİN, E. ABAY, H. SAĞLAMTİMUR, and İ. KAYAN. 2004. *Ulucak Höyük: Excavations Conducted between 1995-2002.* Ancient Near Eastern Studies. Supplement Vol. 15. Louvain-Paris-Dudley, Peeters.

ÇINAR M.E., T. KATAĞAN, and Z. ERGEN. 2002. Zoobenthos-inhabiting *Sarcotragus muscarum* (Porifera: Demospongiae) from the Aegean Sea. *Hydrobiologia* 482: 107-117.

CLAASSEN C. 1983. "Prehistoric shellfishing patterns in N. Carolina." In: C. Grigson and J. Clutton-Brock (Eds.), *Shell Middens, Fishes, and Birds. Animals in Archaeology 2*. B.A.R. International Series 227. Oxford, Oxbow Books: 305-324.

----. 1986. Shellfishing seasons in the prehistoric Southeastern United States. *American Antiquity* 51/1: 21-37.

----. 1988. "Techniques and controls for the determination of seasonality in shellfishing activities." In: R.E. Webb (Ed.), *Recent Developments in Environmental Analysis in Old and New World Archaeology.*

B.A.R. International Series 416. Oxford, BAR Publishing: 51-66.

----. 1993. Problems and choices in shell seasonality studies and their impact on results. *Archaeozoologia* 5/2: 55-76.

----. 1998. *Shells*. Cambridge Manuals in Archaeology. Cambridge, Cambridge University Press.

----. 2000. Quantifying shell: comments on Mason, Peterson and Tiffany. *American Antiquity* 65/2: 415-418.

CLARK II G.R. 1979. Seasonal growth variations in the shells of recent and prehistoric specimens of *Mercenaria Mercenaria* from St. Catherines Island, Georgia. *Anthropological Papers of the American Museum of Natural History* 51: 161-179.

CLARK II G.R. 2005. Daily growth lines in some living Pectens (Mollusca: Bivalvia), and some applications in a fossil relative: time and tide will tell. *Paleogeography, Palaeoclimatology, Palaeoecology* 228: 26-42.

CLASON A.T. and W. PRUMMEL. 1977. Collecting, sieving and archaeozoological research. *Journal of Archaeological Science* 4: 171-175.

COLLEY S.M. 1990. "The analysis and interpretation of archaeological fish remains." In: M. B. Schiffer (Ed.), *Archaeological Method and Theory II.* Tucson, University of Arizona Press: 207-253.

COSMOPOULOS M.B., H.J. GREENFIELD, and D. RUSCILLO. 2003. "Animal and marine remains from the new excavations at Eleusis: an interim report." In: E. Katjabopoulou, Y. Hamilakis, P. Halstead, C. Gamble, and P. Elefanti (Eds.), *Zooarchaeology in Greece, Recent Advances.* British School of Archaeology in Athens. Studies 9: 145-152.

COY J. 1977. "Animal remains (Appendix 4)." In: J.E. Coleman, *Keos I: Kephala: A Late Neolithic Settlement and Cemetery.* Princeton, American School of Classical Studies: 129-133.

CULLEN H.M., P.B. DE MENOCAL, S. HEMMING, G. HEMMING, F.H. BROWN, T. GUILDERSON, and F. SIROCKO. 2000. Climate change and the collapse of the Akkadian Empire: evidence from the deep sea. *Geology* 28: 379-382.

CULLEN H.M., A. KAPLAN, P.A. ARKIN, and P.B. DE MENOCAL. 2002. Impact of the North Atlantic Oscillation on the Middle Eastern climate and streamflow. *Climate Change* 55/3: 315-338.

CUSTER J.F. and K.R. DOMS. 1990. Analysis of microgrowth patterns of the American Oyster (*Crassostrea virginica*) in the Middle Atlantic Region of Eastern North America: Archaeological Applications. *Journal of Archaeological Science* 17: 151-160.

DALFES H.N., G. KUKLA, and H. WEISS. 1997. (Eds.). *Third Millennium BC Climate Change and the Old World Collapse*. Global Environmental Change 49. Berlin, Springer Verlag.

DALL W. 1877. On succession in the shell-heaps of the Aleutian Islands. *United States Geological and Geographic Survey, Contributions to North American Ethnology* 1: 41-91.

DAVIDSON A. 1972. *Akdeniz balık yemekleri*. Translated by T. Odabaşı, 2000. Ankara, Dost Yayınları.

----. 2004. "Mollusks." In: S.H. Katz (Ed.), *Encyclopedia of Food and Culture.* Scribner Library of Daily Life. Vol. 2. New York, Charles Scribner's Sons: 538-543.

DAVIS S. 1987. *The Archaeology of Animals*. London, Batsford.

DE HAAS W and F. KNORR. 1965. *Was Lebt im Meer? Mittelmeer, Atlantik, Nordsee, Ostsee.* Kosmos Naturführer. Stuttgart, Franckhsche Verlagshandlung, W. Keller und Co.

DE MENOCAL P.B. 2001. Cultural responses to climate change during the Late Holocene. *Science* 292: 667-673.

DE NIRO M.J. 1987 Stable isotopy and archaeology. *American Scientist* 75:182-191.

DE SMIT M. and K. BÁBA. 2002. New data to the marine malacofauna from Euboea Island (Greece). *Malacological Newsletter* 20: 83-91.

DEITH M.R. 1983a. Molluscan calendars: the use of growth-line analysis to establish seasonality of shellfish collection at the Mesolithic site of Morton, Fife. *Journal of Archaeological Science* 10: 423-440.

----. 1983b. "Seasonality of shell collecting, determined by oxygen isotope analysis of marine shells from Asturian sites in Cantabria." In: C. Grigson and J. Clutton-Brock (Eds.), *Shell Middens, Fishes, and Birds. Animals and Archaeology 2.* B.A.R. International Series 227. Oxford, Oxbow Books: 67-76.

----. 1985. "Clams and Salmonberries: interpreting seasonality data from shells." In: C. Bonsall (Ed.), *Mesolithic in Europe.* Edinburgh, John Donald: 73-79.

----. 1986. Subsistence strategies at a Mesolithic camp site: evidence from stable isotope analyses of shells. *Journal of Archaeological Science* 13: 61-78.

----. 1988a. "A molluscan perspective on the role of foraging in Neolithic farming economies." In: G. Bailey and J. Parkington (Eds.), *The Archaeology of Prehistoric Coastllines.* New Directions in Archaeology. Cambridge, Cambridge University Press: 116-124.

----. 1988b. "Shell seasonality: an appraisal of the oxygen isotope technique." In: R. E. Webb (Ed.), *Recent Developments in Environmental Analysis in Old and New World Archaeology*. B.A.R. International Series 416. Oxford, Oxbow Books: 37-49.

----. 1989. Shellfish gathering and site function: a case study from Neolithic Apulia. *Archaeozoologica* 3/1-2: 163-176.

DEITH M.R. and N.J. SHACKLETON. 1986. "Seasonal exploitation of marine molluscs: oxygen isotope analysis of shell from La Riera Cave." In: L.G. Straus and G.A. Clark (Eds.), *La Riera cave: Stone Age hunter-gatherer adaptations in northern Spain.* Tucson: Arizona State University Anthropological Research Papers 36: 299-313.

DEITH M.R. and N.J. SHACKLETON. 1988. "Oxygen isotope analysis of marine mollusks from Franchthi

Cave (Part II. Chapter 9).” In: J.C. Shackleton, *Marine Molluscan Remains from Franchthi Cave*. Bloomington, Indiana University Press: 133-157.

DEMİR M. 2003. Shells of mollusca collected from the seas of Turkey. *Turkish Journal of Zoology* 27: 101-140.

DEMİRSOY A. 1996. *Genel ve Türkiye Zoocoğrafyası*. Ankara, Meteksan.

DESSE J. and N. DESSE-BERSET. 2004. “Découverte de chitons (Polyplacophores) sur les sites archéologiques de la côte des Ichthyophages (Makran, Bélouchistan pakistanais).” In : J.-P. Brugal and J. Desse (Eds.), *Petits Animaux et Sociétés Humaines. Du Complément Alimentaire aux Resources Utilitaires. XXIVe Rencontres Internationales d'Archéologie et d'Historie d'Antibes*. Antibes, Éditions APDCA: 59-64.

DETTMANN D.L., A.K. REISCHE, and K.C. LOHMANN. 1999. Controls on the stable isotope composition of seasonal growth bands in aragonitic fresh-water bivalves (Unionidae). *Geochimica et Cosmochimica Acta* 63/7-8: 1049-1057.

DIAMOND J. 1997. *Guns, Germs, and Steel*. New York, W.W. Norton and Company Ltd.

DICKINSON O. 1994. *The Aegean Bronze Age*. Cambridge World Archaeology. Cambridge, Cambridge University Press.

DÖRPFELD W. 1902. Troja und Ilion. Ergebnisse der Ausgrabungen in den vorhistorischen und historischen Schichten von Ilion -1894. Athens.

DOUMET J. 1980. *A study on Ancient Purple and an Attempt to Reproduce the Dyeing Procedure of Tyre as Described by Pliny*. Beirut, Imprinte Catholique.

DUNCA E., B.R. SCHÖNE, and H. MUTVEI. 2005. Freshwater bivalves tell of past climates: but how clearly do shells from polluted rivers speak? *Palaeogeography, Palaeoclimatology, Palaeoecology* 228: 43-58.

EASTON D.F. 2002. *Schliemann's Excavations at Troia 1870-1873*. Studia Troica Monographs 2. Mainz, Philipp von Zabern.

EGEMEN Ö., M. ÖNEN, B. BÜYÜKIŞIK, B. HOŞSUCU, U. SUNLU, Ş. GÖKPINAR, and S. CİRİK. 1999. Güllük Lagünü (Ege Denizi, Türkiye) Ekosistemi. *Turkish Journal of Agriculture and Forestry* 23 (Ek Sayı 3): 927-947.

EMILIANI C. and T. MAYEDA. 1964. Oxygen isotopic analysis of some molluscan shells from fossil littoral deposits of Pleistocene Age. *American Journal of Science* 262: 107-113.

ERLANDSON J.M. 1988. The role of shellfish in prehistoric economies: a protein perspective. *American Antiquity* 53/1: 102-109.

----. 2001. The archaeology of aquatic adaptations: paradigms for a new millennium. *Journal of Archaeological Research* 9/4: 287-350.

ERLANDSON J.M. and S.M. FITZPATRICK. 2006. Oceans, islands, and coasts: Current perspectives on the role of the sea in human prehistory. *Journal of Island and Coastal Archaeology* 1/1: 5-32.

ERLANDSON J.M., C. TORBEN, R.L. VELLANOWETH, and D.J. KENNETT. 1999. Maritime subsistence at a 9300 year old shell midden on Santa Rosa Island, California. *Journal of Field Archaeology* 26: 255-265.

EROL O. 2003. Ceyhan deltasının jeomorfolojik evrimi. *Ege Coğrafya Dergisi (Aegean Geographical Journal)* 12/2: 59-81.

EURONATUR. 2005. *Projekt Zugvögel*. Projekt Bericht 2004-2005.

EUROPEAN ENVIRONMENTAL AGENCY. 1999. *State and Pressures of the Marine and Coastal Mediterranean Environment*. Copenhagen, UNEP.

EVANS J.G. 1972. *Land Snails in Archaeology with Special Reference to the British Isles*. London, Seminar Press.

FABIŠ M. 1996. Archaeofaunal remains from the lower sanctuary: a preliminary report on the 1994 excavations. *Studia Troica* 6: 217-228.

----. 2003. “Troia and fallow deer.” In: G.A. Wagner, E. Pernicka, and H.-P. Uerpmann (Eds.), *Troia and the Troad: Scientific Approaches*. Berlin, Springer Verlag: 264-275.

FACORELLIS Y., Y. MANIATIS, and B. KROMER. 1998. Apparent ^{14}C ages of marine mollusk shells from a Greek island: calculation of the marine reservoir effect in the Aegean Sea. *Radiocarbon* 40/2: 963-973.

FAGAN B. 2004. *The Long Summer: How Climate Changed Civilization*. London, Granta Books.

FALKNER G. 1975. „Systematische Übersicht über die Molluskenfunde von Magula Pevkakia.” In: B. Jordan, *Tierknochenfunde aus der Magula in Thessalien*. München, Institut für Paläoanatomie, Domestikationsforschung und Geschichte der Tiermedizin: 189-90.

FARROW G.E. 1971. Periodicity structures in the Bivalve shell: experiments to establish growth controls in *Cerastoderma edule* from the Thames Estuary. *Palaeontology* 14/4: 571-588.

----. 1972. Periodicity structures in the bivalve shell: analysis of stunting in *Cerastoderma edule* from the Burry Inlet (South Wales). *Palaeontology* 15/1: 61-72.

FİELD C.B., M.J. BEHRENFİELD, J.T. RANDERSON, and P. FALKOWSKİ. 1998. Primary production of biosphere: integrating terrestrial and oceanic components. *Science* 281: 237-240.

FLEMMİNG K., P. JOHNSTON, D. ZWARTZ, Y. YOKOYAMA, K. LAMBECK, and J. CHAPPELL. 1998. Refining the eustatic sea-level curve since the Last Glacial Maximum using far- and intermediate field-sites. *Earth and Planetary Science Letters* 163: 327-342.

FORBES M.H.C. 1976a. “Farming and foraging in prehistoric Greece: a cultural ecological perspective.” In: M. Dimen and E. Friedl (Eds.), *Regional Variation in Modern Greece and Cyprus: Toward a Perspective on the Ethnography of Greece*. Annals of the New York Academy of Science 268. New York: 127-142.

----. 1976b. “Gathering in Argolid: a subsistence subsystem in Greek agricultural community.” In: M. Dimen and E. Friedl (Eds.), *Regional Variation in Modern*

Greece and Cyprus: Toward a Perspective on the Ethnography of Greece. Annals of the New York Academy of Science 268. New York: 251-64.

FORRES R.J. 1987. *Studies in Ancient Technology.* Vol. IV. *Dyes and Dyeing.* Brill Academic Publishers.

FORCHHAMMER G., STEENSTRUP J., and J. WORSAAE. 1851-1857. *Untersogelser i geologisk-antiquarisk Retning.* Copenhagen, Danish National Museum.

FRANK N., A. MANGINI, and M. KORFMANN. 2002. ^{230}Th/U dating of the Trojan 'water quarries'. *Archaeometry* 44/2: 305-314.

GABRIEL U. 2000. Mitteilung zum Stand der Neolithikumsforschung in der Umgebung von Troia (Kumtepe 1993-1995; Beşik-Sivritepe 1983-1984, 1987, 1998-1999). *Studia Troica* 10: 233-238.

GABRIEL U., R. ASLAN, and S.W.E. BLUM. 2004. Alacalıgöl: eine neuentdeckte Siedlung des 5. Jahrtausends v. Chr. in der Troas. *Studia Troica* 14: 121-134.

GAİLLARD J.M. 1987. "Gasteropodes." In: W. Fischer, M.-L. Bauchot and M. Schneider (Eds.), *Fiches FAO d'identification des espèces pour les besoins de la pêche. Méditerranée et mer Noire. Zone de pêche 37.* Vol. I. *Végétaux et Invertébrés.* Rome, FAO: 513-630.

GALILI E., D. ZVIELY, and M. WEINSTEIN-EVRON. 2005. Holocene sea-level changes and landscape evolution on the northern Carmel coast (Israel). *Méditerranée* 1.2: 79-86.

GATES M.-H. 1999. 1997 excavations at Kinet Höyük, Dörtyol, Hatay. *Kazı sonuçları toplantısı* 20/1: 259-281.

GEIST J., K. AUERSWALD, and A. BOOM. 2005. Stable carbon isotopes in freshwater mussel shells: environmental record or marker for metabolic activity? *Geochimica et Cosmochimica Acta* 69/14: 3545-3554.

GEJVALL N.-G. 1969. *Lerna: A Preclassical Site in the Argolid.* Vol. I. *The Fauna.* Princeton, American School of Classical Studies at Athens.

----. 1938. The fauna of Troy: preliminary report. *Kungliga Hujmanistika Vetenskapssamfundet Årsberättesle* 1937-1938: 51-57.

----. 1939. The fauna of Troy: second preliminary report. *Kungliga Hujmanistika Vetenskapssamfundet Årsberättesle* 1938-1939: 1-7.

----. 1946. *The Fauna of Different Settlements at Troy. Part I: Dogs,Horses, and Cattle.*

GENNETT J.A. and J.A. GIFFORD. 1982. "Pollen analysis". In: G.-R. Rapp and J.A. Gifford (Eds.), *Troy: The Archaeological Geology.* Supplementary Monograph 4. Cincinnati, Cincinnati University Press: 105-115.

GIANUZZI-SAVELLI R., F. PUSATERI, A. PALMERI, and C.P.A. EBREO. 1994. (Eds.) *Atlante delle Conchiglie Marine del Mediterraneo.*Vol. 1. *Archaeogastropoda.* Roma, La Conchiglia.

----. 1996. (Eds.) *Atlante delle Conchiglie Marine del Mediterraneo.*Vol. 2. *Caenogastropoda.* Roma, La Conchiglia.

----. 1999. (Eds.) *Atlante delle Conchiglie Marine del Mediterraneo.*Vol. 3.*CaenogastropodaPart 2: Ptenoglossa.* Roma, Evolver.

----. 2003. (Eds.) *Atlante delle Conchiglie Marine del Mediterraneo.*Vol. 4. *Neogastropoda: Muricoidea.* Roma, Evolver.

GIANUZZI-SAVELLI R., F. PUSATERI, A. PALMERI, C.P.A. EBREO, M. COPPINI, A. MARGELLI, and C. BOGI. 2001. (Eds.) *Atlante delle Conchiglie Marine del Mediterraneo.* Vol. 7. *Bivalvia: Protobranchia -Pteriomorpha.* Roma, Evolver.

GLASSOW M.A. and L.R. WILCOXON. 1988. "Coastal adaptations near Point Conception, California, with particular regard to shellfish exploitation." *American Antiquity* 53: 36-51.

GÖBEL J. 2005. *Landschaftrekonstruktion im Siedlungsgebiet von Troia. Geochronologische, geochemische und sedimentologische Untersuchungen.* Hamburg: Verlag Dr. Kovač.

GONTIKAKI E., C. ANTONIADOU and C.C. CHINTIROGLOU. 2003. Population structure of *Cerastoderma glaucum* and *Abra ovata* in Vouliagmeni Lagoon (Attiki). *Journal of Marine Biological Association of U.K.* 83: 1093-1097.

GOODFRIEND G. 1991. Holocene trends in ^{18}O in land snail shells from the Negev Desert and their implications for changes in rainfall source area. *Quaternary Research* 35: 417-426.

GORDON B.C. 1993. Archaeological tooth and bone seasonal increments: the need for standardized terms and techniques. *Archaeozoologia* 5/2: 9-16.

GRAUS R.R. 1974. Latitudinal trends in the shell characteristics of marine gastropods. *Lethaia* 7: 303-314.

GRAYSON D. 1984. *Quantitative Zooarchaeology: Topics in the Analysis of Archaeological Faunas.* London, Academic Press.

GRIFFITHS D. 1977. Caloric variations in Crustacaea and other animals. *Journal of Animal Ecology* 46: 593-605.

GROSSMAN E.L. and T.L. KU. 1986. Oxygen and carbon isotope fractionation in biogenic aragonite: temperature effects. *Chemical Geology* 59: 59-74.

GÜNTHER R.T. 1897. The oyster culture of the ancient Romans. *Journal of the Marine Biological Association of the United Kingdom* 4/4: 360-365.

HARMANKAYA S. 2001. "Gökçeada survey." In: Oktay Belli (Ed.), *Istanbul University's Contributions to Archaeology in Turkey (1932-2000).* Istanbul, Istanbul University Rectorate Publication: 274-278.

HARRISON R.G. and M.A. KATZENBERG. 2003. Paleodiet studies using stable carbon isotopes from bone apatite and collagen: examples from Southern Ontario and San Nicolas Island, California. *Journal of Anthropological Archaeology* 22/3: 227-244.

HAYWARD P.J., G.D. WIGHAM, and N. YONOW. 1995. "Molluscs (Phylum Mollusca)." In: P.J. Hayward and J.S. Ryland (Eds.), *Handbook of the marine fauna of North-West Europe.* Oxford, Oxford University Press: 484-628.

HENSHILWOOD C., P. NILSEN, and J. PARKINGTON. 1994. Mussel drying and food storage in the Late Holocene, SW Cape, S. Africa. *Journal of Field Archaeology* 21/1: 103-109.

HENSHILWOOD C.S., J.C. SEALY, R.J. YATES, K. CRUZ-URIBE, P. GOLDBERG, F.E. GRINE, R.G. KLEIN, C. POGGENPOEL, K. VAN NIEKERK, and I. WATTS. 2001. Blombos Cave, Southern Cape, South Africa: Preliminary report on the 1992-1999 excavations of the Middle Stone Age levels. *Journal of Archaeological Science* 28: 421-448.

HIGGS E.S. and W.J. CLENCH. 1962. Faunal remains of Nea Nikomedia. In: R.J. Rodden. Excavations at the E. Neolithic Site at Nea Nikomedia. *Proceedings of the Prehistoric Society* 11: 273-274.

HITCHNER B.R. 1999. More Italy than Province? Archaeology, texts and culture change in Roman Provence. *Transactions of the American Philological Association* 129: 375-379.

HOEFS J. 2004. *Stable Isotope Geochemistry*. Berlin, Springer Verlag.

HOMER G. 1955. *The Coast Salish of British Columbia*. Westport, Greenwood Press.

HOMER. *Iliad*. Translated by R. Lattimore. Chicago, Perennial Classics.

----. *Odyssey*. Translated by R. Lattimore. Chicago, Perennial Classics.

HOOD S. 1981. *Excavations in Chios 1938-1955. Prehistoric Emporio and Ayio Gala.* British School of Archaeology at Athens. Supplementary Vol. 15. London, Thames and Hudson.

HUBER J.K. and G.J. RAPP. 1987. *Results of a Preliminary Pollen Study From Troy Megaron II A.* Archaeometry Laboratory, University of Minnesota-Duluth.

HÜRYILMAZ H. 1998. Gökçeada Yenibademli Höyük 1996 yılı kurtarma kazısı. *Kazı Sonuçları Toplantısı* 19/1: 357-378.

----. 2000. Gökçeada Yenibademli Höyük 1998 yılı kazıları. *Kazı Sonuçları Toplantısı* 21/1: 229-238.

----. 2001. Gökçeada Yenibademli Höyük 1999 yılı kazıları. *Kazı Sonuçları Toplantısı* 22/1: 247-258.

----. 2002. 2000 Yılı Yenibademli Höyüğü kazıları. *Kazı Sonuçları Toplantısı* 23/1: 295-304

----. 2002. Yenibademli Höyük: Kuzeydoğu Ege Denizi'nde bir Erken Tunç Çağı yerleşmesi. *Hacettepe Üniversitesi Edebiyat Fakültesi Dergisi* 19: 27-44.

----. 2004. Gökçeada'ya hayat veren su kaynaklarının kullanımları ve korunmaları. *Çanakkale-Troas Arkeoloji Buluşması III: Kent ve Su.* 30 Ağustos-3 Eylül 2004. Çanakkale, Çanakkale Belediyesi.

İLHAN R. and H. GÜLYAVUZ. 2004. Çanakkale'de bulunan bazı bivalvia türlerinin kimyasal kompozisyonlarının karşılaştırılması. *Türk Sucul Yaşam Dergisi* 2: 199-204.

İLHAN R., M. GÖKOĞLU, and H. GÜLYAVUZ. 2004. Nehir midyesi (*Unio crasssus* Philipson, 1788)'nin et verimi ve kimyasal kompozisyonu. *Türk Sucul Yaşam Dergisi* 2: 205-210.

ISSAR A.S. and M. ZOHAR. 2004. *Climate Change – Environment and Civilization in the Middle East.* Berlin, Springer Verlag.

IVELL R. 1979a. The biology and ecology of a brackish lagoon bivalve, *Cerastoderma glaucum* Bruguière, in Lago Lungo, Italy. *Journal of Molluscan Studies* 45: 364-382.

----. 1979b. The biology and ecology of a brackish lagoon bivalve, *Cerastoderma glaucum* Bruguiere, in an English Lagoon, The Widewater, Sussex. *Journal of Molluscan Studies* 45: 383-400.

JABLONKA P. 2006. "Leben außerhalb der Burg –Die Unterstadt von Troia." In: M.O. Korfmann (Ed.), *Troia: Archäologie eines Siedlungshügels und seiner Landschaft.* Mainz, Philipp von Zabern: 167-180.

JACKSON J.B.C., M.X. KIRBY, WOLFGANG H. BERGER, K.A. BJORNDAL, L.W. BOTSFORD, B.J. BOURQUE, R.H. BRADBURY, R. COOKE, J. ERLANDSON, J.A. ESTES, T.P. HUGHES, S. KIDWELL, C. B. LANGE, H.S. LENIHAN, J.M. PANDOL, C. H. PETERSON, R.S. STENECK, M.J. TEGNER, and R.R. WARNER. 2001. Historical overfishing and the recent collapse of coastal ecosystems. *Science* 293: 629-638.

JERARDINO A. 1997. Changes in shellfish species composition and mean shell size from a Late Holocene record of the west coast of Southern Africa. *Journal of Archaeological Science* 24: 1031-1044.

JONES D.S. and I.R. QUITMYER. 1996. Marking time with bivalve shells: oxygen isotopes and season of annual increment formation. *Palaios* 11: 340-346.

JONES D.S., I.R. QUITMYER, and C.F.T. ANDRUS. 2005. Oxygen isotopic evidence for greater seasonality in Holocene shells of *Donax variabilis* from Florida. *Palaeogeography, Palaeoclimatology, Palaeoecology* 228: 96-108.

JÖNSSON L. 2002. SeaWIFs satellite data analysis of Black Sea water discharge pattern into the Aegean Sea. *Water Science and Technology* 46/8: 195-202.

KANG K.H. and J.M. KIM. 2004. The predation of trumpet shell, *Charonia* sp., on eight different marine invertebrate species. *Aquaculture Research* 35/13: 1202-1206.

KARALI L. 1999. *Shells in Aegean Prehistory*. B.A.R. International Series 761. Oxford, BAR Publishing.

----. 2005a. "Shells from prehistoric sites of Northern Greece." In: D. Bar-Yosef Mayer, *Archaeomalacology: Molluscs in Former Environments of Human Behaviour. Proceedings of the 9th ICAZ Conference, Durham 2002.* Oxford, Oxbow Books: 91-98.

----. 2005b. Archaeomalacological remains from Ulucak, Western Turkey (İzmir Region): a preliminary report. *Mediterranean Archaeology and Archaeometry* 5/3 (Special Issue): 23-26.

KAYAN İ. 1988. Late Holocene sea-level changes on the Western Anatolian coast. *Palaeogeography, Palaeoclimatology, Palaeoecology* 68/2-4: 205-218.

----. 1991. Holocene geomorphic evolution of the Beşik Plain and the changing environment of ancient man. *Studia Troica* 1: 79-92.

----. 1995. Troia and the supposed harbor sites. *Studia Troica* 5: 211-235.

----. 1996. Holocene stratigraphy of the lower Karamenderes-Dümrek plain and the archaeological material in the alluvial sediments to the north of the Troia Ridge. *Studia Troica* 6: 239-249.

----. 1997a. Geomorphological evolution of the Çıplak Valley and archaeological material in the alluvial sediments to the south of the lower city of Troia. *Studia Troica* 7: 489-507.

----. 1997b. "Bronze Age regression and chage of sedimentation on the Aegean coastal plains of Anatolia (Turkey)." In: N.H. Dalfes, G. Kukla, and H. Weiss (Eds.), *Third Millennium BC Climate Change and the Old World Collapse*. Global Environmental Change 49. Berlin, Springer Verlag: 431-450.

----. 1999. "Kemalpaşa çevresinde geçmişten günümüze arazi kullanımı ve günümüzdeki sorunlar." In: *Kemalpaşa Kültür ve Çevre Sempozyumu*. 3-5 Haziran 1999. Kemalpaşa-İzmir, Kemalpaşa Kaymakamlığı ve Ege Üniversitesi İzmir Araştırma ve Uygulama Merkezi Bildiriler Kitabı: 1-16.

----. 2000. "İzmir çevresinin morfotektonik birimleri ve alüvyal jeomorfolojisi." *Batı Anadolu'nun Depremselliği Sempozyumu Bildiriler Kitabı*. İzmir, İzmir Valiliği: 103-111.

KAYAN İ., E. ÖNER, L. UNCU, B. HOCAOĞLU, and S. VARDAR. 2003. "Geoarchaeological interpretations of the Troian Bay." In: G.A. Wagner, E. Pernicka, and H.-P. Uerpmann (Eds.), *Troia and the Troad: Scientific Approaches*. Berlin, Springer Verlag: 379-401.

KEEGAN W.F. and J.M. DIAMOND. 1987. "Colonization of islands by humans: a biogeographical perspective." In: M. Schiffer (Ed.), *Archaeological Method and Theory 10*. New York, Academic Press: 49-91.

KELLER N., D. DEL PIERO, and A. LONGINELLI. 2002. Isotopic composition, growth rates, and biological behavior of *Chamelea galena* and *Callista chione* from the Gulf of Trieste (Italy). *Marine Biology* 140: 9-15.

KELLETAT D. 2005. A Holocene sea-level curve for the Eastern Mediterranean from multiple indicators. *Zeitschrift für Geomorphologie*. Supplement Vol. 137: 1-9.

KENNEDY H., C.A. RICHARDSON, C.M. DUARTE, and D.P. KENNEDY. 2001. Oxygen and carbon stable isotopic profiles of the fan mussel, *Pinna nobilis*, and reconstruction of sea surface temperatures in the Mediterranean. *Marine Biology* 139: 1115-1124.

KENNETT D. and B. VOORHIES. 1995. Middle Holocene periodicities in rainfall inferred from oxygen and carbon isotopic fluctuations in prehistoric tropical estuarine mollusc shells. *Archaeometry* 37/1: 157-170.

KENT B. 1988. *Making Dead Oysters Talk*. St. Mary's City, Maryland Historical Trust.

KILLINGLEY J.S. 1981. Seasonality of mollusk collecting determined from O-18 profiles of midden shells. *American Antiquity* 46/1: 152-158.

KOIKE H. 1980. *Seasonal Dating by Growth-Line Counting of the Clam* Meretrix lusoria. Tokyo, University of Tokyo Museum. Bulletin no. 18.

KORFMANN M. 1984. Beşik-Tepe. Vorbericht über die Ergebnisse der Grabung von 1982. Die Hafenbucht von "Troja" (Hisarlik), Grabungen am Beşik-Yassıtepe. *Archäologischer Anzeiger*: 165-195.

----. 1985. Vorbericht über die Ergebnisse der Grabung von 1983. Grabungen am Beşik-Yassıtepe und Beşik-Sivritepe. *Archäologischer Anzeiger*: 157-194.

----. 1986. "Troy: topography and navigation." In: M.J. Mellink (Ed.), *Troy and the Trojan War.A Symposium Held at Bryn Mawr College, October 1984*. Pennsylvania, Bryn Mawr College: 1-16.

----. 1986. Beşik-Tepe. Vorbericht über die Ergebnisse der Grabung von 1984. Grabungen am Beşik-Yassıtepe, Beşik-Sivritepe und Beşik-Gräberfeld. *Archäologischer Anzeiger*: 303-363.

----. 1988. Beşik-Tepe. Vorbericht über die Ergebnisse der Grabung von 1985 und 1986. Grabungen am Beşik-Yassıtepe und im Beşik-Gräberfeld. *Archäologischer Anzeiger*: 391-404.

----. 1992. Troia-Ausgrabungen 1990 und 1991. *Studia Troica* 2: 1-41.

----. 1993. Troia-Ausgrabungen 1992. *Studia Troica* 3: 1-38.

----. 1994. Troia-Ausgrabungen 1993. *Studia Troica* 4: 1-49.

----. 1995. Troia-Ausgrabungen 1994. *Studia Troica* 5: 1-38.

----. 1996. Troia-Ausgabungen 1995. *Studia Troica* 6: 1-63.

----. 1998. Troia-Ausgrabungen 1997. *Studia Troica* 8: 1-69.

----. 2000. Troia –Ausgrabungen 1999. *Studia Troica* 10: 1-52.

----. 2003. *Troia in the Light of New Research*. Keynote Lecture, English Edition. Universität Trier.

KOUTSOUBAS D., C. ARVANİTİDİS, C. DOUNAS, and L. DRUMMOND. 2000. Community structure and dynamics of the Molluscan fauna in a Mediterranean lagoon (Gialova lagoon, SW Greece). *Belgian Journal of Zoology* 130. Supplement Vol. 1: 131-138.

KRAFT J.C., G.R. RAPP, I. KAYAN, and J.V. LUCE. 2003a. Harbor areas at ancient Troy: sedimentology and geomorphology complement Homer's *Illiad*. *Geology* 31/2: 163-166.

KRAFT J.C., I. KAYAN, H. BRÜCKNER, and G. RAPP 2003b. "Sedimentary facies patterns and the interpretation of palaeogeographies of ancient Troia." In: G.A. Wagner, E. Pernicka, and H.-P. Uerpmann (Eds.), *Troia and the Troad: Scientific Approaches*. Berlin, Springer Verlag: 361-378.

KRAFT J.C., İ. KAYAN, and O. EROL. 1980. Geomorphic reconstructions in the environs of ancient Troy. *Science* 209: 776-782.

KROMER B., M. KORFMANN, and P. JABLONKA. 2003. "Heidelberg radiocarbon dates for Troia I to VIII and Kumtepe." In: G.A. Wagner, E. Pernicka, and H.-P. Uerpmann (Eds.), *Troia and the Troad:*

Scientific Approaches. Berlin, Springer Verlag: 44-54.

KRÖNNECK P. 1996. Vogelknochen aus Troia: ein Beitrag zur Umweltrekonstruktion. *Studia Troica* 6: 229-238.

KRÖNNECK P. 2003. "Troian bird remains: environment and hunting." In: G.A. Wagner, E. Pernicka, and H.-P. Uerpmann (Eds.), *Troia and the Troad: Scientific Approaches*. Berlin: Springer Verlag: 277-283.

KÜÇÜK F., Z. YILDIRIM, H. GÜLYAVUZ, and M. ÜNLÜSAYIN. 1996. "Deniz salyangozlarından *Patella caerula* L. 1758 ve *Monodonta articulata* (Lamarck, 1822) (Prosobranchia: Gastropoda)'nın Antalya Körfezi'nde yayılışları, et verimi ve etlerinin kimyasal yapısı üzerine bir araştırma." In: *II. International Symposium on Aquatic Products. September 21-23, 1996. Istanbul-Türkiye. Organized by Faculty of Aquatic Products, University of Istanbul*: 1-17.

KUMLU M. 2001. *Karides, İstakoz ve Midye Yetiştiriciliği.* Çukurova Universitesi Su Ürünleri Fakültesi Yayınları No: 6. Adana.

LACOMBE P. 2000. Anthropology of shellfish-gathering. *Dialectical Anthropology* 25: 161-187.

LAMB W. 1936. *Excavations at Thermi in Lesbos.* Cambridge, Cambridge University Press.

LAMPRELL K. 1987. *Spiny oysters of the world:* Spondylus. Leiden, Brill and Backhuys Publ.

LAUFER B. 1915. The story of the Pinna and the Syrian Lamb. *Journal of American Folklore* 28/108: 103-128.

LEAKE W. Sir. 1830. *Travels in the Morea.* London.

LEONTARAKIS P.K. and C.A. RICHARDSON. 2005. Growth of the smooth clam, *Callista chione* (Linnaeus, 1758) (Bivalvia: Veneridae) from the Thracian Sea, northeastern Mediterranean. *Journal of Molluscan Studies* 71: 189-192.

LIGHT J. 2005. "Marine mussel shells –Wear is the evidence." In: D.E. Bar-Yosef Mayer (Ed.), *Archaeomalacology: Molluscs in Former Environments of Human Behaviour. Proceedings of the 9th ICAZ Conference, Durham 2002.* Oxford, Oxbow Books: 56-62.

LITTLE C. 1996. *The Biology of Rocky Shores.* Oxford, Oxford University Press.

LUBELL D. 2004. "Prehistoric edible land snails in the circum-Mediterranean: the archaeological evidence." In: J.-P. Brugal and J. Desse (Eds.), *Petits Animaux et Sociétés Humaines. Du Complément Alimentaire aux Resources Utilitaires. XXIVe Rencontres Internationales d'Archéologie et d'Historie d'Antibes*. Antibes, Éditions APDCA: 77-98.

LYMAN R.L. 1994. *Vertebrate Taphonomy.* Cambridge Manuals in Archaeology. Cambridge, Cambridge University Press.

MANGOLD K. and S.V. BOLETZKY. 1987. "Cephalopodes." In: Fischer, W., M.-L. Bauchot, and M. Schneider (Eds.), Fiches FAO d'identification des espèces pour les besoins de la pêche. Méditerranée et mer Noire. Zone de pêche 37. Vol. I. Végétaux et Invertébrés. Rome, FAO: 634-714.

MANNING S.W., B. KROMER, P. KUNIHOLM, and M. NEWTON. 2001. Anatolian tree rings and a new chronology for the Eastern Mediterranean Bronze-Iron Ages. *Science* 294: 2532-2535.

MANNING S.W., M. BARBETTI, B. KROMER, P.I. KUNIHOLM, I. LEVIN, N.W. NEWTON, and P.I. REINER. 2002. No systematic early bias to Mediterranean ^{14}C ages: radiocarbon measurements from tree-ring and air samples provide tight limits to age offsets. *Radiocarbon* 44/3: 739-754.

MANNINO M.A. and K.D. THOMAS. 2001. Intensive Mesolithic exploitation of coastal resources? Evidence from a shell deposit on the Isle of Portland (Southern England) for the impact of human foraging on populations of intertidal rocky shore molluscs. *Journal of Archaeological Science* 28: 1101-1114.

----. 2002. Depletion of a resource? The impact of prehistoric human foraging on intertidal mollusc communities and its significance for human settlement, mobility and dispersal. *World Archaeology* 33/3: 452-474.

MANUEL J.L., S.M. GALLAGER, C.M. PEARCE, D.A. MANNING, and R.K. O'DOR. 1996. Veligers from different populations of sea scallop *Placopecten magellanicus* have different vertical migration patterns. *Marine Ecology Progress Series* 142: 147-163.

MARELLI D.C. and W.S. ARNOLD. 2001. Shell morphologies of bay scallops, *Argopecten irradians,* from extant and prehistoric populations from the Florida Gulf coast: Implications for the biology of past and present metapopulations. *Journal of Archaeological Science* 28: 577-586.

MARIANI S., F. PICCARI, and E. DE MATTHEIS. 2002. Shell Morphology in *Cerastoderma* spp. (Bivalvia: Cardiidae) and its significance for adaptation to tidal and non-tidal coastal habits. *Journal of Marine Biological Association of U.K.* 82: 483-490.

MASON R.D., M.L. PETERSON, and J.A. TIFFANY. 1998. Weighing vs. counting: measurement reliability and the California school of midden analysis. *American Antiquity* 63/2: 303-324.

MAYEWSKI P.A., E.J. ROHLING, J.C. STAGER, W. KARLEN, K.A. MAASCH, L.D. MEEKER, E.A. MEYERSON, F. GASSE, S. VAN KREVELD, K. HOLMGREN, J. LEE-THORP, G. ROSQVIST, F. RACK, M. STAUBWASSER, R. R. SCHNEIDER, and E. STEIG. 2004. Holocene climate variability. *Quaternary Research* 62/3: 243-255.

MCCONNAUGHEY T. 1989a. ^{13}C and ^{18}O disequilibrium in biological carbonates: I. Patterns. *Geochimica et Cosmochimica Acta* 53: 151-162.

----. 1989b. ^{13}C and ^{18}O disequilibrium in biological carbonates: II. *In vitro* simulation of kinetic isotope effects. *Geochimica et Cosmochimica Acta* 53: 163-171.

MCKILLOP H. 1984. Prehistoric Maya reliance on marine resources: analysis of a midden from Moho Cay, Belize. *Journal of Field Archaeology* 11/1: 25-35.

MEEHAN B. 1977. "Man does not live by calories alone: the role of shellfish in a coastal cuisine." In: J. Allen, J. Golson, and R. Jones (Eds.), *Sunda and Sahul: Prehistoric Studies in Southeast Asia, Melanesia and Australia*. London, New York, San Francisco, Academic Press: 493-532.

----. 1982. *Shell Bed to Shell Midden*. Canberra, Australian Institute for Aboriginal Studies.

----. 1983. "A Matter of choice? Some thoughts on shell gathering strategies in northern Australia." In: C. Grigson and J.Clutton-Brock (Eds.), *Animals and Archaeology 2: Shell Middens, Fishes and Birds*. B.A.R. International Series 227. Oxford, Oxbow Books: 3-17.

MEGALOUDI F. 2005. Archaeobotanical finds from Ulucak, Western Turkey (İzmir Region): a preliminary study. *Mediterranean Archaeology and Archaeometry* 5/3 (Special Issue): 27-32.

MEIGHAN C.W. 1969. "Molluscs as food remains in archaeological sites." In: D. Brothwell and E.S. Higgs (Eds.), *Science in Archaeology*, 2nd Edition. Thames and Hudson, London: 415-422.

MELLARS P.A. and S. PAYNE 1971. Excavation of two Mesolithic shell middens on the island of Oronsay (Inner Hebrides). *Nature* 231: 397-8.

MIENIS H.K. 2000. A blister pearl in *Arca noae* Linnaeus, 1758. *Of Sea and Shore* 23/3: 175.

MILLER M.A. 1996. The manufacture of cockle shell beads at E. Neolithic Franchthi Cave, Greece: a case of craft specialization? *Journal of Mediterranean Archaeology* 9/1:7-37.

MILNER N. 2001. At the cutting edge: using thin sectioning to determine season of death of the European Oyster, *Ostrea edulis*. *Journal of Archaeological Science* 28: 861-873.

MINNITI C. 2005. "Shells at the Bronze Age settlement of Coppa Nevigata (Apulia, Italy)." In: D.E. Bar-Yosef (Ed.), *Archaeomalacology: Molluscs in Former Environments of Human Behaviour*. Proceedings of the 9th ICAZ Conference, Durham 2002. Oxford, Oxbow Books: 71-81.

MITCHELL D.R. and M.S. FOSTER. 2000. Hohokam shell middens along the Sea of Cortez, Puerto Penasco, Sonora, Mexico. *Journal of Field Archaeology* 27: 27-41.

MONKS G.G. and R. JOHNSTON. 1993. Estimating season of death from growth increment data: a critical review. *Archaeozoologia* 5/2: 17-40.

MORALES-MUNIZ A. 1993. "Where are the tunas? Ancient Iberian fishing industries from an archaeozoological perspective." In: In A.T. Clason, S. Payne, and H.-P. Uerpmann (Eds.), *Skeletons in her cupboard. Festschrift for Juliet Clutton-Brock*. Oxford, Oxbow Monograph 34: 135-141.

MOSS M.L. 1993. Shellfish, gender, and status on the Northwest Coast. Reconciling archaeological ethnographic, and ethnohistorical records of the Tlingit. *American Anthropologist* 95/3: 631-652.

NASH C.E. 2000. "Aquatic animals." In: K.F. Kiple and K.C. Ornelas (Eds.), *The Cambridge World History of Food*. Vol. 1: 456-467.

NEUMANN J. 1991. Number of days that Black Sea bound sailing ships were delayed by winds at the entrance to the Dardanelles near Troy's site. *Studia Troica* 1:93-100.

NIKULA R. and R. VÄINÖLÄ 2003. Phylogeography of *Cerastoderma glaucum* (Bivalvia: Cardiidae) across Europe: a major break in the Eastern Mediterranean. *Marine Biology* 114: 339-350.

NISHIDA A.K., N. NORDI, and R. ALVES. 2006. Molluscs production associated to lunar-tide cycle: a case study in Paraiba State under ethnoecology viewpoint. *Journal of Ethnobiology and Ethnomedicine* 2: 2-8.

NOLL E. 2002. *Ethnoarchäologische Studien an Muschelhaufen*. Tübinger Schriften zur Ur- und Frühgeschichtlichen Archäologie 7. Münster, Waxman.

NORDSIECK F. 1969. *Die europäischen Meeresmuscheln (Bivalvia) vom Eismeer bis Kapverden, Mittelmeer und Schwarzes Meer*. Stuttgart, Fischer.

----. 1982. *Die europäischen Meeres-Gehäuseschnecken (Prosobranchia) vom Eismeer bis Kapverden, Mittelmeer und Schwarzes Meer*. Stuttgart-New York, Fischer.

ÖNER E. 2000. Geoarchäologische und paläogeographische Forschungen auf der Insel Gökçeada (Imbros): Siedlungshügel von Yenibademli, Nordwestanatolien-Nordostägäisches Meer-Türkei. *Bremer Beiträge zur Geographie und Raumplanung* 36: 23-33.

----. 2001. "Gökçeada kıyılarında Holosen deniz seviyesi ve kıyı çizgisi değişmeleri." In: E. Özhan and Y. Yüksel (Eds.), *Türkiye'nin Kıyı ve Deniz Alanları III. Ulusal Konferansı Bildiriler Kitabı*. Ankara: 779-790.

OYBAK-DÖNMEZ E. 2005. Early Bronze Age crop plants from Yenibademli Höyük (Gökçeada), Western Turkey. *Environmental Archaeology* 10/1: 39-49.

ÖZCAN H. 2005. "Monitoring and evaluation of the spatial and temporal changes of pH, spcond, DO, ORP and TDS in the waters of Troia (Turkey) by GIS." In: S. Erasmi, B. Cyffka, M. Kappas (Eds.), *Remote Sensing and GIS for Environmental Studies*. Göttingen, Göttinger Geographische Abhandlungen 113: 175-182.

ÖZDOĞAN M. 1999. "Northwest Turkey: Neolithic cultures in between the Balkans and Anatolia." In: M. Özdoğan (Ed.), *Neolithic in Turkey*. Istanbul, Arkeoloji ve Sanat Yayınları: 203-224.

ÖZSOY E., M.A. LATİF, H.I. SUR, and Y. GORYACHKİN 1996. "A review of the exchange of flow regimes and mixing in the Bosphorus Strait." In: Briand, F. (Ed.), *Mediterranean Tributary Seas. Bulletin de l'Institut Océanographique*, Monaco, Special No. 17, CIESME Science Series 2: 187-204.

ÖZSOY E., M.A. LATIF, S. TUĞRUL, and Ü. ÜNLÜATA 1995. "Exchanges with the Mediterranean, fluxes and boundary mixing processes in the Black Sea." In: F. Briand (Ed.), *Mediterranean Tributary Seas. Bulletin de l'Institut Océanographique*, Monaco, Special No. 15, CIESME Science Series 1: 1-25.

ÖZTÜRK B. and Z. ERGEN. 1999. Saros Körfezi'nde (Kuzey Ege Denizi) dağılım gösteren *Patella* (Archaeogastropoda) türleri. *Turkish Journal of Zoology* 23: Ek Sayı 2: 513-519.

PALAZ M. 2005. The dynamics of the benthic molluscs in the inner bay of İzmir. *Journal of Applied Sciences* 5/1: 44-46.

PALAZ M. and S. BERBER. 2005. The bivalve species of the Dardanelles. *Journal of the Marine Biological Association of the UK* 85: 357-358.

PANNELLA G. and C. MACCLINTOCK. 1968. Biological and environmental rhythms reflected in molluscan shell growth. *Journal of Palaeontology* 42: 64-80.

PAPATHANASIOU A., C.S. LARSEN, and L. NORR. 2000. Bioarchaeological inferences from a Neolithic Ossuary from Alepotrypa Cave, Diros, Greece. *International Journal of Osteoarchaeology* 10: 210-228.

PARMALEE P.W. and W.E. KLIPPEL. 1974. Freshwater mussels as a prehistoric food source. *American Antiquity* 39/3: 421-434.

PAVUK P. 2006. *Die Troia VI - Früh - Keramik und Kleinfunde.* Unpublished PhD thesis. Tübingen University.

PEHARDA M., C.A. RICHARDSON, V. ONOFRI, A. BRATOŠ, and M. CRNČEVĆ. 2002. Age and growth of the bivalve *Arca Noae* L. in the Croatian Adriatic Sea. *Journal of Molluscan Studies* 68: 307-310.

PETERS J. 1983. "Archaic Milet: daily life and religious customs from an archaezoological perspective." In: H. Buitenhuis and A.T. Clason (Eds.), *Archaeozoology of the Near East. Proceedings of the First International Symposium on the Archaeozoology of Southwestern Asia and Adjacent Areas.* Leiden, Universal Book Services/Dr. W. Backhuys: 88-96.

PETERSEN K.S. 2004. Shellfish from the excavations at Aetolian Chalkis. *Proceedings of the Danish Institute at Athens* 4. 215-227.

PLINY. *Natural History.* English translation. Perseus ebooks.

POPPE T.G. and Y. GOTO. 2000. *European Seashells.* Hackenheim, Conchbooks.

POUTIERS J. M. 1987. "Bivalvia." In: Fischer, W., M.-L. Bauchot, and M. Schneider (Eds.), *Fiches FAO d'identification des espèces pour les besoins de la pêche. Méditerranée et mer Noire. Zone de pêche 37.* Vol. I. *Végétaux et Invertébrés.* Rome, FAO: 369-500.

PRIEUR A. 2005. Les Coquillages du Paléolithique à l'âge du Bronze au Moyen-Orient et en Méditerranée Orientale : interprétations environnementales et utilisation humaine. *Paléorient* 31/1: 158-168.

PRUMMEL W. 2005. "Molluscs from a Middle Bronze Age site and two Hellenistic sites in Thessaly, Greece." In: D.E. Bar-Yosef (Ed.), *Archaeomalacology: Molluscs in former environments of human behavior. Proceedings of the 9th ICAZ Conference in Durham, 2002.* Oxbow, Oxbow Books: 107-121.

PURCELL N. 1996. "Eating fish: the paradoxes of seafood." In: J. Wilkins, D. Harvey, and M. Dobson (Eds.), *Food in Antiquity.* Exeter, University of Exeter Press: 132-149.

QUITMYER I. and K. SZABO. 2008. *Proceedings of the 1st Meeting of Archaeomalacology Working Group of ICAZ in Florida, 2004. Archaeofauna* 18.

QUITMYER I.R., D.S. JONES, and W.S. ARNOLD. 1997. The sclerochronology of Hard Clams, *Mercenaria* spp., from the South-EasternU.S.A.: a method of elucidating the zooarchaeological records of seasonal resource procurement and seasonality in prehistoric shell middens. *Journal of Archaeological Science* 24: 825-840.

RAAB M.L. 1992. An optimal foraging analysis of prehistoric shellfish collecting on San Clemente Island, California. *Journal of Ethnobiology* 12/1: 63-80.

RAPP G.-R. and J.A. GIFFORD. 1982. *Troy: The Archaeological Geology.* Supplementary Monograph 4. Cincinnati, Cincinnati University Press.

RAPP G.-R. and J.C. KRAFT. 1979. "Agean sea level changes in the Bronze Age." In: C. Doumas (Ed.), *Thera and the Aegean World. Acts of the 2nd International Science Congress on the Volcano of Thera.* Thera Foundation. London, Bishopgate: 183-193.

REESE D.S. 1980. Industrial exploitation of murex shells: Purple-dye and lime production at Sidi Khrebish, Benghazi (Berenice). *Libyan Studies* 11: 79-93.

----. 1982. "The molluscs from Bronze Age to Post-Geometric Asine in the Argolid, Greece (Appendix 2)." In: S. Dietz (Ed.), *Asine II: Results of the Excavations East of the Acropolis 1970-74. Fasc. 1.* Stockholm, Swedish Institute of Archaeology in Athens.

----. 1983. The use of cone shells in Neolithic and Bronze Age Greece. *Annual of the British School of Archaeology at Athens* 78: 353-357.

----. 1984. Topshell rings in the Aegean Bronze Age. *Annual of the British School of Archaeology in Athens* 79: 237-238.

----. 1985. "The Late Bronze Age to Geometric shells from Kition (Appendix VIII, A)." In V. Karageorghis, *Excavations at Kition* V/II. Nichosia, Department of Antiquities: 340-371.

----. 1986. "Shells at Aphrodisias." In: M. Joukowsky (Ed.), *Prehistoric Aphrodisias: An Account of the excavations and artifact studies.* Vol. 1. Louvain, Providence: 191-196.

----. 1987a. Palaikastro shells and Bronze Age purple-dye production in the Mediterranean Basin. *Annual of the British School of Archaeology at Athens* 82: 201-206.

----. 1987b. The EM IIA Shells from Knossos, with comments on Neolithic to EM III shell utilization. *The Annual of the British School of Archaeology at Athens* 82: 207-211.

----. 1987c. "Marine and fresh-water molluscs." In: P. Hellström (Ed.), *Paradeisos, A Late Neolithic*

Settlement in Aegean Thrace. Memoir 7. Stockholm: Medelhavsmuseet, 119-134.

----. 1989. On cassid lips and helmet shells. *Bulletin of the American Schools of Resarch* 275: 33-39.

----. 2000. "Iron Age shell purple-dye production in the Aegean (Appendix 6.1)." In: J.W. and M.C. Shaw (Eds.). *Kommos IV The Greek Sanctuary 1.* Princeton, Princeton University Press.

----. 2005. Whale bones and shell purple-dye at Motya (Western Sicily, Italy). *Oxford Journal of Archaeology* 24/2: 107-114.

----. 2006. "The exploitation of aquatic resources at Bakla Tepe, Liman Tepe and Panaztepe." In: B. Avunç (Ed.), *Studies in Honor of Hayat Erkanal: Cultural Reflections.* Istanbul, Homer Kitabevi.

REESE D.S. and ROSE M. J. 2002. "Fish and shells from Sarepta." Unpublished manuscript.

REITZ E. and E. WING. 1999. *Zooarchaeology.* Cambridge Manuals in Archaeology. Cambridge, Cambridge University Press.

RENFREW C. and P. BAHN. 1996. *Archaeology: Theories, Methods and Practice.* 2nd Edition. London, Thames and Hudson.

RHOADS D.C. and R.A. LUTZ. 1980. "Skeletal records of environmental change." In: D.C. Rhoads and R.A. Lutz (Eds.), *Shell Growth of Aquatic Organisms: Biological Records of Environmental Change.* New York, Plenum Press: 1-19.

RICHARDS M.P. and P.A. MELLARS. 1998. Stable isotopes and the seasonality of the Oronsay middens. *Antiquity* 72:178-184.

RICHARDS M.P. and R.E.M. HEDGES. 1999. Stable isotope evidence for similarities in the types of the marine foods used by late Mesolithic humans at sites along the Atlantic coast of Europe. *Journal of Archaeological Science* 26: 717-722.

RICHARDSON C.A. 1988. Exogenous and endogenous rhythms of band formation in the shell of the clam *Tapes phillippinarum* (Adams et Reeve, 1850). *Journal of Experimental Marine Biology and Ecology* 122: 105-126.

RICHARDSON C.A., H. KENNEDY, C.M. DUARTE, D.P. KENNEDY, and S.V. PROUD. 1999. Age and growth of the fan mussel *Pinna nobilis* from south-east Spanish Mediterranean seagrass (*Posidonia oceanica*) meadows. *Marine Biology* 133: 205-212.

RICHARDSON C.A. 2001. Molluscs as archives of environmental change. *Oceanography and Marine Biology - an annual review* 39: 103-164.

RICHARDSON C.A., D.J. CRISP, and N.W. RUNHAM. 1979. Tidally deposited growth bands in the shell of the common cockle *Cerastoderma edule* (L.). *Malacologia* 18: 277-290.

RICHARDSON C.A., D.J. CRISP, N.W. RUNHAM, and LL. D. GRUFFYDD. 1980. The use of tidal growth bands in the shell of *Cerastoderma edule* to measure seasonal growth rates under cool temperate and sub-arctic conditions. *Journal of the Marine Biological Association of the United Kingdom* 60: 977-989.

RICKEN W., T. STEUBER, H. FREITAG, M. HIRSCHFELD, and B. NIEDENZU. 2003. Recent and historical discharge of a large European river system –oxygen isotopic composition of river water and skeletal aragonite of Unionidae in the Rhine. *Palaeogeography, Palaeoclimatology, Palaeoecology* 193: 73-86.

RIDOUT-SHARPE J. 1998. "The molllusca (Chapter 24)." In: E. Pelterburg (Ed.), *Lemba Archaeological Project, Cyprus. Vol II. 1B. Excavations at Kissonerga-Mosphilia, 1979, 1992.* University of Edinburgh, Department of Archaeology: 338-348.

RIEHL S. 1999. *Bronze Age Environment and Economy in the Troad. The Archaeobotany of Kumtepe and Troy.* BioArchaeologica 2. Tübingen, Mo Vince Verlag.

RIEHL S. and E. MARINOVA. 2008. Mid-Holocene vegetation change in the Troad (W Anatolia): man-made or natural? *Vegetation History and Archaeobotany* 17: 297-312.

RILOV G., G. AVITAL, and Y. BENAYAHU. 2002. Effect of an exotic prey on the feeding pattern of a predatory snail. *Marine Environmental Research* 54: 85-98.

RILOV G., Y. BENAYAHU, and A. GASITH. 2004. Life on the edge: Do biomechanical and behavioral adaptations to wave-exposure correlate with habitat partitioning in predatory whelks? *Marine Ecology Progress Series* 282: 193-194.

ROBERTS N. 1991. *The Holocene: An Environmental History.* Oxford, Cambridge, Basil Blackwell.

----. 1993. Sub-Milankovich palaeoclimatic events: their recognition and correlation. *Climatic Change* 24: 175-178.

ROBERTS N., T. STEVENSON, D. BASIL, R. CHEDDADI, S. BREWSTER, and A. ROSEN. 2004. "Holocene climate, environment and cultural change in the Circum-Mediterranean Region." In: R.W. Battarbee, F. Gasse, and C. Stickley (Eds.), *Past Climate Variability through Europe and Africa.* Spinger, Dordrecht: 343-362, (Figures): 627-638.

ROBERTS N., J.M. REED, M.J. LENG, C. KUZUCUOĞLU, M. FONTUGNE, J. BERTAUX, H. WOLDRING, S. BOTTEMA, S. BLACK, E. HUNT, and M. KARABIYIKOĞLU. 2001. The tempo of Holocene climatic change in the eastern Mediterranean region: new high-resolution creater-lake sediment data from central Turkey. *The Holocene* 11/6: 721-736.

ROHLING E.J., P.A. MAYEWSKI, A. HAYES, R.H. ABU-ZIED, and J.S.L. CASFORD. 2002. Holocene atmosphere-ocean interactions: records from Greenland and the Aegean Sea. *Climate Dynamics* 18: 587-593.

ROSE C. 2006. "Auf mythengetränkem Boden –Ilion in griechisher, römischer und byzantinisher Zeit." In: M.O. Korfmann (Ed.), *Troia: Archäologie eines Siedlungshügels und seiner Landschaft.* Mainz, Philipp von Zabern: 189-198.

ROSE M.J. 1994. *With Line and Glittering Bronze Hook: Fishing in the Aegean Bronze Age.* Unpublished PhD thesis. Indiana University.

ROSENBERG G.D. 1975. "A comment on terminology: the increment and the series." In: G.D. Rosenberg and S.K. Runcorn (Eds.). *Growth Rhythms and the History of the Earth's Rotation.* London, John Wiley and Sons Ltd.: 1-8.

RUSCILLO D. 2005. "Reconstructing Murex royal purple and Biblical blue in the Aegean." In: D. Bar-Yosef (Ed.), *Archaeomalacology: Molluscs in Former Environments of Human Behaviour.* Proceedings of the 9th ICAZ Conference, Durham 2002. Oxford, Oxbow Books: 99-106.

RUSSO A.R. 1997. Epifauna living on sublittoral seaweeds around Cyprus. *Hydrobiologia* 344/1-3: 169-179.

SABELLI B. 1979. *Simon and Schuster's Guide to Shells.* New York, Simon and Schuster Inc.

SABELLI B., R. GIANNUZZI-SAVELLI, and D. BEDULLI. 1990. *Annotated check-list of Mediterranean marine mollusks.* Bologna, Liberia Naturalistica Bolognese.

SAMPSON A. 1998. The Neolithic and Mesolithic occupation of the Cave of Cyclope, Youra, Alonnessos, Greece. *Annual of the British School at Athens* 93: 1-22.

SANTORO C.M., .T. ARRIAZA, V.G. STANDEN, and P.A. MARQUET. 2005. "People of the coastal Atacama Desert: living between sand dunes and waves of the Pacific Ocean." In: P. Veth, M. Smith, P. Hiscock (Eds.), *Desert Peoples. Archaeological Perspectives.* London, Blackwell Publishers: 243-260.

SARI H.M., S. BALIK, M.R. USTAOĞLU, and A. İLHAN. 2005. Distribution and ecology of freshwater ichthyofauna of the Biga Peninsula, North-western Anatolia, Turkey. *Turkish Journal of Zoology* 29: 1-11.

SAYIN E. 2003. Physical features of the İzmir Bay. *Continental Shelf Research* 23: 957-970.

SAZCI G. 2003. *Der Schatzfundhorizont in Troia –seine Einordnung innerhalb des Grabungsplatzes und der bronzezeitlichen Welt.* Unpublished PhD thesis. Tübingen University.

----. 2005. Troia I-III, die maritime Troia-Kultur und Troia IV-V, die anatolische Troia-Kultur: eine Untersuchung der Funde und Befunde im mittleren Schliemanngraben (D07, D08). *Studia Troica* 15: 35-98.

SCHACHNER A. 1999. Der Hanay Tepe und seine Bedeutung für die bronzezeitliche Topographie der Troas: die prähistorische Funde der Grabungen von Frank Calvert im Berliner Museum für Vor- und Frühgeschichte. *Acta Praehistorica et Archaeologica* 31: 7-47.

SCHLIEMANN H. 1881. *Ilios: Stadt und Land der Trojaner.* Leipzig, F.A. Brockhaus.

SCHÖNE B.R., D.L. RODLAND, D.M. SURGE, J. FIEBIG, D.P. GILLIKIN, S.M. BAIER, and A. GOEWERT. 2006. Comment on "Stable carbon isotopes in freshwater mussel shells: environmental record or marker for metabolic activity?" by Geist et al. 2005. *Geochimica et Cosmochimica Acta* 70: 2658-2661.

SCHÖNE B.R., E. DUNCA, J. FIEBIG, and M. PFEIFFER. 2005. Mutvei's solution: an ideal agent for resolving microgrowth structures of biogenic carbonates. *Palaeogeography, Palaeoclimatology, Palaeoecology* 228: 149-166.

SCHÖNE B.R. and D. SURGE. 2005. Editorial: looking back over skeletal diaries –high-resolution environmental reconstructions from accretionary hard parts of aquatic organisms. *Palaeogeography, Palaeoclimatology, Palaeoecology* 228: 1-3.

SCOTT W. Sir. 1972. *The Journal.* Oxford, Oxford University Press.

SEÇMEN Ö. 1977. *Gökçeada ve Bozcaada Adalarının Flora ve Vejetasyonu.* TBAG 211. Ankara, Türkiye Bilimsel ve Teknik Araştırma Kurumu.

SEED R. 1980. "Shell growth and form in bivalvia." In: D.C. Rhoads and R.A. Lutz (Eds.), *Skeletal Growth in Aquatic Organisms.* New York, Plenum: 23-67.

SEEHER J. 1983. Vorläufiger bericht über die Keramik des Beşik-Sivritepe. *Archäologischer Anzeiger* 172-182.

SHACKLETON J.C. 1983. "An approach to determining prehistoric shellfish collecting patterns." In: C. Grigson and J. Clutton-Brock (Eds.), *Shell Middens, fishes and Birds. Animals in Archaeology 2.* B.A.R. International Series 227. Oxford, Oxbow Books: 77-85.

----. 1988a. *Marine Molluscan Remains from Franchthi Cave.* Bloomington: Indiana University Press.

----. 1988b. "Reconstructing past shorelines as an approach to determining factors affecting shellfish collecting in the prehistoric past." In: G. Bailey and J. Parkington (Eds.), *The Archaeology of Prehistoric Coastlines.* Cambridge, Cambridge University Press: 11-21.

SHACKLETON J.C. and H. ELDERFIELD. 1980. Strontium isotope dating of the source of Neolithic European *Spondylus* artifacts. *Antiquity* 64/243: 312-315.

SHACKLETON N.J. 1968. "The Mollusca, the Crustacea, the Echinodermata (Appendix IX)." In: J.D. Evans and C. Renfrew, *Excavations at Saliagos near Antiparos.* The British School of Archaeology at Athens. London, Thames and Hudson: 122-138.

----. 1969. "Marine mollusca in archaeology." In: D. Brothwell and E.S. Higgs (Eds.), *Science in Archaeology.* London, Thames and Hudson: 407-414.

----. 1970. Stable isotope study of the palaeoenvironment of Nea Nikomedeia, Greece. *Nature*, 227, 943-944.

----. 1972 "The shells (Appendix VII)." In: P.M. Warren, *Myrtos: An Early Bronze Age Settlement in Crete.* British School of Archaeology at Athens, Supplementary Vol. 7: 321-325.

----. 1973. Oxygen isotope analysis as a means of determining season of occupation of prehistoric midden sites. *Archaeometry* 15: 133-141.

----. 1974. Oxygen isotopic demonstration of winter seasonal occupation. (Contribution to) Results of recent investigations at Tamar Hat. E.C. Saxon (Ed.), *Libyca* 22: 49-91.

SHACKLETON N.J. and M. DEITH. 1988. "Oxygen isotope analyses of marine molluscs from Franchthi Cave", In: J. Shackleton, *Excavations at Franchthi Cave IV: Molluscan Remains from Franchthi Cave*. Bloomington, Indiana University Press: 189-194.

SHACKLETON N.J. and C. RENFREW. 1970 Neolithic trade routes re-aligned by oxygen isotope analyses. *Nature* 228: 1062-1065.

SHACKLEY M. 1981. *Environmental Archaeology*. London, George Allen and Unwin.

SHELDON R.W. 1967. Relationship between shell-weight and age in certain molluscs. *Journal of Fisheries Research Board of Canada* 24/5: 1165-1171.

SORRENTINO C. 1997. "Poliochni: Il Materiale Faunistico." In: C. G. Doumas and V. La Rosa, *Poliochni: E l'Antica Eta del Bronzo Nell'Egeo Settentrionale*. Atene, Scuola Archeologica Italiana di Atene: 157-167.

SOUCI S.W., W. FACHMANN, and H. KRAUT. 2000. *Food Composition and Nutrition Tables*. Boca Raton, London, New York, and Washington D.C., Medpharm, Scientific Publishers Stuttgart.

STEININGER F. and E. MAUSER. 2002. *Die Molluskenfauna aus den "shell-middens" der präkolumbianischen Siedlung von Pointe de Caille, St. Lucia, West Indies: Bedeutung für Ernährungsstrategien und Werkzeugherstellung*. Abhandlungen der Senckenbergischen naturforschenden Gesellschaft. Frankfurt am Main. 556. Stuttgart, E. Schweizerbart'sche Verlagsbuchhandlung (Nägele und Obermiller).

STIEGLITZ R.R. 1994. The Minoan origin of Tryian Purple. *Biblical Archaeologist* 57/1: 46-54.

STRABO. *Geography*. Translated by H.L. Jones. 1917-1932. Harvard University Press and Heinemann.

STRATOULI G. 1996. Die Fischerei in der Ägäis während des Neolithikums: zur Technik und zum potentiellen Ertrag. *Prähistorische Zeitschrift* 71/1: 1-27.

TEKİNAY A.A., M. ALPARSLAN, Ö. ÖZEN, P. AKYÜZ, and D. GÜROY. 2002. 1996-2001 yılları arasında Çanakkale Balık Hali'nde pazarlanan su ürünleri ve Çanakkale bölgesi üretim miktarlarının karşılaştırılması. *Ege University Journal of Fisheries and Aquatic Sciences* 19: 455-463.

TEVESZ M. and J. CARTER. 1980. "Environmental relationship of shell form and structure of Unionacean Bivalves." In: D.C. Rhoads and R.A. Lutz (Eds.), *Shell Growth of Aquatic Organisms: Biological Records of Environmental Change*. New York, Plenum Press: 295-322.

THOMAS F.R. 2002. An evaluation of central-place foraging among mollusk gatherers in Western Kirpati, Micronesia: linking behavioral ecology with ethnoarchaeology. *World Archaeology: Archaeology and Evolutionary Ecology* 34/1: 182-208.

TRANTALİDOU K. 2005. Faunal exploitation at the Ulucak Höyük area: preliminary report. *Mediterranean Archaeology and Archaeometry* 5/3 (Special Issue): 47-61.

TREGANZA A.E. and S.F. COOK. 1950. *The Topanga Culture: First Season's Excavation of the Tank Site, 1947. University of California Anthropological Record 12/4*. California, Coyote Press.

TUNÇER S. and Ç. ERDEMİR. 2001. "Dünya kültür mirası olan Troia'nın geleceği üzerine araştırmalar." In: E. Özhan and Y. Yüksel (Eds.), *Türkiye Kıyıları '01. Türkiye'nin Kıyı ve Deniz Alanları III. Ulusal Konferansı Bildiriler Kitabı*. Kıyı Alanları Yönetimi Türkiye Milli Komitesi: 173-178.

TÜRKEŞ M. 2003. "Spatial and temporal variations in precipitation and aridity index series of Turkey." In: H.-J. Bolle (Ed.), *Mediterranean Climate: Variability and Trends*. Berlin, Springer Verlag: 181-212.

UÇKAÇ Ş. 2005 Sea surface variability in the Aegean Sea. *Ege University Journal of Fisheries and Aquatic Sciences* 22/1-2: 129-135.

UERPMANN H.-P. 1973. Animal bone finds and economic archaeology: a critical study of 'osteoarchaeological method'. *World Archaeology* 4/3: 307-322.

----. 1992. Stand und Konzeption der Archäo-Biologischen Begleitforschung zu den neuen Grabungen in Troia. *Mitteilungsblatt der archaeologica Venatoria e.v.*: 28-35.

----. 1993. "Proposal for a separate nomenclature of domestic animals." In: A.T. Clason, S. Payne, and H.-P. Uerpmann (Eds.), *Skeletons in her cupboard. Festschrift for Juliet Clutton-Brock*. Oxford, Oxbow Monograph 34: 239-241.

----. 1999. "Der Rückzucht-Auerochse und sein ausgestorbenes Vorbild." In: G.-C. Weniger (Ed.), *Archäologie und Biologie des Aurochsen.* Mettman, Neanderthal-Museum: 93-102.

----. 2003. "Environmental aspects of economic changes in Troia." In: G.A. Wagner, E. Pernicka, and H.-P. Uerpmann (Eds.), *Troia and the Troad: Scientific Approaches*. Berlin, Springer Verlag: 251-262.

UERPMANN H.-P. and M. Uerpmann. 2001 "Leben in Troia: die Pflanzen- und Tierwelt in Troia." *Traum und Wirklichkeit. Ausstellungskatalog*: 315-318.

UERPMANN M. 2006. "Von adler bis Zahnbrassen –Der Beitrag der Archäozoologie zur Erforschung Troias." In: M.O. Korfmann (Ed.), *Troia: Archäologie eines Siedlungshügels und seiner Landschaft*. Mainz, Philipp von Zabern: 283-296.

UERPMANN M. and W. VAN NEER. 1998. "Fish remains from the new excavations at Troy." In: H. Buitenhuis, L. Bartosiewicz, and A. M. Choyke (Eds.), *Archaeozoology of the Near East 3: Proceedings of the third international symposium on the archaeozoology of Southwest Asia and the adjacent areas*. Groningen: ARC-Publicatie 18: 243-254.

UERPMANN M. and W. VAN NEER. 2000. Fischreste aus den neuen Grabungen in Troia (1989-1999). *Studia Troica* 10: 145-179.

ÜNLÜSOY S. 2006. "Vom Reihenhaus zum Megaron – Troia I bis Troia III." In: M.O. Korfmann (Ed.), *Troia: Archäologie eines Siedlungshügels und seiner Landschaft.* Mainz, Philipp von Zabern: 133-144.

VAN ANDEL T.H. and J. SHACKLETON. 1982. Late Palaeolithic and Mesolithic coastlines in Greece and the Aegean. *Journal of Field Archaeology* 9: 445-454.

VAN NEER W., I. ZOHAR, and O. LERNAU. 2005. The emergence of fishing communities in the Eastern Mediterranean region: a survey of evidence from pre- and protohistoric periods. *Paléorient* 31/1: 131-157.

VAN ZEIST W. and S. BOTTEMA. 1991. *Late Quaternary Vegetation of the Near East.* Beihefte zum Tübinger Atlas des Vorderen Orients (TAVO). Reihe A (Naturwissenschaften) No. 18. Wiesbaden.

VERMEIJ G.J. 1978. *Biogeography and Adaptation: Patterns of Marine Life.* Cambridge: Harvard University Press.

----. 1980. "Gastropod shell growth rate, allometry, and adult size: environmental implications." In: D.C. Rhoads und R.A. Lutz (Eds.), *Skeletal Growth in Aquatic Organisms*. New York, Plenum: 379-394.

----. 1993. *A Natural History of Shells.* Princeton, Princeton University Press.

VICENTI H. 2005. Unusual spurting behavior of the freshwater mussel *Unio crassus. Journal of Molluscan Studies* 71: 409-410.

VİRCHOW R. 1879. Beiträge zur Landeskunde der Troas. *Abhandlungen der Königlichen Akademie der Wissenschaften Berlin, Physikalische Klasse* 3:1-190.

----. 1881. "Fauna der Troas." In: H. Schliemann, *Ilios: Stadt und Land der Trojaner*. Leipzig, F.A. Brockhaus: 130-135.

VON DEN DRIESCH A. 1999. Archäozoologische Untersuchungen an Tierknochen aus dem dritten und ersten vorchristlichen Jahrtausend vom Beşik-Yassıtepe, Westtürkei. *Studia Troica* 9: 439-474.

VON MARTENS E. 1879. Conchylien, welche Geh. Rath. Professor Virchow von Troas mitgebracht hat. *Sitzungsberichte der Gesellschaft Naturforschender Freunde zu Berlin* 1879: 86-93.

----. 1880. Eine Anzahl Conchylienschalen, welche Geh. Rath Prof. Virchow bei seiner Ausgrabung des Hanai-Tepeh gefunden hat. *Sitzungsberichte der Gesellschaft Naturforschender Freunde zu Berlin* 1880: 63-64.

WALTER R.C., R.T., BUFFLER, J.H. BRUGGEMANN, M.M., GUILLAUME, S.M. BERHE, B. NEGASSI, Y. LIBSEKAL, H. CHENG, R.L. EDWARDS, R. VON COSEL, D. NERAUDEAU, M. GAGNON. 2000. Early human occupation of the Red Sea coast of Eritrea during the last interglacial. *Nature* 405: 65-69.

WASELKOV G. 1987. "Shellfish gathering and shell midden archaeology." In: M. Schiffer (Ed.), *Archaeological Method and Theory 10.* New York, Academic Press: 93-210.

WHITEHEAD P.J., M.L. BAUCHOT, J.C. HUREAU, J. NIELSEN, and E. TORTONESE. 1987. *Fishes of the North-Eastern Atlantic and the Mediterranean* [CD-ROM]. Paris, Unesco Publications.

WIDMER R. 1989. Archaeological research strategies in the investigation of shell-bearing sites, a Florida perspective. Paper delivered at the Annual Meeting of the Society of American Archaeology, Atlanta.

WİLSON J.G. and B. ELKAİM. 1997. Seasonal and geographical differences in oxygen consumption with temperature of *Cerastoderma glaucum* (Poiret) and a comparison with *C. edule* (L.). *Estuarine, Coastal and Shelf Science* 45: 571-577.

WİTTER-BACKOFEN U. 2003. "Regional palaeodemographic aspects of Troia and its ecosystem." In: G.A. Wagner, E. Pernicka, and H.-P. Uerpmann (Eds.), *Troia and the Troad: Scientific Approaches*. Berlin, Springer Verlag: 303-315.

WOLKERSDORFER C. 2006. "Wasser; Quelle des Lebens –Hydrogeologische Untersuchungen in Troia." In: M.O. Korfmann (Ed.), *Troia: Archäologie eines Siedlungshügels und seiner Landschaft.* Mainz, Philipp von Zabern: 329-336.

WOLKERSDORFER C., J. GÖBEL, C. BLUME, and C. WEBER. 2004. Hydrogeologische Probenahmestellen in der Troas. *Studia Troica 14:* 169-200.

YANKSON K. 1986. Precocious sexual maturity in *Cerastoderma glaucum* (Bruguière) reared in the laboratory. *Journal of Molluscan Studies* 52: 79-80.

YESNER D.R. 1984. Population pressure in coastal environments: an archaeological test. *World Archaeology (Coastal Archaeology)* 16/1: 108-127.

YESNER D.R. 1988. "Island biogeography and prehistoric human adaptation on the southern coast of Maine (USA)." In: G. Bailey and J. Parkington (Eds.), *The Archaeology of Prehistoric Coastlines.* New Directions in Archaeology. Cambridge, Cambridge University Press: 53-63.

YİGIT N., A. DEMİRSOY, A. KARATAŞ, Ş. ÖZKURT, AND E. ÇOLAK. 2006. Notes on the mammals found in Kazdağı National Park and its environs. *Turkish Journal of Zoology* 30: 73-82.

YILDIZ H. AND A. LÖK. 2005. Çanakkale Boğazı'nda farklı boy gruplarındaki midyelerin (*Mytilus galloprovincialis* Lamarck, 1819) iki değişik sistemde büyüme ve yaşama performansları. *Ege Üniversitesi Su Ürünleri Dergisi* 22/1-2: 69-74.

ZANGGER E. 2003. "Some open questions about the plain of Troia." In: G.A. Wagner, E. Pernicka, and H.-P. Uerpmann (Eds.), *Troia and the Troad: Scientific Approaches*. Berlin, Springer Verlag: 317-324.

ZAOUALI J. 1979. Contribution à l'étude malacologique de la mer de Bou Grara (Tunisie méridionale) : Première données biométrique sur *Cerastoderma glaucum* Poirèt, 1789 (Bivalvia, Eulamallibranchia, Cardiidae). *Archives Institute Pasteur Tunis* 58: 403-424.

----. 1980. Sexual cycle of *Cerastoderma glaucum* Poirèt, 1789 (Bivalvia) in Lake Tunis and in the Bou Grara Sea (Tunisia). *Archives Institute Pasteur Tunis* 57/3: 281-295.

ZENETOS A. 1996. *Marine Bivalvia of Greece.* Fauna Graeciae 7.

ZIDERMAN I.I. 1990. Seashells and ancient purple dyeing. *Biblical Archaeologist* 53/2: 98-103.

ZODIATIS S., S. ALEXANDRI, P. PAVLAKIS, L. JÖNSSON, G. KALLOS, A. DEMETROPOULOS, G. GEORGIOU, A. THEODOROU, and E. BALOPOULOS. 1997. Tentative study of flow patterns in the North Aegean Sea using NOA-AVHRR images and 2D model simulation. *Earth and Environmental Science* 14: 1221-1231.

ZUSCHIN M., M. STACHOWITSCH M., and R.J. STANTON Jr. 2003. Patterns and processes of shell fragmentation in modern and ancient marine environments. *Earth-Science Reviews* 63: 33-82.

Appendix: Tables

Table 13. Temperature and precipitation at the nearest stations to Troia, Yenibademli, and Ulucak (After www.weatherreports.com).

Site	Troia		Yenibademli		Ulucak	
Station location	40° 8' N 26° 24' E		40° 11' N 25° 54' E		38° 24' N 27° 18' E	
Station altitude	8 m		72 m		25 m	
Over years	18	40	11	7	148	132
Months	Mean T	Mean P	Mean T	Mean no. of days with P	Mean T	Mean P
January	6 ° C	107	7 ° C	12	8 ° C	124
February	7 ° C	69	7 ° C	13	9 ° C	94
March	9 ° C	69	9 ° C	13	11 ° C	79
April	13 ° C	46	14 ° C	9	15 ° C	43
May	17 ° C	30	18 ° C	6	20 ° C	33
June	22 ° C	23	22 ° C	6	24 ° C	10
July	25 ° C	13	24 ° C	3	27 ° C	3
August	24 ° C	8	24 ° C	2	26 ° C	3
September	21 ° C	25	22 ° C	3	23 ° C	13
October	16 ° C	48	17 ° C	8	18 ° C	41
November	12 ° C	91	12 ° C	12	13 ° C	89
December	9 ° C	107	9 ° C	15	10 ° C	137
Annual	15 ° C	648	16 ° C	102	17 ° C	663

T = Temperature. P = Precipitation.

Table 14. Sea surface temperatures off the Çanakkale coast (after Anonymous 1982: 1809, Tables 2 and 3).

Months	Mean (° C)	Maximum (° C)	Minimum (° C)
January	9.2	13.8	0.3
February	8.4	15.3	0.6
March	8.5	12.8	5
April	10.9	16	7.5
May	15.4	19.5	11.0
June	19.8	25.2	16.2
July	22.7	25.9	19.3
August	23.6	25.6	21.0
September	21.8	24.9	17.2
October	18.3	22.5	14.4
November	15	19	9.8
December	11.9	16	6.8
Annual	15.5	25.9	0.3

Table 15. Systematic list of the mollusk taxa represented in the archaeomalacological assemblages.

Kingdom: Animalia	
Phylum: Mollusca	
Class: Gastropoda	
Family: Patellidae	*Patella caerulea*
	Patella rustica
Family: Haliotidae	*Haliotis tuberculata*
Family: Trochidae	
Subfamily: Monodontinae	
Family: Cerithidae	*Cerithium* spp.
Family: Cypraeidae	
Family: Cassidae	*Phalium granulatum*
Family: Ranellidae	
Subfamily: Cymatiinae	*Charonia tritonis*
Family: Muricidae	
Subfamily: Muricinae	*Bolinus brandaris*
	Hexaplex trunculus
Subfamily: Fasciolariinae	*Fasciolaria lignaria*
Family: Columbellidae	*Columbella rustica*
Family: Conidae	*Conus mediterraneus*
Class: Bivalvia	
Family: Arcidae	
Subfamily: Arcinae	*Arca noae*
Family: Glycymerididae	
Subfamily: Glycymeridinae	
Family: Mytilidae	
Subfamily: Mytilinae	*Mytilus galloprovincialis*
Family: Pinnidae	*Pinna* spp.
Family: Pectenidae	
Subfamily: Pectininae	Pectenidae
Family: Spondylidae	*Spondylus gaederopus*
Family: Ostridae	
Subfamily: Ostreinae	*Ostrea edulis*
Family: Cardiidae	
Subfamily: Cardiinae	*Acanthocardia tuberculatum*
Subfamily: Laevicardinae	*Cerastoderma glaucum*
Family: Solenidae	
Subfamily: Soleninae	*Solen marginatus*
Family: Donacidae	*Donax* spp.
Family: Veneridae	
Subfamily: Tapetinae	*Tapes decussatus*
Family: Unionidae	
Subfamily: Unioninae	*Unio* spp.

Table 16. Scientific and common names of the mollusk taxa represented in the archaeomalacological assemblages.

Scientific names	**English common names**	**Deutsche Volksnamen**
Gastropoda	Snails	Schnecken
Patella caerulea	Mediterranean limpet	Blaue Napfschnecke
Patella rustica	Portuguese limpet	Portuegesische Napfschnecke
Haliotis tuberculata	Abalone	Meerohr
Monodontinae	Topshells	Kreiselschnecken
Cerithium spp.	Hornshells	Seenadel
Cypraeidae	Cowries	Porzellanschnecken
Phalium granulatum	Lipped helmet shell	Helmschnecke
Charonia tritonis	Triton shell	Tritonhorn
Bolinus brandaris	Spiny murex	Herkuleskeule
Hexaplex trunculus	Murex	Purpurschnecke
Fasciolaria lignaria	Tulip shell	Bündelhorn
Columbella rustica	Rustic dove shell	Schlichte Täubchenschnecke
Conus mediterraneus	Mediterranean cone shell	Mittelmeerkegel
Bivalvia	Clams	Muscheln
Arca noae	Noah's ark shell	Arche Noah
Glycymerididae	Dog cockles	Pastetenmuscheln
Mytilus galloprovincialis	Mediterranean mussel	Mittelmeer-Miesmuscheln
Pinna spp.	Fan mussels	Steckmuschel
Pectenidae	Scallops	Kammusmuscheln
Spondylus gaederopus	Thorny oyster	Stachelauster
Ostrea edulis	European oyster	Europäische Auster
Acanthocardia tuberculatum	Red cockle	Knotige Herzmuschel
Cerastoderma glaucum	Lagoon cockle	Lagune Herzmuchel
Solen marginatus	European razorclam	Schwertmuschel
Donax spp.	Wedge shell	Stumpfmuscheln
Tapes decussatus	Carpetshell	Teppichmuscheln
Unio spp.	Freshwater clams	Flussmuscheln

Table 17. Morphology of modern *C. glaucum* populations in the Troad.

Measurements	Station	No. of specimens	Maximum	Upper quartile	Median	Lower quartile	Minimum	Mean	Standard deviation	Standard error of mean
Valve height (mm)	1	86	42.8	39.9	38.2	35.5	27.8	37.5	3.2	0.3
	2	101	45.0	39.8	35.8	32.5	13.4	34.8	6.4	0.6
	3	226	41.9	36.0	32.8	26.6	11.4	31.0	6.8	0.5
	All	413	45.0	37.9	35.0	30.5	11.4	33.2	6.7	0.3
Valve length (mm)	1	86	50.1	46.0	43.1	46.0	32.0	42.8	4.1	0.4
	2	101	50.4	44.4	41.3	36.1	14.6	39.2	7.5	0.8
	3	226	48.9	40.3	37.4	29.4	11.3	34.8	7.9	0.5
	All	413	50.4	43.0	39.1	34.1	11.3	37.5	7.9	0.4
Breadth (mm)	1	86	17.3	15.5	14.8	13.9	11.3	14.6	1.3	0.1
	2	101	17.6	15.7	14.4	12.9	5.4	13.8	2.6	0.3
	3	226	16.6	14.3	13.2	10.8	4.1	12.3	2.7	0.2
	All	413	17.6	14.9	13.8	12.0	4.1	13.1	2.6	0.1
Weight (g)	1	86	10.2	7.0	6.2	5.3	3.4	6.1	1.3	0.1
	2	101	10.6	7.6	6.0	4.3	0.4	5.7	2.3	0.2
	3	226	8.8	5.2	4.3	2.7	0.2	4.0	1.8	0.1
	All	413	10.6	6.3	5.0	3.7	0.2	4.9	2.1	0.1
Solidity	1	86	1.11	0.93	0.86	0.76	0.61	0.86	0.11	0.01
	2	101	1.37	1.01	0.88	0.77	0.44	0.88	0.18	0.01
	3	226	1.54	0.87	0.76	0.67	0.29	0.78	0.17	0.01
	All	413	1.5	0.9	0.8	0.7	0.3	0.8	0.17	0.00
Curve	1	86	0.79	0.71	0.68	0.65	0.61	0.69	0.04	0.00
	2	101	0.80	0.72	0.70	0.68	0.63	0.70	0.04	0.00
	3	226	0.80	0.73	0.71	0.69	0.57	0.71	0.04	0.00
	All	413	0.80	0.72	0.70	0.68	0.57	0.70	0.04	0.00

APPENDIX

Table 18. Faunal assemblage of Troia (All occupational phases).

	Troia			
Taxa	**NIS**	**% NIS**	**WIS (g)**	**% WIS**
Unidentified mammals	34238	21.1	88850.1	100
Domestic animals	62664	49.1	602685.4	66.9
Domestic or wild animals	407	0.3	6308.4	0.7
Wild mammals	7449	5.8	93517.6	10.4
Wild birds	1306	1	1668.2	0.2
Reptiles and amphibians	538	0.4	1357.7	0.2
Fish	533	0.4	1389.1	0.2
Crustaceans	6	0	2.9	0
Mollusks	54548	33.7	192735.3	19.5
Identified specimens	127682	78.9	900854.8	91
Unidentified bones	34238	21.1	88850.1	9
All specimens	161920	100	989704.9	100

Table 19. Faunal assemblages of the Early Bronze Age Troia.

	Troia I				Troia II				Troia III				Troia I/II/III				Early Bronze Age			
Taxa	**NIS**	**% NIS**	**WIS (g)**	**% WIS**	**NIS**	**% NIS**	**WIS (g)**	**% WIS**	**NIS**	**% NIS**	**WIS (g)**	**% WIS**	**NIS**	**% NIS**	**WIS (g)**	**% WIS**	**NIS**	**% NIS**	**WIS (g)**	**% WIS**
Unidentified mammals	6680	100	15192.3	100	2077	100	3842.5	1000	846	100	1994.1	100	939	100	1923.5	100	10542	100	22952.4	100
Domestic animals	12923	49.7	101443.5	67.2	3907	25.7	25687	40.8	1949	83.1	14370.1	76.3	1780	25.5	12828.1	44.4	20559	33.6	154328.7	54.2
Domestic or wild animals	74	0.3	1359.7	0.9	15	0.1	271.6	0.4	14	0.6	146.5	0.8	16	0.2	160	0.6	119	0.2	1937.8	0.7
Wild mammals	438	1.7	6754	4.5	276	1.8	3092.7	4.9	275	11.7	3754.8	19.9	187	2.7	1487	5.2	1176	1.9	15088.5	5.3
Wild birds	109	0.4	139.7	0.1	107	0.7	142.4	0.2	28	1.2	46.4	0.2	36	0.5	51.7	0.2	280	0.5	380.2	0.1
Reptiles and amphibians	30	0.1	57.8	0	11	0.1	81	0.1	3	0.1	19.9	0.1	13	0.2	42.1	0.1	57	0.1	200.8	0.1
Fish	41	0.2	121.1	0.1	166	1.1	67.3	0.1	8	0.3	15.4	0.1	30	0.4	58.3	0.2	245	0.4	262.1	0.1
Crustaceans	1	0	1	0	0	0	0	0	0	0	0	0	0	0	0	0	1	0	1	0
Mollusks	12398	37.9	41115.9	24.7	10735	62.1	33674.9	50.4	68	2.1	458.5	2.2	4928	62.2	14238	46.2	28129	46.0	89487.3	31.4
Identified specimens	26019	79.6	151020.8	90.9	15223	88	63030.9	94.3	2346	73.5	18828.6	90.4	6990	88.2	28865.2	93.8	50578	82.8	261745.5	91.9
Unidentified bones	6680	20.4	15192.3	9.1	2077	12	3842.5	5.7	846	26.5	1994.1	9.6	939	11.8	1923.5	6.2	10542	17.2	22952.4	8.1
All specimens	32699	100	166213.1	100	17300	100	66873.4	100	3192	100	20822.7	100	7929	100	30788.7	100	61120	100	284697.9	100

Table 20. Faunal assemblages of the Middle Bronze Age Troia.

	Troia IV				Troia V				Troia IV/V				Middle Bronze Age			
Taxa	**NIS**	**% NIS**	**WIS (g)**	**% WIS**	**NIS**	**% NIS**	**WIS (g)**	**% WIS**	**NIS**	**% NIS**	**WIS (g)**	**% WIS**	**NIS**	**% NIS**	**WIS (g)**	**% WIS**
Unidentified mammals	684	100	1730.3	100	2875	100	7323.5	100	65	100	162.7	100	3624	28.8	9216.5	12.5
Domestic animals	1207	66.9	10519.9	68.1	3662	53.1	35924.5	80.8	166	65.1	2952.7	65.2	5035	40.0	49397.1	67.1
Domestic or wild animals	5	0.3	217.2	1.4	52	0.8	731.1	1.6	4	1.6	119	2.6	61	0.5	1067.3	1.4
Wild mammals	273	15.1	3573.6	23.1	282	4.1	4590.1	10.3	55	21.6	1269.8	28.1	610	4.8	9433.5	12.8
Wild birds	59	3.3	59.4	0.4	79	1.1	107.9	0.2	23	9	47.9	1.1	161	1.3	215.2	0.3
Reptiles and amphibians	7	0.4	12.3	0.1	25	0.4	66.5	0.1	2	0.8	5.6	0.1	34	0.3	84.4	0.1
Fish	8	0.4	4.8	0	20	0.3	21.9	0	0	0	0	0	28	0.2	26.7	0.0
Crustaceans	0	0	0	0	0	0	0	0	0	0	0	0	0	0	0	0
Mollusks	245	9.8	1067.5	6.2	2773	28.4	2977.5	5.8	4	1.3	30.3	0.6	3022	24.0	4075.3	5.5
Identified specimens	1804	72.5	15454.7	89.9	6898	70.6	44446.2	85.9	255	79.7	4526.3	96.5	8957	71.2	64427.2	87.5
Unidentified bones	684	27.5	1730.3	10.1	2875	29.4	7323.5	14.1	65	20.3	162.7	3.5	3624	28.8	9216.5	12.5
All specimens	2488	100	17185	100	9773	100	51769.7	100	320	100	4689	100	12581	100	73643.7	100

Table 21. Faunal assemblages of the Late Bronze Age Troia.

	Troia VI Early				Troia VI Middle				Troia VI Late			
Taxa	**NIS**	**% NIS**	**WIS (g)**	**% WIS**	**NIS**	**% NIS**	**WIS (g)**	**% WIS**	**NIS**	**% NIS**	**WIS (g)**	**% WIS**
Unidentified mammals	2590	100	6524.6	100	3351	100	6772.8	100	2151	100	5852.2	100
Domestic animals	4475	36	43277.1	53.2	4557	54.8	44983.7	68.5	3908	52.9	39222	68.2
Domestic or wild animals	42	0.3	379.1	0.5	58	0.7	495.7	0.8	22	0.3	261.1	0.5
Wild mammals	552	4.4	7108.8	8.7	511	6.1	5606.8	8.5	308	4.2	4226.5	7.3
Wild birds	38	0.3	38.9	0	83	1	102.2	0.2	80	1.1	89.8	0.2
Reptiles and amphibians	30	0.2	62.4	0.1	49	0.6	90.5	0.1	34	0.5	79.3	0.1
Fish	26	0.2	41.8	0.1	27	0.3	74.9	0.1	9	0.1	13.2	0
Crustaceans	3	0	1	0	1	0	0.8	0	0	0	0	0
Mollusks	7167	47.7	30215.8	34.4	3005	25.7	14113.6	19.5	3010	31.6	13609.6	21.5
Identified specimens	12438	82.8	81329.8	92.6	8323	71.3	65686.1	90.7	7386	77.4	57519.5	90.8
Unidentified bones	2590	17.2	6524.6	7.4	3351	28.7	6772.8	9.3	2151	22.6	5852.2	9.2
All specimens	15028	100	87854.4	100	11674	100	72458.9	100	9537	100	63371.7	100

Table 21. Faunal assemblages of the Late Bronze Age Troia (Continued).

	Troia VI Early/Middle/Late				Troia VIIa				Late Bronze Age Troia			
Taxa	**NIS**	**% NIS**	**WIS (g)**	**% WIS**	**NIS**	**% NIS**	**WIS (g)**	**% WIS**	**NIS**	**% NIS**	**WIS (g)**	**% WIS**
Unidentified mammals	2665	100	9064.3	100	4195	100	11572	100	14952	23.5	39785.9	9.5
Domestic animals	4795	64.5	59068.6	84	6664	50.7	65444.1	63.7	24399	38.3	251995.5	60.4
Domestic or wild animals	19	0.3	142.3	0.2	29	0.2	599	0.6	170	0.3	1877.2	0.4
Wild mammals	279	3.8	3176.4	4.5	1322	10.1	17993.9	17.5	2972	4.7	38112.4	9.1
Wild birds	32	0.4	34.3	0	279	2.1	319.8	0.3	512	0.8	585	0.1
Reptiles and amphibians	14	0.2	25.6	0	28	0.2	108.3	0.1	155	0.2	366.1	0.1
Fish	14	0.2	19.8	0	84	0.6	110	0.1	160	0.3	259.7	0.1
Crustaceans	0	0	0	0	1	0	0.1	0	5	0	1.9	0
Mollusks	2254	22.3	7545.3	9.5	4738	27.3	18062.3	15.8	20174	31.7	83546.6	20.0
Identified specimens	7438	73.6	70307.6	88.6	13153	75.8	102664.6	89.9	48738	76.5	377507.6	90.5
Unidentified bones	2665	26.4	9064.3	11.4	4195	24.2	11572	10.1	14952	23.5	39785.9	9.5
All specimens	10103	100	79371.9	100	17348	100	114236.6	100	63590	100	417293.5	100

Table 22. Faunal assemblages of the Troia VIIb and VIIa/b periods.

	Troia VIIb				Troia VIIa/b			
Taxa	**NIS**	**% NIS**	**WIS (g)**	**% WIS**	**NIS**	**% NIS**	**WIS (g)**	**% WIS**
Unidentified mammals	1155	100	3648.9	100	522	100	1287.6	100
Domestic animals	3772	60.4	33668.7	60.9	1240	55.6	9794.1	58.2
Domestic or wild animals	6	0.1	49	0.1	3	0.1	103	0.6
Wild mammals	1367	21.9	16937.3	30.6	432	19.4	5110.5	30.4
Wild birds	64	1	134.9	0.2	15	0.7	34.4	0.2
Reptiles and amphibians	12	0.2	63.6	0.1	12	0.5	41	0.2
Fish	50	0.8	724	1.3	11	0.5	9.2	0.1
Crustaceans	0	0	0	0	0	0	0	0
Mollusks	966	15.5	3700	6.7	519	23.3	1742.7	10.4
Identified specimens	6240	84.4	55306.7	93.8	2232	81	16834.9	92.9
Unidentified bones	1155	15.6	3648.9	6.2	522	19	1287.6	7.1
All specimens	7395	100	58955.6	100	2754	100	18122.5	100

Table 23. Faunal assemblages of the Post-Bronze Age Troia.

	Troia VIII				Troia IX				Troia X			
Taxa	**NIS**	**% NIS**	**WIS (g)**	**% WIS**	**NIS**	**% NIS**	**WIS (g)**	**% WIS**	**NIS**	**% NIS**	**WIS (g)**	**% WIS**
Unidentified mammals	1609	100	6516	100	998	100	2332	100	826	100	3083	100
Domestic animals	4028	70.5	58238	80.3	1760	73.6	22889	89.6	1798	73.4	21216	83.9
Domestic or wild animals	40	0.7	1258	1.7	8	0.3	16	0.1	0	0	0	0
Wild mammals	734	12.9	7574	10.4	105	4.4	654.2	2.6	52	2.1	600.4	2.4
Wild birds	158	2.8	150.8	0.2	89	3.7	128.3	0.5	27	1.1	39.4	0.2
Reptiles and amphibians	71	1.2	34.2	0	96	4	245.8	1	101	4.1	321.8	1.3
Fish	17	0.3	44.3	0.1	12	0.5	48.4	0.2	10	0.4	14.7	0.1
Crustaceans	0	0	0	0	0	0	0	0	0	0	0	0
Mollusks	660	9	5085	6.4	321	9.5	1573	5.6	448	13.7	2990	10.5
Identified specimens	5712	78	72482	91.8	2391	70.6	25555	91.6	2451	74.8	25295	89.1
Unidentified bones	1609	22	6516	8.2	998	29.4	2332	8.4	826	25.2	3083	10.9
All specimens	7321	100	78998	100	3389	100	27887	100	3277	100	28378	100

Table 23. Faunal assemblages of the Post-Bronze Age Troia (Continued).

	Troia VII/IX/X				Post-Bronze Age Troia			
Taxa	**NIS**	**% NIS**	**WIS (g)**	**% WIS**	**NIS**	**% NIS**	**WIS (g)**	**% WIS**
Unidentified mammals	10	100	28.1	100	3443	23.9	11958.8	8.7
Domestic animals	73	19.1	1159	68.1	7659	53.3	103501.7	75.6
Domestic or wild animals	1	0.3	6.7	0.4	49	0.3	1280.8	0.9
Wild mammals	0	0	0	0	891	6.2	8828.8	6.4
Wild birds	0	0	0	0	274	1.9	318.5	0.2
Reptiles and amphibians	0	0	0	0	268	1.9	601.8	0.4
Fish	0	0	0	0	39	0.3	107.4	0.1
Crustaceans	0	0	0	0	0	0	0	0
Mollusks	309	78.6	536.6	31	1738	12.1	10184.2	7.4
Identified specimens	383	97.5	1702	98.4	10937	76.1	125033.5	91.3
Unidentified bones	10	2.5	28.1	1.6	3443	23.9	11958.8	8.7
All specimens	393	100	1730	100	14380	100	136992.3	100

Table 24. Mollusk assemblages of Troia (All occupational phases).

	Troia			
Taxa	**NIS**	**% NIS**	**WIS (g)**	**% WIS**
Patella spp.	648	1.2	1687.1	0.9
Monodontinae	175	0.3	567.1	0.3
Cerithium spp.	11	0.0	38.2	0.0
Hexaplex trunculus	3517	6.4	16919.5	8.8
Arca noae	8	0.0	45	0.0
Glycymerididae	14	0.0	134.4	0.1
Mytilus galloprovincialis	2140	3.9	4574.3	2.4
Pinna spp.	8	0.0	21.3	0.0
Pectenidae	218	0.4	924.2	0.5
Spondylus gaederopus	12	0.0	440.9	0.2
Ostrea edulis	1775	3.3	20243.3	10.5
Cerastoderma glaucum	40455	74.2	136980.6	71.1
Solen marginatus	3764	6.9	5846.2	3.0
Tapes decussatus	1140	2.1	3310.8	1.7
Unio spp.	207	0.4	505.3	0.3
Unidentified bivalves	72	0.1	110.7	0.1
Unidentified marine gastropods	10	0.0	60.7	0.0
Terrestrial gastropods	209	0.4	146.7	0.1
Unidentified mollusks	165	0.3	179	0.1
Sum of mollusks	54548	33.7	192735.3	19.5

Table 25. Mollusk assemblages of the Early Bronze Age Troia.

	Troia I				Troia II				Troia III				Troia I/II/III				Maritime Troia (All)			
Taxa	**NIS**	**% NIS**	**WIS (g)**	**% WIS**	**NIS**	**% NIS**	**WIS (g)**	**% WIS**	**NIS**	**% NIS**	**WIS (g)**	**% WIS**	**NIS**	**% NIS**	**WIS (g)**	**% WIS**	**NIS**	**% NIS**	**WIS (g)**	**% WIS**
Patella spp.	14	0.1	18.8	0	79	0.7	146.5	0.4	1	1.5	18	3.7	26	0.5	45	0.3	120	0.4	228.3	0.3
Monodontinae	24	0.2	43	0.1	6	0.1	8.5	0	1	1.5	0.9	0.2	1	0	1	0	32	0.1	53.4	0.1
Cerithium spp.	2	0	7.2	0	1	0	0.9	0	0	0	0	0	0	0	0	0	3	0.0	8.1	0.0
Hexaplex trunculus	11	0.1	45.9	0.1	14	0.1	87	0.3	0	0	0	0	4	0.1	19	0.1	29	0.1	151.9	0.2
Arca noae	6	0	40	0.1	2	0	5	0	0	0	0	0	0	0	0	0	8	0.0	45	0.1
Glycymerididae	0	0	0	0	0	0	0	0	0	0	0	0	0	0	0	0	0	0.0	0	0.0
Mytilus galloprovincialis	1109	8.9	1536.7	3.7	186	1.7	567.3	1.7	14	20.6	46.2	9.5	44	0.9	131	0.9	1353	4.8	2281.2	2.5
Pinna spp.	0	0	0	0	0	0	0	0	0	0	0	0	0	0	0	0	0	0.0	0	0.0
Pectenidae	5	0	23.5	0.1	1	0	12	0	2	2.9	13	2.7	0	0	0	0	8	0.0	48.5	0.1
Spondylus gaederopus	0	0	0	0	0	0	0	0	1	1.5	32	6.6	0	0	0	0	1	0.0	32	0.0
Ostrea edulis	806	6.5	7791.3	18.9	127	1.2	1283	3.8	12	17.6	221	45.4	71	1.4	815	5.7	1016	3.6	10109.8	11.3
Cerastoderma glaucum	8939	72.1	28020.1	68.1	10237	95.4	31463	93.5	34	50	123	25.3	4694	95.3	13171	92.5	23904	85.0	72776.9	81.3
Solen marginatus	72	0.6	68.9	0.2	19	0.2	6.5	0	1	1.5	0.5	0.1	28	0.6	14	0.1	120	0.4	89.9	0.1
Tapes decussatus	1102	8.9	3183	7.7	25	0.2	56	0.2	0	0	0	0	0	0	0	0	1127	4.0	3239	3.6
Unio spp.	89	0.7	179.1	0.4	1	0	1.9	0	1	1.5	3.4	0.7	0	0	0	0	91	0.3	184.4	0.2
Unidentified bivalves	0	0	0	0	8	0.1	13.4	0	0	0	0	0	3	0.1	9	0.1	11	0.0	22.4	0.0
Unidentified marine gastropods	16	0.1	26.7	0.1	1	0	4.1	0	0	0	0	0	0	0	0	0	17	0.1	30.8	0.0
Terrestrial gastropods	71	0.6	44.8	0.1	26	0.2	13	0	1	1.5	1	0.2	56	1.1	31	0.2	154	0.5	89.8	0.1
Unidentified mollusks	132	1.1	86.9	0.2	2	0	7	0	0	0	0	0	1	0	2	0	135	0.5	95.9	0.1
Sum of mollusks	12398	100	41115.9	100	10735	100	33675	100	68	100	459	100	4928	100	14238	100	28129	100	89487.3	100

Table 26. Mollusk assemblages of the Middle Bronze Age Troia.

	Troia IV				Troia V				Troia IV/V				Middle Bronze Age Troia			
Taxa	**NIS**	**% NIS**	**WIS (g)**	**% WIS**	**NIS**	**% NIS**	**WIS (g)**	**% WIS**	**NIS**	**% NIS**	**WIS (g)**	**% WIS**	**NIS**	**% NIS**	**WIS (g)**	**% WIS**
Patella spp.	1	0.4	3	0.3	15	0.5	62.8	0.5	2	50	2.1	6.9	18	0.6	67.9	1.7
Monodontinae	0	0	0	0	0	0	0	0	0	0	0	0	0	0	0	0
Cerithium spp.	0	0	0	0	1	0	4.4	0	0	0	0	0	1	0	4.4	0.1
Hexaplex trunculus	4	1.6	46.7	4.4	14	0.5	161	0.5	1	25	27	89.1	19	0.6	234.7	5.8
Arca noae	0	0	0	0	0	0	0	0	0	0	0	0	0	0	0	0
Glycymerididae	0	0	0	0	0	0	0	0	0	0	0	0	0	0	0	0
Mytilus galloprovincialis	3	1.2	15.5	1.5	29	1.0	117.3	1.0	1	25	1.2	4.0	33	1.1	134.0	3.3
Pinna spp.	0	0	0	0	0	0	0	0	0	0	0	0	0	0	0	0
Pectenidae	2	0.8	12.6	1.2	2	0.1	12.8	0.1	0	0	0	0	4	0.1	25.4	0.6
Spondylus gaederopus	0	0	0	0	0	0	0	0	0	0	0	0	0	0	0	0
Ostrea edulis	9	3.7	213	19.9	54	1.9	1221	1.8	0	0	0	0	63	2.1	1433.6	35.2
Cerastoderma glaucum	223	91.0	770	72.1	394	14.2	1220	13.2	0	0	0	0	617	20.4	1989.3	48.8
Solen marginatus	1	0.4	0.4	0	2260	81.5	175.5	75.9	0	0	0	0	2261	74.8	175.9	4.3
Tapes decussatus	1	0.4	3.7	0.3	0	0	0	0	0	0	0	0	1	0	3.7	0.1
Unio spp.	0	0	0	0	3	0.1	2.1	0.1	0	0	0	0	3	0.1	2.1	0.1
Unidentified bivalves	1	0.4	3.1	0.3	1	0	1.2	0	0	0	0	0	2	0.1	4.3	0.1
Unidentified marine gastropods	0	0	0	0	0	0	0	0	0	0	0	0	0	0	0	0
Terrestrial gastropods	0	0	0	0	0	0	0	0	0	0	0	0	0	0	0	0
Unidentified mollusks	0	0	0	0	0	0	0	0	0	0	0	0	0	0	0	0
Sum of mollusks	245	100	1068	100	2773	100	2973	100	4	100	30.3	100	3022	100	4075.3	100

Table 27. Mollusk assemblages of the Late Bronze Age Troia.

	Troia VI Early				Troia VI Middle				Troia VI Late				Troia VI Early/Middle/Late				Troia VIIa				Late Bronze Age Troia			
Taxa	NIS	% NIS	WIS (g)	% WIS	NIS	% NIS	WIS (g)	% WIS	NIS	% NIS	WIS (g)	% WIS	NIS	% NIS	WIS (g)	% WIS	NIS	% NIS	WIS (g)	% WIS	NIS	% NIS	WIS (g)	% WIS
Patella spp.	396	5.5	1143	3.8	5	0.2	10.3	0.1	6	0.2	13.2	0.1	7	0.3	18.2	0.2	33	0.7	82	0.5	447	2.5	1267.1	1.5
Monodontinae	139	1.9	508	1.7	0	0	0	0	0	0	0	0	0	0	0	0	4	0.1	5.7	0	143	0.8	513.7	0.6
Cerithium spp.	3	0	12.2	0	0	0	0	0	1	0	2	0	1	0	4	0.1	2	0	7.5	0	7	0	25.7	0
Hexaplex trunculus	586	8.2	2247	7.4	328	10.9	1372	9.7	324	10.8	1125	8.3	416	18.5	1530	20.3	1695	35.8	9635	53.3	3349	18.9	15909.1	19.0
Arca noae	0	0	0	0	0	0	0	0	0	0	0	0	0	0	0	0	0	0	0	0	0	0	0	0
Glycymerididae	3	0	44	0.1	3	0.1	12.6	0.1	1	0	11	0.1	2	0.1	15.7	0.2	4	0.1	42.9	0.2	13	0.1	126.2	0.2
Mytilus galloprovincialis	108	1.5	366.6	1.2	181	6.0	364.2	2.6	172	5.7	430.3	3.2	75	3.3	250.6	3.3	106	2.2	300.4	1.7	642	3.6	1712.1	2.0
Pinna spp.	0	0	0	0	0	0	0	0	0	0	0	0	0	0	0	0	3	0.1	11	0.1	3	0	11	0
Pectenidae	0	0	0	0	1	0	13	0.1	1	0	3	0	6	0.3	29.6	0.4	4	0.1	10.3	0.1	12	0.1	55.9	0.1
Spondylus gaederopus	0	0	0	0	0	0	0	0	2	0.1	1.5	0	0	0	0	0	5	0.1	397	2.2	7	0	398.5	0.5
Ostrea edulis	41	0.6	617.8	2.0	9	0.3	174.7	1.2	8	0.3	327.2	2.4	27	1.2	343.5	4.6	26	0.5	441.4	2.4	111	0.6	1904.6	2.3
Cerastoderma glaucum	5215	72.7	20088	66.5	2273	75.6	12001	85.0	2437	81.0	11580	85.1	1427	63.3	5199	68.9	2775	58.6	7024	38.9	14127	79.9	55891	66.9
Solen marginatus	619	8.6	5027	16.6	144	4.8	54.9	0.4	16	0.5	8.9	0.1	274	12.2	123.9	1.6	54	1.1	33	0.2	1107	6.3	5247.3	6.3
Tapes decussatus	1	0	2.4	0	0	0	0	0	2	0.1	4.6	0	0	0	0	0	1	0	1.6	0	4	0	8.6	0
Unio spp.	35	0.5	137.5	0.5	25	0.8	54.4	0.4	30	1.0	74.8	0.5	6	0.3	13.8	0.2	10	0.2	18.1	0.1	106	0.6	298.6	0.4
Unidentified bivalves	1	0	1.3	0	24	0.8	14.4	0.1	0	0	0	0	1	0	0.4	0	4	0.1	12.7	0.1	30	0.2	28.8	0
Unidentified marine gastropods	1	0	8.1	0	0	0	0	0	0	0	0	0	0	0	0	0	4	0.1	29.7	0.2	5	0	37.8	0
Terrestrial gastropods	11	0.2	8.1	0	8	0.3	7.5	0.1	4	0.1	4.6	0	8	0.4	4.6	0.1	4	0.1	3.5	0	35	0.2	28.3	0
Unidentified mollusks	8	0.1	5.3	0	4	0.1	33.9	0.2	6	0.2	23.5	0.2	4	0.2	12.1	0.2	4	0.1	6.7	0	26	0.1	81.5	0.1
Sum of mollusks	7167	100	30216	100	518	100	14114	100	3010	100	13610	100	2254	100	7545	100	4738	100	18062	100	17687	100	83546.6	100

Table 28. Mollusk assemblages of Troia VIIb and VIIa/b.

	Troia VIIb				Troia VIIa/b			
Taxa	**NIS**	**% NIS**	**WIS (g)**	**% WIS**	**NIS**	**% NIS**	**WIS (g)**	**% WIS**
Patella spp.	26	2.7	65.5	1.8	12	2.3	29.5	1.7
Monodontinae	0	0	0	0	0	0	0	0
Cerithium spp.	0	0	0	0	0	0	0	0
Hexaplex trunculus	43	4.5	209	5.6	21	4.0	114.8	6.6
Arca noae	0	0	0	0	0	0	0	0
Glycymerididae	1	0.1	8.2	0.2	0	0	0	0
Mytilus galloprovincialis	69	7.1	343.8	9.4	13	2.5	31	1.8
Pinna spp.	5	0.5	10.3	0.3	0	0.0	0	0.0
Pectenidae	3	0.3	7.1	0.2	1	0.2	4	0.2
Spondylus gaederopus	0	0	0	0	0	0	0	0
Ostrea edulis	22	2.3	321.6	8.7	9	1.7	83.2	4.8
Cerastoderma glaucum	776	80.3	2692.8	72.8	460	88.6	1514.8	86.9
Solen marginatus	0	0	0	0	1	0.2	1.1	0.1
Tapes decussatus	0	0	0	0	0	0	0	0
Unio spp.	2	0.2	10.7	0.3	0	0	0	0
Unidentified bivalves	3	0.3	5.4	0.1	8	1.5	22.6	1.3
Unidentified marine gastropods	0	0	0	0	1	0.2	10	0.6
Terrestrial gastropods	0	0	0	0	0	0	0	0
Unidentified mollusks	2	0.2	1	0.0	0	0	0	0
Sum of mollusks	966	100	3700	100	519	100	1742.7	100

Table 29. Mollusk assemblages of the Post-Bronze Age Troia.

	Troia VIII				**Troia IX**				**Troia X**			
Taxa	**NIS**	**% NIS**	**WIS (g)**	**% WIS**	**NIS**	**% NIS**	**WIS (g)**	**% WIS**	**NIS**	**% NIS**	**WIS (g)**	**% WIS**
Patella spp.	1	0.2	1	0.0	24	7.5	27.8	1.8	0	0	0	0
Monodontinae	0	0	0	0	0	0	0	0	0	0	0	0
Cerithium spp.	0	0	0	0	0	0	0	0	0	0	0	0
Hexaplex trunculus	32	4.8	123	2.4	14	4.4	120	7.6	13	2.9	111.4	3.7
Arca noae	0	0	0	0	0	0	0	0	0	0	0	0
Glycymerididae	0	0	0	0	0	0	0	0	0	0	0	0
Mytilus galloprovincialis	10	1.5	18.8	0.4	1	0.3	2	0.1	15	3.3	44.2	1.5
Pinna spp.	0	0	0	0	0	0	0	0	0	0	0	0
Pectenidae	98	14.8	467	9.2	88	27.4	305	19.4	2	0.4	7.6	0.3
Spondylus gaederopus	0	0	0	0	0	0	0	0	0	0	0	0
Ostrea edulis	187	28.3	3415	67.2	103	32.1	873	55.5	260	58.0	2033	68.0
Cerastoderma glaucum	310	47.0	1029	20.2	87	27.1	243	15.4	143	31.9	721.6	24.1
Solen marginatus	0	0.0	0	0	2	0.6	0	0	0	0.0	0	0
Tapes decussatus	0	0	0	0	0	0	0	0	8	1.8	59.5	2
Unio spp.	1	0.2	1.9	0	1	0.3	2	0.1	2	0.4	3.4	0.1
Unidentified bivalves	0	0	0	0	0	0	0	0	2	0.4	0.5	0.0
Unidentified marine gastropods	0	0	0	0	0	0	0	0	3	0.7	8.8	0.3
Terrestrial gastropods	20	3.0	28.6	0.6	0	0	0	0	0	0	0	0
Unidentified mollusks	1	0.2	0.1	0.0	1	0.3	0.5	0.0	0	0	0	0
Sum of mollusks	660	100	5085	100	321	100	1573	100	448	100	2990	100

Table 29. Mollusk assemblages of the Post-Bronze Age Troia (Continued).

	Troia VIII/IX/X				Post-Bronze Age Troia (All)			
Taxa	**NIS**	**% NIS**	**WIS (g)**	**% WIS**	**NIS**	**% NIS**	**WIS (g)**	**% WIS**
Patella spp.	0	0	0	0	25	1.4	28.8	25
Monodontinae	0	0	0	0	0	0.0	0	0
Cerithium spp.	0	0	0	0	0	0.0	0	0
Hexaplex trunculus	4	1.3	13.7	0.7	63	3.6	368.3	63
Arca noae	0	0	0	0	0	0.0	0	0
Glycymerididae	0	0	0	0	0	0.0	0	0
Mytilus galloprovincialis	4	1.3	2.2	0.7	30	1.7	67.2	30
Pinna spp.	0	0	0	0	0	0.0	0	0
Pectenidae	2	0.6	3.6	0.4	190	10.9	783.3	190
Spondylus gaederopus	0	0	0	0	0	0.0	0	0
Ostrea edulis	4	1.3	70	0.7	554	31.9	6390.5	554
Cerastoderma glaucum	31	10	122.1	5.8	571	32.9	2115.8	571
Solen marginatus	263	85.1	322.8	49	265	15.2	322.8	265
Tapes decussatus	0	0.0	0	0	8	0.5	59.5	8
Unio spp.	1	0.3	2.2	0.2	5	0.3	9.5	5
Unidentified bivalves	0	0	0	0	2	0.1	0.5	2
Unidentified marine gastropods	0	0	0	0	3	0.2	8.8	3
Terrestrial gastropods	0	0	0	0	20	1.2	28.6	20
Unidentified mollusks	0	0	0	0	2	0.1	0.6	2
Sum of mollusks	309	100	536.6	100	1738	100	10184.2	100

Table 30. Morphology of *C. glaucum* populations from Troia.

Measurement	Phase	No. of specimens	Maximum	Upper quartile	Median	Mean	Lower quartile	Minimum	Standard deviation	Standard error of mean
Valve height (mm)	I	878	39.0	27.6	24.6	24.9	22.1	13.2	4.3	1.5
	II	496	35.4	26.4	23.6	23.6	21.1	9.9	4.1	1.8
	III	9	34.5	30.0	26.0	27.0	23.5	21.4	4.2	13.8
	I/II/III	16	38.9	30.8	24.6	26.9	23.5	21.7	5.1	12.7
	V	30	30.1	26.0	23.0	23.2	20.0	15.6	3.4	6.2
	VI E	1143	43.1	28.1	25.8	25.8	23.2	14.2	3.8	1.1
	VI M	401	44.6	28.9	26.6	26.7	24.0	14.3	4.0	2.0
	VI L	578	45.4	31.3	27.4	28.0	24.3	15.0	5.0	2.1
	VI E/M/L	459	43.4	28.2	25.2	25.6	41.7	15.3	4.2	1.9
	VIIa	161	36.2	28.0	25.4	25.4	23.0	17.4	3.7	2.9
	VIIb	54	38.8	31.1	26.5	27.2	22.8	16.9	5.5	7.5
	EBA	1399	39.0	27.1	24.2	24.5	21.7	9.9	4.3	1.2
	LBA	2742	45.4	39.0	26.1	26.3	23.4	14.2	4.3	0.8
	All	4225	45.4	28.3	25.5	25.7	22.7	9.9	4.4	0.7
Valve length (mm)	I	843	42.0	29.5	26.3	26.7	23.5	14.1	46.7	1.6
	II	475	39.5	28.1	25.2	25.2	22.3	10.6	44.4	2.0
	III	9	39.5	31.9	28.5	29.7	26.7	23.4	46.1	1.5
	I/II/III	11	36.8	28.6	26.5	27.4	24.5	23.4	38.7	11.7
	V	25	29.4	27.3	24.1	24.1	21.2	17.0	33.2	6.6
	VI E	934	47.0	30.1	27.3	27.6	24.7	15.8	42.0	1.4
	VI M	326	48.0	31.4	29.1	28.9	25.7	16.0	45.5	2.5
	VI L	484	43.7	34.2	29.1	30.1	25.7	16.2	57.2	2.6
	VI E/M/L	395	41.5	30.5	27.1	27.4	24.1	16.7	45.3	2.3
	VIIa	144	42.4	30.4	27.2	27.8	25.1	18.7	43.9	3.7
	VIIb	51	40.5	34.0	28.3	29.7	25.2	18.9	60.7	8.5
	EBA	1338	42.0	29.0	26.1	26.2	23.1	10.6	46.5	1.3
	LBA	2283	48.0	31.0	27.8	28.3	25.0	15.8	47.9	1.0
	All	3697	48.0	30.4	27.1	27.6	24.3	10.6	48.6	0.8

Table 30. Morphology of *C. glaucum* populations from Troia (Continued).

Measurement	Phase	No. of specimens	Maximum	Upper quartile	Median	Mean	Lower quartile	Minimum	Standard deviation	Standard error of mean
Breadth (mm)	I	881	17.6	12.0	10.7	10.8	9.4	5.3	2.0	0.7
	II	496	15.6	11.6	10.3	10.3	9.0	4.0	2.0	0.9
	III	9	17.2	14.0	12.9	12.6	10.3	9.9	2.4	0.8
	I/II/III	16	16.7	13.7	10.7	11.7	10.0	9.3	2.4	0.6
	V	30	13.1	11.2	10.1	10.1	8.8	7.0	1.6	2.9
	VI E	1140	18.9	12.6	11.4	11.5	10.4	6.6	1.7	0.5
	VI M	402	19.1	12.5	11.5	11.5	10.3	5.8	1.7	0.9
	VI L	579	18.3	13.9	12.0	12.2	10.6	6.4	2.3	0.9
	VI E/M/L	459	19.2	12.3	11.0	11.1	9.7	6.9	1.9	0.9
	VIIa	161	15.4	12.1	10.9	10.0	9.9	7.2	1.7	1.3
	VIIb	54	17.4	13.4	11.5	11.7	9.7	6.5	2.5	3.5
	EBA	1402	17.6	11.9	10.6	10.7	9.3	4.0	2.0	0.5
	LBA	2741	19.2	12.7	11.4	11.5	10.2	5.8	1.9	0.4
	All	4227	19.2	12.5	11.2	11.2	9.9	4.0	2.0	0.3
Weight (g)	I	838	11.1	4.0	2.8	3.2	1.9	0.4	1.8	0.1
	II	472	9.9	3.8	2.6	2.9	1.7	0.1	1.7	0.1
	III	9	10.0	6.4	4.9	4.9	2.6	2.3	2.5	0.8
	I/II/III	12	7.3	4.0	2.6	3.3	2.7	1.9	1.6	0.5
	V	23	3.8	3.2	2.5	2.4	1.7	0.9	0.9	0.2
	VI E	904	12.9	4.4	3.4	3.7	2.6	0.5	1.7	0.1
	VI M	315	21.0	4.3	3.2	3.4	2.3	0.6	1.8	0.1
	VI L	473	14.7	6.7	4.1	5.0	2.8	0.6	2.9	0.1
	VI E/M/L	384	11.2	4.3	2.9	3.3	2.0	0.6	1.6	0.1
	VIIa	144	8.9	3.8	2.8	2.9	1.9	0.7	1.4	0.1
	VIIb	50	11.1	6.3	3.5	4.2	1.8	0.4	2.8	0.4
	EBA	1331	11.1	3.9	2.7	31.3	1.9	0.1	1.9	0.0
	LBA	2220	21.0	4.6	3.4	3.8	2.4	0.5	2.1	0.0
	All	3624	21.0	4.4	3.1	3.6	2.2	0.1	2.0	0.0

Table 30. Morphology of *C. glaucum* populations from Troia (Continued).

Measurement	Phase	No. of specimens	Maximum	Upper quartile	Median	Mean	Lower quartile	Minimum	Standard deviation	Standard error of mean
Solidity	I	838	1.74	1.10	0.93	0.96	0.79	0.35	0.25	0.00
	II	472	1.96	1.11	0.94	0.95	0.76	0.20	0.28	0.01
	III	9	1.68	1.48	1.31	1.26	1.00	0.89	0.27	0.09
	I/II/III	12	1.37	1.18	0.95	1.00	0.85	0.63	0.20	0.05
	V	23	1.35	1.03	0.89	0.93	0.77	0.58	0.19	0.04
	VI E	904	1.85	1.23	1.06	10.7	0.89	0.36	0.23	0.01
	VI M	315	2.11	1.09	0.89	0.92	0.75	0.40	0.25	0.01
	VI L	473	2.21	1.42	1.15	1.15	0.88	0.40	0.34	0.02
	VI E/M/L	384	1.81	1.12	0.91	0.94	0.75	0.40	0.25	0.01
	VIIa	144	1.73	1.01	0.84	0.87	0.70	0.45	0.23	0.02
	VIIb	50	1.67	1.24	0.99	0.99	0.70	0.27	0.35	0.05
	EBA	1331	1.96	1.11	0.94	0.96	0.79	0.20	0.26	0.00
	LBA	2220	2.20	1.21	1.00	1.03	0.82	0.36	0.28	0.01
	All	3624	2.20	1.17	0.98	1.00	0.81	0.20	0.27	0.00
Curve	I	843	0.95	0.83	0.80	0.80	0.77	0.50	0.04	0.00
	II	475	0.96	0.84	0.81	0.81	0.78	0.67	0.05	0.00
	III	9	0.96	0.90	0.85	0.85	0.78	0.74	0.07	0.02
	I/II/III	11	0.88	0.84	0.80	0.81	0.77	0.73	0.05	0.01
	V	25	0.90	0.84	0.81	0.81	0.77	0.74	0.05	0.01
	VI E	929	1.03	0.86	0.83	0.83	0.79	0.65	0.05	0.00
	VI M	325	0.90	0.83	0.79	0.79	0.75	0.58	0.05	0.00
	VI L	483	1.04	0.85	0.81	0.82	0.78	0.55	0.05	0.00
	VI E/M/L	392	0.97	0.84	0.80	0.80	0.77	0.64	0.05	0.00
	VIIa	144	0.94	0.83	0.78	0.79	0.75	0.58	0.06	0.00
	VIIb	51	0.89	0.82	0.79	0.79	0.76	0.68	0.05	0.00
	EBA	1338	0.96	0.83	0.81	0.81	0.77	0.50	0.05	0.00
	LBA	2273	1.05	0.85	0.81	0.81	0.77	0.55	0.06	0.00
	All	3687	1.05	0.84	0.81	0.81	0.77	0.50	0.06	0.00

Table 31. Faunal assemblages of Yenibademli.

	Yenibademli Early				Yenibademli Late				Yenibademli (All)			
Taxa	**NIS**	**% NIS**	**WIS (g)**	**% WIS**	**NIS**	**% NIS**	**WIS (g)**	**% WIS**	**NIS**	**% NIS**	**WIS (g)**	**% WIS**
Unidentified mammals	450	100	1710.1	100	1507	100	5614.2	100	2081	100	7873	100
Domestic animals	1640	38.8	21802.4	60.6	4032	51.3	50844.0	72.7	6000	47.4	75670.5	68.6
Domestic or wild animals	9	0.2	614.9	1.7	30	0.4	1245.3	1.8	39	0.3	1860.2	1.7
Wild mammals	268	6.3	4696.6	13	417	5.3	5613.0	8.0	722	5.7	10727.6	9.7
Wild birds	1	0	0.5	0	7	0.1	12.7	0	8	0.1	13.2	0
Reptiles and amphibians	7	0.2	23.4	0.1	16	0.2	96.1	0.1	24	0.2	120.7	0.1
Fish	2	0	9.8	0	2	0	2.6	0	5	0	15.3	0
Mollusks	2300	54.4	8855.6	24.6	3348	42.6	12080.4	17.3	5849	39.7	21976.7	18.6
Identified specimens	4227	90.4	36003.2	95.5	7852	83.9	69894.1	92.6	12647	85.9	110384.2	93.3
Unidentified bones	450	9.6	1710.1	4.5	1507	16.1	5614.2	7.4	2081	14.1	7873	6.7
All specimens	4677	100	37713.3	100	9359	100	75508.3	100	14728	100	118257.2	100

(This listing excludes a few non-stratified units.)

Table 32. Mollusk assemblages of Yenibademli.

Taxa	Yenibademli Early				Yenibademli Late				Yenibademli (All)			
	NIS	% NIS	WIS (g)	% WIS	NIS	% NIS	WIS (g)	% WIS	NIS	% NIS	WIS (g)	% WIS
Patella caerulea	1828	79.5	4230.4	47.8	2469	73.0	5013.9	41.5	4407	75.3	9499.5	43.2
Patella rustica	21	0.9	36.3	0.4	24	0.7	40.6	0.3	45	0.8	76.9	0.3
Haliotis spp.	0	0	0	0	1	0	1.7	0	1	0	1.7	0
Monodontinae	8	0.3	28.3	0.3	22	0.7	68	0.6	34	0.6	104.5	0.5
Cerithium spp.	15	0.7	56.6	0.6	47	1.4	327.1	2.7	67	1.1	402.5	1.8
Cypraeidae	1	0	16	0.2	2	0.1	27.1	0.2	4	0.1	51.5	0.2
Phalium granulatum	1	0	17	0.2	2	0.1	10.1	0.1	3	0.1	27.1	0.1
Charonia tritonis	0	0	0	0	0	0	0	0	1	0	11	0.1
Bolinus brandaris	1	0	14	0.2	1	0	12	0.1	2	0	26	0.1
Hexaplex trunculus	7	0.3	44.2	0.5	46	1.4	199.5	1.7	57	1.0	279.6	1.3
Fasciolaria lignaria	0	0	0	0	1	0	3.9	0	1	0	3.9	0
Conus mediterraneus	0	0	0	0	1	0	3.8	0	2	0	13.6	0.1
Arca noae	10	0.4	52.8	0.6	19	0.6	118.9	1.0	33	0.6	199.7	0.9
Glycymerididae	1	0	30	0.3	5	0.1	47.9	0.4	6	0.1	77.9	0.4
Mytilus galloprovincialis	6	0.3	20	0.2	7	0.2	20.8	0.2	12	0.2	31	0.1
Pinna spp.	0	0	0	0	1	0	1.8	0	1	0	1.8	0
Pectenidae	0	0	0	0	1	0	3.8	0	1	0	3.8	0
Spondylus gaederopus	9	0.4	454.3	5.1	11	0.3	621.1	5.1	22	0.4	1111.4	5.1
Ostrea edulis	235	10.2	3021	34.1	358	10.6	4228.7	35.0	621	10.6	7613.7	34.6
Acanthocardium tuberculatum	10	0.4	85.9	1.0	12	0.4	52.9	0.4	25	0.4	152.7	0.7
Cerastoderma glaucum	138	6.0	732.9	8.3	290	8.6	1192.9	9.9	466	8.0	2177.7	9.9
Solen marginatus	1	0	0.7	0	4	0.1	12.1	0.1	5	0.1	12.8	0.1
Donax trunculus	2	0.1	2.4	0	5	0.1	3.3	0	5	0.1	3.6	0
Tapes decussatus	4	0.2	6.6	0.1	6	0.2	7.5	0.1	10	0.2	14.1	0.1
Unidentified marine gastropods	0	0	0	0	1	0	11	0.1	4	0.1	9.9	0
Unidentified bivalves	2	0.1	6.2	0.1	3	0.1	3.6	0	5	0.1	20.3	0.1
Terrestrial gastropods	0	0	0	0	9	0.3	46.4	0.4	7	0.1	46.4	0.2
Sum of mollusks	2300	100	8855.6	100	3348	100	12080.4	100	5849	100	21976.7	100

Table 33. Morphology of *P. caerulea* populations from Yenibademli.

Measurement	Phase	No. of specimens	Maximum	Upper quartile	Median	Mean	Lower quartile	Minimum	Standard deviation	Standard error of mean
Length (mm)	Early	234	58.0	39.7	34.9	35.0	28.9	19.1	8.0	0.5
	Late	353	61.7	37.1	31.7	32.7	27.0	15.5	8.3	0.4
	All	587	61.7	38.2	32.6	33.6	28.0	19.1	8.1	0.3
Height (mm)	Early	234	14.1	9.0	7.5	8.0	6.4	4.3	2.1	0.1
	Late	353	16.1	8.4	7.0	7.4	6.0	3.7	2.1	0.1
	All	587	16.1	8.7	7.3	7.6	8.7	3.9	2.1	0.1
Width (mm)	Early	234	51.5	33.5	28.2	28.9	23.1	15.5	7.5	0.5
	Late	353	55.7	30.6	25.7	26.9	21.8	14.5	7.2	0.4
	All	587	55.7	31.9	26.5	27.7	22.2	14.5	7.4	0.3
Height/Length	Early	234	0.38	0.25	0.22	0.22	0.20	0.12	0.03	0.00
	Late	353	0.37	0.25	0.22	0.23	0.20	0.12	0.04	0.00
	All	587	0.38	0.25	0.22	0.23	0.20	0.12	0.04	0.00
Length/Width	Early	234	2.38	1.26	1.21	1.22	1.17	0.96	0.09	0.00
	Late	353	1.61	1.27	1.22	1.23	1.18	0.78	0.08	0.00
	All	587	1.61	1.27	1.22	1.22	1.17	0.78	0.08	0.00

Table 34. Morphology of *C. glaucum* populations from Yenibademli.

Measurement	Phase	No. of specimens	Maximum	Upper quartile	Median	Mean	Lower quartile	Minimum	Standard deviation	Standard error of mean
Valve height (mm)	Early	33	37.2	33.1	27.6	28.7	25.4	19.6	4.7	0.8
	Late	55	51.2	32.2	26.1	27.6	23.0	15.7	7.0	1.0
	All	88	51.2	32.4	26.9	28.0	23.8	15.7	6.3	0.7
Valve length (mm)	Early	33	43.1	35.8	29.6	31.1	26.8	21.3	5.4	0.9
	Late	55	47.7	34.5	27.7	29.2	24.9	12.8	7.0	0.9
	All	88	47.7	34.7	29.4	29.9	34.7	12.8	6.5	0.7
Breadth (mm)	Early	33	18.2	15.2	12.5	13.0	11.5	8.2	2.5	0.4
	Late	55	22.0	15.0	11.2	12.1	9.9	6.5	3.3	0.4
	All	88	22.0	15.1	12.1	12.4	10.1	6.5	3.1	0.3
Weight (g)	Early	33	13.3	8.5	4.9	6.1	3.7	1.3	3.3	0.6
	Late	55	15.3	8.2	3.7	5.1	2.2	0.7	3.7	0.5
	All	88	15.3	8.4	4.3	5.5	2.5	0.7	3.4	0.4
Solidity	Early	33	1.97	1.63	1.41	1.34	1.06	0.50	0.38	0.07
	Late	55	1.91	1.51	1.05	1.14	0.78	0.49	0.42	0.06
	All	88	1.97	1.55	1.16	1.21	0.84	0.49	0.42	0.04
Curve	Early	33	0.92	0.88	0.84	0.83	0.81	0.57	0.07	0.01
	Late	55	1.09	0.87	0.84	0.83	0.77	0.63	0.08	0.01
	All	88	1.09	0.87	0.84	0.83	0.78	0.57	0.07	0.01

Table 35. Faunal assemblages of Ulucak.

	Chalcolithic Ulucak				Early Bronze Age Ulucak				Ulucak (All)			
Taxa	**NIS**	**% NIS**	**WIS (g)**	**% WIS**	**NIS**	**% NIS**	**WIS (g)**	**% WIS**	**NIS**	**% NIS**	**WIS (g)**	**% WIS**
Unidentified mammals	1130	100	4380	100	443	100	2295.7	100	1573	28.6	6675.8	13.7
Domestic animals	1678	81.7	17131	78.5	930	49.4	13316.1	66	2608	47.4	30447.2	62.6
Domestic or wild animals	22	1.1	556	2.5	12	0.6	331.5	1.6	34	0.6	887.5	1.8
Wild mammals	296	14.4	3961	18.2	254	13.5	4999.2	24.8	550	10	8960.1	18.4
Wild birds	2	0.1	5.3	0	1	0.1	0.5	0	3	0.1	5.8	0
Reptiles and amphibians	5	0.2	16.5	0.1	3	0.2	7	0	8	0.1	23.5	0
Fish	2	0.1	4.1	0	0	0	0	0	2	0	4.1	0
Mollusks	48	1.5	138.6	0.5	681	29.3	1524.7	6.8	729	13.2	1663.3	3.4
Identified specimens	2053	64.5	21813	83.3	1881	80.9	20179	89.8	3934	71.4	41991.5	86.3
Unidentified bones	1130	35.5	4380	16.7	443	19.1	2295.7	10.2	1573	28.6	6675.8	13.7
All specimens	3183	100	26193	100	2324	100	22474.7	100	5507	100	48667.3	100

Table 36. Mollusk assemblages of Ulucak.

	Chalcolithic Ulucak				Early Bronze Age Ulucak				Ulucak (All)			
Taxa	**NIS**	**% NIS**	**WIS (g)**	**% WIS**	**NIS**	**% NIS**	**WIS (g)**	**% WIS**	**NIS**	**% NIS**	**WIS (g)**	**% WIS**
Monodontinae	2	4.2	6.5	4.7	1	0.1	0.5	0.0	3	0.4	7	0.4
Cerithium spp.	0	0	0	0	1	0.1	1.7	0.1	1	0.1	1.7	0.1
Charonia tritonis	0	0	0	0	1	0.1	13	0.9	1	0.1	13	0.8
Bolinus brandaris	1	2.1	5.3	3.8	0	0	0	0	1	0.1	5.3	0.3
Columbella rustica	1	2.1	0.8	0.6	0	0	0	0	1	0.1	0.8	0.0
Arca noae	4	8.3	13.2	9.5	6	0.9	52.9	3.5	10	1.4	66.1	4.0
Pinna spp.	1	2.1	2.7	1.9	1	0.1	1.2	0.1	2	0.3	3.9	0.2
Ostrea edulis	1	2.1	6.5	4.7	2	0.3	37	2.4	3	0.4	43.5	2.6
Cerastoderma glaucum	37	77.1	100.8	72.7	668	98.1	1416.1	92.9	705	96.7	1516.9	91.2
Solen marginatus	0	0	0	0	1	0.1	2.3	0.2	1	0.1	2.3	0.1
Unio spp.	1	2.1	2.8	2.0	0	0	0	0	1	0.1	2.8	0.2
Unidentified marine gastropods	0	0	0	0	0	0	0	0	0	0	0	0
Unidentified bivalves	0	0	0	0	0	0	0	0	0	0	0	0
Terrestrial gastropods	0	0	0	0	0	0	0	0	0	0	0	0
Sum of mollusks	48	1.5	138.6	0.5	681	100	1524.7	100	729	100	1663.3	100

Table 37. Morphology of *C. glaucum* assemblages from Ulucak.

Measurement	Phase	No. of specimens	Maximum	Upper quartile	Median	Mean	Lower quartile	Minimum	Standard deviation	Standard error of mean
Valve height (mm)	Chalcolithic	7	31.3	26.0	24.1	24.7	22.5	22.3	3.2	1.2
	EBA	189	36.4	23.7	21	21.8	19.3	15.1	3.6	0.3
	All	196	36.4	23.8	21.2	21.9	19.3	15.1	3.6	0.3
Valve length (mm)	Chalcolithic	7	32.1	26.9	26.7	26.3	23.3	23.3	3	1.1
	EBA	189	37.5	25.5	23.0	23.6	20.9	16.4	3.7	0.3
	All	196	37.5	25.6	23.0	23.7	20.9	16.4	3.7	0.3
Breadth (mm)	Chalcolithic	7	14.3	11.3	10.2	10.7	9.2	9.1	1.8	0.7
	EBA	189	14.8	10.3	9.1	9.3	8.1	6.4	1.6	0.1
	All	196	14.8	10.3	9.1	9.3	8.1	6.4	1.7	0.1
Weight (g)	Chalcolithic	7	5.8	4.1	3.1	3.2	2	1.7	1.4	0.5
	EBA	189	7.2	2.3	1.6	1.1	1.1	0.4	1.1	0.1
	All	196	7.2	2.4	1.7	1.9	1.2	0.4	1.2	0.1
Solidity	Chalcolithic	7	1.21	1.18	1.06	1	0.79	0.68	0.2	0.01
	EBA	189	1.4	0.83	0.73	0.73	0.62	0.35	0.19	0.01
	All	196	1.4	0.84	0.73	0.74	0.6	0.33	0.44	0.05
Curve	Chalcolithic	7	0.89	0.86	0.81	0.81	0.76	0.72	0.06	0.02
	EBA	189	1.05	0.81	0.78	0.79	0.75	0.67	0.05	0
	All	196	1.05	0.81	0.78	0.79	0.75	0.67	0.05	0

www.ingramcontent.com/pod-product-compliance
Lightning Source LLC
LaVergne TN
LVHW071207210926
848482LV00015B/744